S0-CUN-194

Advances in Industrial Control

Springer
London
Berlin
Heidelberg
New York
Barcelona
Hong Kong
Milan
Paris
Santa Clara
Singapore
Tokyo

Other titles published in this Series:

Control of Modern Integrated Power Systems
E. Mariani and S.S. Murthy

Advanced Load Dispatch for Power Systems: Principles, Practices and Economies
E. Mariani and S.S. Murthy

Supervision and Control for Industrial Processes
Björn Sohlberg

Modelling and Simulation of Human Behaviour in System Control
Pietro Carlo Cacciabue

Modelling and Identification in Robotics
Krzysztof Kozlowski

Spacecraft Navigation and Guidance
Maxwell Noton

Robust Estimation and Failure Detection
Rami Mangoubi

Adaptive Internal Model Control
Aniruddha Datta

Price-Based Commitment Decisions in the Electricity Market
Eric Allen and Marija Ilić

Compressor Surge and Rotating Stall: Modeling and Control
Jan Tommy Gravdahl and Olav Egeland

Radiotheraphy Treatment Planning: New System Approaches
Olivier Haas

Feedback Control Theory for Dynamic Traffic Assignment
Pushkin Kachroo and Kaan Özbay

Control and Instrumentation for Wastewater Treatment Plants
Reza Katebi, Michael A. Johnson & Jacqueline Wilkie

Autotuning of PID Controllers
Cheng-Ching Yu

Robust Aeroservoelastic Stability Analysis
Rick Lind & Marty Brenner

Performance Assessment of Control Loops:Theory and Applications
Biao Huang & Sirish L. Shah

Data Mining and Knowledge Discovery for Process Monitoring and Control
Xue Z. Wang

Advances in PID Control
Tan Kok Kiong, Wang Quing-Guo & Hang Chang Chieh with Tore J. Hägglund

Advanced Control with Recurrent High-order Neural Networks: Theory and Industrial Applications
George A. Rovithakis & Manolis A. Christodoulou

Aniruddha Datta, Ming-Tzu Ho
and Shankar P. Bhattacharyya

Structure and Synthesis of PID Controllers

With 105 Figures

Springer

Aniruddha Datta, PhD
Department of Electrical Engineering, Texas A&M University, College Station, TX 77843-3128, USA

Ming-Tzu Ho, PhD
Engineering Science Department, National Cheung Kung university, 1 University Road, Tainan 701, Taiwan, Republic of China

Shankar P. Bhattacharyya, PhD
Department of Electrical Engineering, Texas A&M University, College Station, TX 77843-3128, USA

ISBN 1-85233-614-5 Springer-Verlag London Berlin Heidelberg

British Library Cataloguing in Publication Data
Datta, Aniruddha, 1963-
Structure and synthesis of PID controllers. - (Advances in industrial control)
1.PID controllers
I.Title II.Ho, Ming-Tzu III.Bhattacharyya, S.P.
629.8
ISBN 1852336145

Library of Congress Cataloging-in-Publication Data
A catalog record for this book is available from the Library of Congress

Printed in Great Britain

Typesetting: Camera ready by authors
Printed and bound by Athenæum Press Ltd., Gateshead, Tyne & Wear
69/3830-543210 Printed on acid-free paper SPIN 10712015

Advances in Industrial Control

Professor Dr -Ing M. Thoma
Institut für Regelungstechnik
Universität Hannover
Appelstr. 11
30167 Hannover
Germany

Professor H. Kimura
Department of Mathematical Engineering and Information Physics
Faculty of Engineering
The University of Tokyo
7-3-1 Hongo
Bunkyo Ku
Tokyo 113
Japan

Professor A.J. Laub
College of Engineering - Dean's Office
University of California
One Shields Avenue
Davis
California 95616-5294
United States of America

Professor J.B. Moore
Department of Systems Engineering
The Australian National University
Research School of Physical Sciences
GPO Box 4
Canberra
ACT 2601
Australia

Dr M.K. Masten
Texas Instruments
2309 Northcrest
Plano
TX 75075
United States of America

Professor Ton Backx
AspenTech Europe B.V.
De Waal 32
NL-5684 PH Best
The Netherlands

THIS BOOK IS DEDICATED TO

My Parents Akhil Chandra Datta and Sheela Datta
A. Datta

My Parents Mei-Ni Wang and Chen-Cheng Ho
M. T. Ho

J. Boyd Pearson
Friend, advisor, mentor and
Outstanding control theorist of the 20th century.
S. P. Bhattacharyya

SERIES EDITORS' FOREWORD

The series *Advances in Industrial Control* aims to report and encourage technology transfer in control engineering. The rapid development of control technology has an impact on all areas of the control discipline. New theory, new controllers, actuators, sensors, new industrial processes, computer methods, new applications, new philosophies..., new challenges. Much of this development work resides in industrial reports, feasibility study papers and the reports of advanced collaborative projects. The series offers an opportunity for researchers to present an extended exposition of such new work in all aspects of industrial control for wider and rapid dissemination.

It has taken a surprisingly long time for control theorists to turn their attention to the paradoxical success of three-term control in industrial applications but some small groups of researchers in Singapore, Sweden, Taiwan and elsewhere have continued to investigate fundamental issues concerned with PID and the autotune culture. In this new text, A. Datta, M.T. Ho and S. P. Bhattacharyya report on their fundamental investigations into three-term control. They concentrate on determining the set of all stabilising PID controllers for a given single-input, single-output plant and provide efficient computational routes for depicting this set. The examples displayed are instructive and insightful, showing that the topology of the stabilising set can be exceedingly complicated.

The importance of this PID work is that it creates a framework for determining the limits of performance of PID control to satisfy *other* design specifications. In one sense this research activity in PID control is only just beginning and there are many unanswered questions remaining; however, this volume complements that by C.–C. Yu on *Autotuning of PID Controllers: Relay Feedback Approach* and the new volume by K. K. Tan, Q.–G. Wang, C.–C. Hang and T. J. Hägglund on *Advances in PID Control* which are both published in the Advances in Industrial Control series. Industrial control engineers will be pleased to see the spotlight turned on to the ubiquitous PID controller and will welcome the insight that these investigations provide. In the academic community these volumes will be read to see the hidden theoretical depth and subtly that exists with this commonest form of controller.

M.J. Grimble and M.A. Johnson
Industrial Control Centre
Glasgow, Scotland, UK

PREFACE

In this monograph we describe some new results on the structure and design of Proportional-Integral-Derivative (PID) controllers. We believe that these results could form the beginnings of an attempt to close the significant gap that has arisen between control theory and control engineering practice since the late 1950's. This gap may be described as follows.

Kalman's pathbreaking work on the linear quadratic regulator problem in 1960, ushered in the era of state space based optimal control as the design method of choice for linear multivariable feedback control systems. Almost overnight, research on classical control design methods, which were largely of an adhoc nature, were abandoned in favor of the new analytically based optimal designs. This trend has persisted until now with the added element that the H_∞ and l^1 optimal control problems assess performance based on worst case inputs.

Over the last 40 years, many feedback design problems have all been elegantly formulated and solved within the broad framework of *optimal* control theory. In each case the resulting feedback controller is of high dynamic order, typically comparable to that of the plant. From the early days these high order controllers have been justified by control theorists on the grounds that they could be easily implemented using digital computers, since the cost of memory was decreasing, and the speed of computing increasing, at exponential rates.

In the control industry, however, the situation is quite different. The majority of control systems in the world continue to be operated by PID controllers, just as they were in the pre 1960 period. The net effect of computer technology has been that today's PID's are efficiently implemented digitally by microprocessors, A/D and D/A converters, and special purpose signal processing chips. There is a great deal of current research and an extensive literature on the design of PID controllers. However, modern optimal control methods are conspicuously absent from this literature. This is the situation prevalent in most industries including process control, motion control and aerospace industries.

This severe gap is not healthy for theory or for practice and needs to be bridged if control theory is to remain a useful or even relevant engineering discipline. Although it is generally argued that the gap is due to practice lagging theory, our opinion is exactly the opposite. As it turns out despite extensive progress, control theory has been unable to solve the robust or optimal PID control problem in any degree of generality. This means that there are no known procedures for designing PID controllers using the most successful modern theories, namely H_2, H_∞, l^1 or μ optimal control.

In this environment it is indeed unrealistic to expect an industry to overhaul its existing PID hardware in favor of a complex and high order controller. Even if optimality under a worst case scenario is accepted as a worthwhile design goal, unless it is shown that the best performance attainable by the PID controller is indeed inadequate, the argument in favor of a high order optimal controller is questionable. This is the reason that the aforementioned "gap" persists till today, several decades after the initial introduction of optimal control theory.

At this point it is important to clarify, that although modern controls researchers have, by and large, ignored the PID controller, an important exception is the work of a small group of researchers prominent among which is Astrom and his associates. This group and others have developed extensive results and expertise on the theory, autotuning, adaptation and antiwindup techniques for the design of PID controllers, based mostly on experimental methods.

A major obstacle to designing the "best" PID controller, in any sense whatsoever, has been the difficulty in characterizing the entire set of stabilizing PID controllers. Indeed this difficulty holds for every fixed and low order controller. The solution of this problem is a necessary first step to optimal, or at least, rational design based on achievable performance with fixed order and structure, and thus the first step in closing the gap. The absence of results of this type is a significant bottleneck in applying control theory to practical problems.

The main contribution of the present monograph is an effective solution of the above problem, in the case of PID control. Specifically, we give a constructive and computationally efficient characterization of the set of all stabilizing PID controllers for a given but arbitrary, single input single output linear time invariant (LTI) plant. This is accomplished by first generalizing a classical stability result developed in the last century, the Hermite-Biehler Theorem. This generalization is used to reduce the PID stabilization problem to a one parameter family of linear programming problems. Thus a definite yes or no answer to the question of existence of stabilizing PID controllers is given, and in case stabilization is possible, the entire set of stabilizing PID controllers is found. This result is applicable to all LTI plants regardless of order, and regardless of whether it is open loop stable or not.

In general we see, through the various examples presented in the book, that the set of PID controllers stabilizing a given system has a very complicated topology. For example the set is not convex nor is it even connected. This reinforces our belief that optimum design with prescribed structure and order is a very difficult problem. Nevertheless, we believe that without explicit knowledge of the stabilizing set the problem is orders of magnitude more difficult.

Our solution, although numerically based, does expose some hitherto unknown structure; for instance we show that with k_p, the proportional gain, fixed, the set of stabilizing gains in the k_i-k_d plane is a union of convex sets. We give various examples to demonstrate, that with the set of stabilizing PID's in hand we can carry out optimum H_2, H_∞ designs, minimize overshoot or settling time in response to step inputs. Similarly the time delay tolerance and gain and phase margins can be maximized over the stabilizing set. In addition we show how we can robustify the PID controller to plant parameter uncertainty and avoid fragility, that is, extreme sensitivity of the closed loop system stability, to controller parameter perturbations.

The last two aspects are significant since they show that we have 1) extended the capability of the parametric approach to robust control, to the domain of robust synthesis and 2) effectively addressed the issue of fragility, which has recently been proved to be a significant difficulty in many optimal control designs.

The results given here will need to be extended in many directions. In particular, the extension to multivariable systems and the connection with the autotuning results of Astrom and his associates would be useful from the point of view of both theory and applications.

Several people made contributions to this project in many different ways. A.D. and S.P.B. would like to thank R. Kishan Baheti, Director of the Knowledge Modelling and Computational Intelligence Program at the National Science Foundation, for supporting their research program. M.T.H. would like to thank Ming-Yish Chen and Yu-Chuan Lin of Ritek Corporation in Taiwan for supporting his research while he was a researcher there. The authors would also like to thank Guillermo Silva who contributed in a major way towards the writing of Chapters 2 and 7. In fact, the results reported in Chapter 7 were obtained by Guillermo in the course of his doctoral research. Last but not the least, A.D. wouuld like to thank his wife Anindita and his daughters Aparna and Anisha without whose sacrifices this project would not have been possible.

A. Datta
M. T. Ho
S. P. Bhattacharyya

June 23, 1999
College Station, Texas
USA

TABLE OF CONTENTS

CHAPTER 1
OVERVIEW OF CONTROL SYSTEMS

In this chapter we give a quick overview of control theory, explaining why integral feedback control works, describing PID controllers and summarizing modern optimal and robust control theory. This background will serve, as an introduction to control for the nonspecialist and also to motivate our results on PID control, presented in subsequent chapters.

1.1 Introduction to Control

Control theory and control engineering deal with dynamic systems such as aircraft, spacecraft, ships, trains and automobiles, chemical and industrial processes such as distillation columns and rolling mills, electrical systems such as motors, generators and power systems, machines such as numerically controlled lathes and robots. In each case the *setting* of the control problem is:

1. There are certain dependent variables, called *outputs* to be controlled, which must be made to behave in a prescribed way. For instance it may be necessary to *assign* the temperature and pressure at various points in a process, or the position and velocity of a vehicle, or the voltage and frequency in a power system, to given desired fixed values, despite uncontrolled and unknown variations at other points in the system.
2. Certain independent variables called *inputs*, such as voltage applied to the motor terminals, or valve position, are available to regulate and control the behaviour of the system. Other dependent variables, such as position, velocity or temperature are accessible as dynamic *measurements* on the system.
3. There are unknown and unpredictable *disturbances* impacting the system. These could be, for example, the fluctuations of load in a power system, disturbances such as wind gusts acting on a vehicle, external weather conditions acting on an airconditioning plant or the fluctuating load torque on an elevator motor, as passengers enter and exit.
4. The equations describing the plant dynamics, and the parameters contained in these equations, are not known at all or at best known imprecisely. This uncertainty can arise, even when the physical laws and

equations governing a process are known well, for instance, because these equations were obtained by linearizing a nonlinear system about an operating point. As the operating point changes so do the system parameters.

These considerations suggest the following general representation of the *plant*, or system to be controlled:

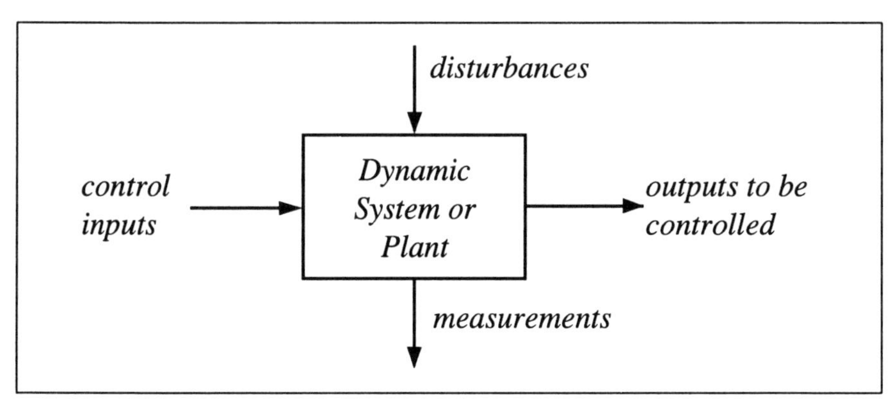

Fig. 1.1. A general plant.

In the above diagram the inputs or outputs shown could actually be representing a vector of signals. In such cases the plant is said to be a *multivariable plant* as opposed to the case where the signals are scalar in which case the plant is said to be a *scalar or monovariable plant.*

Control is exercised by feedback, which means that the corrective control input to the plant is generated by a device which is driven by the available measurements. Thus the controlled system can be represented by the following *feedback* or *closed loop system*:

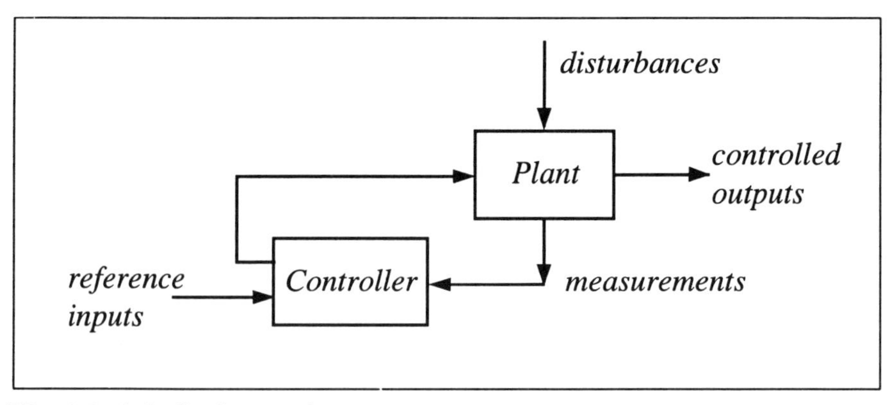

Fig. 1.2. A feedback control system.

The control design problem is to determine the characteristics of the controller so that the controlled outputs can be:

1. Set to prescribed values called references.
2. Maintained at the reference values despite the unknown disturbances.
3. Conditions 1) and 2) are met despite the inherent uncertainties and changes in the plant dynamic characteristics.

The first condition above is called *tracking*, the second, *disturbance rejection* and the third *robustness* of the system. The simultaneous satisfaction of 1),2) and 3) is called *robust tracking and disturbance rejection* and control systems designed to achieve this are called *robust servomechanisms.*

In the next section we discuss how integral and PID control are useful in the design of robust servomechanisms.

1.2 The Magic of Integral Control

Integral control is used almost universally in the control industry to design robust servomechanisms. Integral action is most easily implemented by computer control. It turns out that hydraulic, pneumatic, electronic and mechanical integrators are also commonly used elements in control systems. In this section we explain how integral control works in general to achieve robust tracking and disturbance rejection.

Let us first consider an integrator:

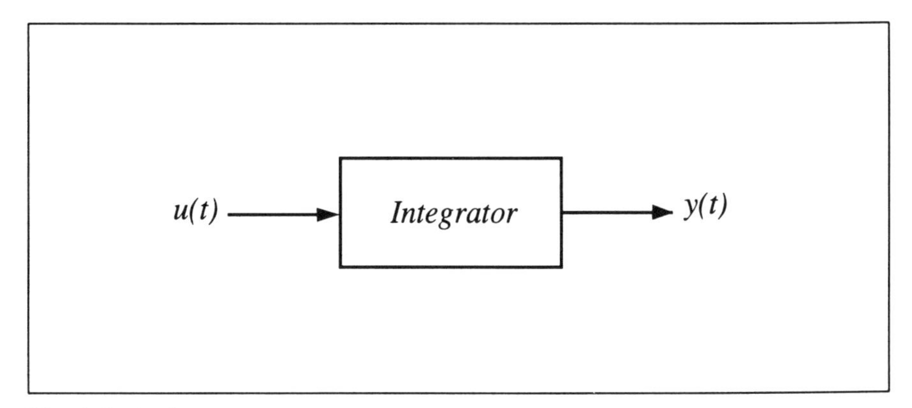

Fig. 1.3. An integrator.

The input-output relationship is:

$$y(t) = K \int_0^t u(\tau) d\tau + y(0) \tag{1.1}$$

or

$$\frac{dy}{dt} = Ku(t) \tag{1.2}$$

where K is the integrator gain.

Now *suppose* that the output $y(t)$ is a *constant*. It follows from eq. (1.2) that

$$\frac{dy}{dt} = 0 = Ku(t) \; \forall \, t > 0. \tag{1.3}$$

Equation 1.3 proves the following important facts about the operation of an integrator:

1) If the output of an integrator is constant over a segment of time, then the input must be identically zero over that same segment.
2) The output of an integrator changes as long as the input is nonzero.

The simple fact stated above suggests how an integrator can be used to solve the servomechanism problem. If a plant output $y(t)$ is to track a constant reference value r, despite the presence of unknown constant disturbances, it is enough to:

a) attach an integrator to the plant and make the error

$$e(t) = r - y(t)$$

the input to the integrator
b) ensure that the closed loop system is asymptotically stable so that under constant reference and disturbance inputs, all signals, including the integrator output, reach constant steady state values.

This is depicted in the block diagram below:

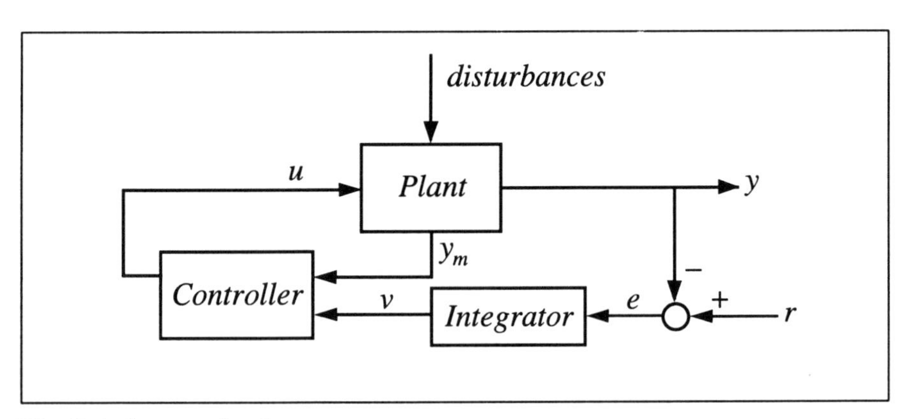

Fig. 1.4. Servomechanism.

If the feedback system shown in the block diagram above is asymptotically stable, and the inputs r and d are constant, it follows that all signals in the closed loop will tend to constant values. In particular the integrator output $v(t)$ tends to a constant value. Therefore by the fundamental fact about the operation of an integrator established above, it follows that the integrator input tends to zero. Since we have arranged that this input is the tracking error it follows that $e(t) = r - y(t)$ tends to zero and hence $y(t)$ tracks r as $t \to \infty$.

We emphasize that the steady state tracking property established above is *very robust*. It holds as long as the closed loop is asymptotically stable and 1) is independent of the particular values of the constant disturbances or references, 2) is independent of the initial conditions of the plant and controller and 3) is independent of whether the plant and controller are linear or nonlinear. Thus the tracking problem is reduced to guaranteeing that stability is assured. In many practical systems stability of the closed loop system can even be ensured without detailed and exact knowledge of the plant characteristics and parameters, and this is known as *robust stability*.

We next discuss how several plant outputs $y_1(t), y_2(t), \cdots, y_m(t)$ can be pinned down to prescribed but arbitrary constant reference values $r_1, r_2, \cdots$, r_m in the presence of unknown but constant disturbances $d_1, d_2, \cdots, d_q$. The previous argument can be extended to this multivariable case by attaching m integrators to the plant, and driving each integrator with its corresponding error input $e_i(t) = r_i - y_i(t), i = 1, \cdots m$. This is shown in the configuration below:

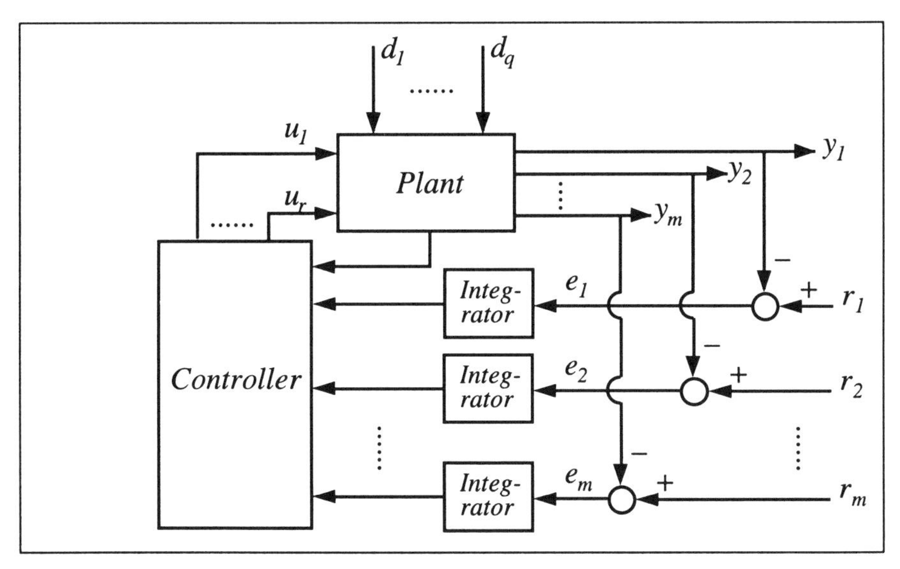

Fig. 1.5. Multivariable servomechanism.

Once again it follows that as long as the closed loop system is stable, all signals in the system must tend to constant values and integral action forces the $e_i(t), i = 1, \cdots, m$ to tend to zero asymptotically, regardless of the actual values of the disturbances $d_j, j = 1, \cdots, q$. The existence of steady state inputs $u_1, u_2, \cdots, u_r$ that make $y_i = r_i, i = 1, \cdots, m$ for arbitrary $r_i, i = 1, \cdots, m$ requires that the plant equations relating $y_i, i = 1, \cdots, m$ to $u_j, j = 1, \cdots, r$ be invertible for constant inputs. In the case of linear time invariant systems this is equivalent to the requirement that the corresponding transfer matrix have rank equal to m at $s = 0$. Sometimes this is restated as the two conditions 1) $r \geq m$ or at least as many control inputs as outputs to be controlled and 2) $G(s)$ has no transmission zero at $s = 0$.

In general, the addition of an integrator to the plant tends to make the system less stable. This is because the integrator is an inherently unstable device; for instance, its response to a step input, a bounded signal, is a ramp, an unbounded signal. Therefore the problem of stabilizing the closed loop becomes a critical issue even when the plant is stable to begin with.

Since integral action and thus the attainment of zero steady state error is *independent* of the particular value of the integrator gain K, we can see that this gain can be used to try to stabilize the system. This single degree of freedom is sometimes insufficient for attaining stability and an acceptable transient response, and additional gains are introduced as explained in the next section. This leads naturally to the proportional integral derivative (PID) controller structure commonly used in industry.

1.3 PID Controllers

In the last section we have seen that when an integrator is part of an asymptotically stable system and constant inputs are applied to the system, the integrator input is forced to become zero. This simple and powerful principle is the basis for the design of linear, nonlinear, single-input single- output and multivariable servomechanisms. All we have to do is: 1) Attach as many integrators as outputs to be regulated 2) drive the integrators with the tracking errors required to be zeroed and 3) stabilize the closed loop system by using any adjustable parameters.

As argued in the last section the input zeroing property is independent of the gain cascaded to the integrator. Therefore this gain can be freely used to attempt to stabilize the closed loop system. Additional free parameters for stabilization can be obtained, without destroying the input zeroing property, by adding parallel branches to the controller, processing in addition to the integral of the error, the error itself and its derivative, when it can be obtained. This leads to the Proportional-Integral-Derivative controller structure shown in Fig. 1.6.

As long as the closed loop is stable it is clear that the input to the integrator will be driven to zero independent of the values of the gains. Thus

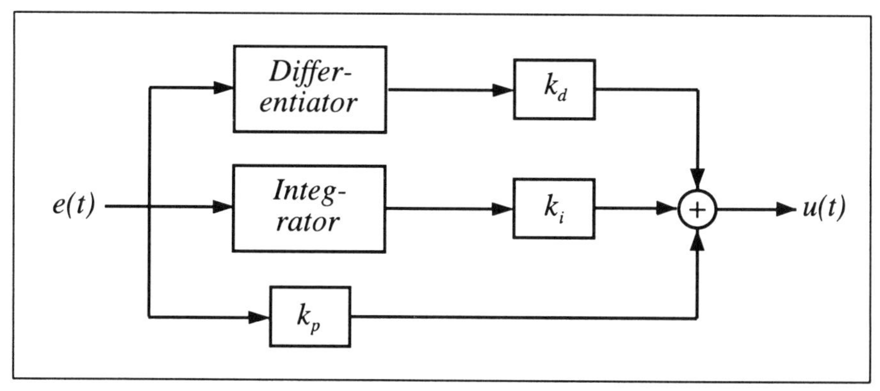

Fig. 1.6. PID controller.

the function of the gains k_p, k_i and k_d is to stabilize the closed loop system if possible, and to adjust the transient response of the system.

In general the derivative can be computed or obtained if the error is varying slowly. Since the response of the derivative to high frequency inputs is much higher than its response to slowly varying signals (see Fig. 1.7), the derivative term is usually omitted, if the error signal is corrupted by high frequency noise.

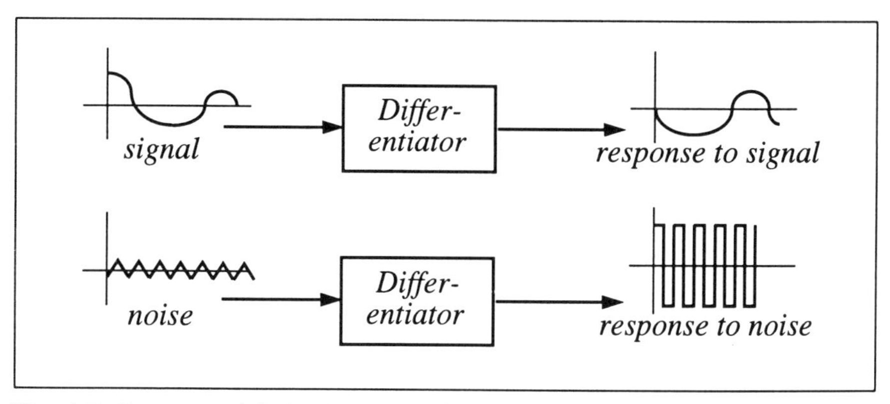

Fig. 1.7. Response of derivative to signal and noise.

In such cases the derivative gain k_d is set to zero or equivalently the differentiator is switched off and the controller is a Proportional-Integral or PI controller. Such controllers are most common in industry.

In subsequent chapters of the book we solve the problem of stabilization of a linear time invariant plant by a PID controller. If fact our solution will uncover the entire set of stabilizing controllers in a computationally efficient way.

In the rest of this introductory chapter we discuss informally the currently available techniques for stabilization. These are based mainly on arbitrary pole assignment or optimal control. These techniques which invariably produce high order controllers are generally inapplicable to the design of PID controllers.

1.4 Feedback Stabilization of Linear Systems

We have seen that steady state tracking and disturbance rejection can be accomplished by attaching integrators to a plant provided the closed loop is stable. Since an integrator is an unstable device the problem of closed loop stability is a nontrivial one even when the open loop plant is stable. In this section we discuss informally the results and techniques available for feedback stabilization of linear time invariant multivariable systems.

1.4.1 The Characteristic Polynomial

For linear time invariant control systems, stability is characterized by the root locations of the closed loop characteristic polynomial. Stability corresponds to these roots lying in the open left half plane (LHP).

Consider the standard feedback control system shown in Figure 1.8. If

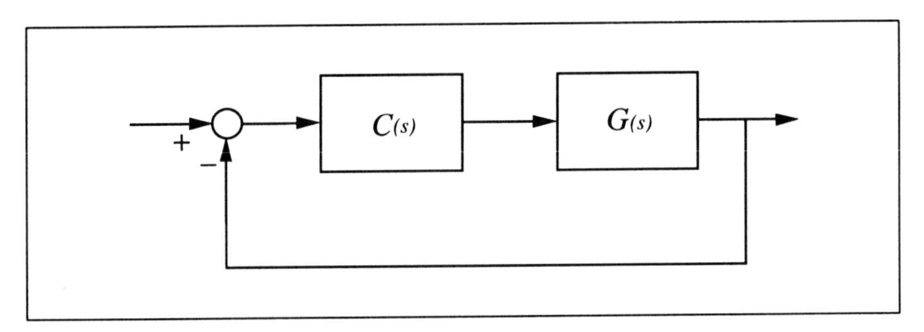

Fig. 1.8. Standard feedback system.

the plant and controller are linear, time invariant dynamic systems they can be described by their respective real rational transfer function matrices $G(s)$ and $C(s)$. We can carry out a polynomial factorization and write

$$C(s) = N_c(s)D_c^{-1}(s) \quad \text{and} \quad G(s) = D_p^{-1}(s)N_p(s)$$

where N_c, D_c, N_p and D_p are polynomial matrices in the complex variable s. The **characteristic polynomial** of the closed loop control system is given by

$$\delta(s) = \det\left[D_p(s)D_c(s) + N_p(s)N_c(s)\right].$$

If a state space model of a strictly proper plant and proper controller are employed, we have, setting the external input to zero, the following set of differential equations describing the system in the time domain:

$$\text{Plant}: \quad \begin{aligned} \dot{x}_p &= A_p x_p + B_p u \\ y &= C_p x_p \end{aligned}$$

$$\text{Controller}: \quad \begin{aligned} \dot{x}_c &= A_c x_c + B_c y \\ u &= C_c x_c + D_c y \end{aligned}$$

$$\text{Closed loop}: \quad \begin{bmatrix} \dot{x}_p \\ \dot{x}_c \end{bmatrix} = \underbrace{\begin{bmatrix} A_p + B_p D_c C_p & B_p C_c \\ B_c C_p & A_c \end{bmatrix}}_{A_{cl}} \begin{bmatrix} x_p \\ x_c \end{bmatrix}$$

and the closed loop characteristic polynomial is given by

$$\delta(s) = \det\left[sI - A_{cl}\right].$$

In the above, the size n of the state vector x_p represents the dynamic order of the plant, the size n_c of x_c represents the dynamic order of the controller and $n + n_c$ is the dynamic order of the closed loop system.

In each of the above cases system stability is equivalent to the condition that the closed loop characteristic polynomial have all its roots in the open left half plane of the complex plane. We describe below some of the available techniques for stabilization.

1.4.2 Stabilization by observer based state feedback

A fundamental result in linear system theory states that for a given pair of real matrices A, B of size $n \times n$ and $n \times r$ respectively, the n eigenvalues of the real matrix $A + BF$ can be arbitrarily assigned by choice of a suitable real matrix F, provided that the pair (A, B) satisfies the *controllability* condition:

$$rank[B, AB, A^2B, \cdots, A^{n-1}B] = n.$$

The dual result is that for a given pair of real matrices C, A of size $m \times n$ and $n \times n$ respectively, the n eigenvalues of the real matrix $A - LC$ can be arbitrarily assigned by choice of a suitable real matrix L provided the pair (C, A) satisfies the *observability* condition:

$$rank[C', A'C', A'^2C', \cdots, A'^{n-1}C'] = n.$$

The above result can be used to easily find a stabilizing compensator of order $n_c = n$ when the plant A_p, B_p, C_p is controllable and observable. The steps to be followed are:

1. Choose a set Λ_1 of n complex numbers in the left half plane and determine F to assign them as eigenvalues of $A_p + B_p F$.
2. Choose another set Λ_2 of n complex numbers in the LHP and determine L to assign them as eigenvalues of $A_p - LC_p$.
3. The stabilizing compensator is given by:

$$\{A_c, B_c, C_c, D_c\} = \{A_p + B_p F - LC_p, L, F, 0\}.$$

4. The closed loop characteristic polynomial can be shown to factor as:

$$det\,[sI - A_{cl}] = det\,[sI - A_p - B_p F]\,det\,[sI - A_p + LC_p]$$

and this proves that the $2n$ characteristic roots of the closed loop system are just the union of the assigned sets Λ_1 and Λ_2.

The implementation of the controller in 3) above is as follows:

$$\text{Controller}: \quad \begin{aligned} \dot{x}_c &= A_p x_c + B_c u + L(y - C_p x_c) \\ u &= F x_c \end{aligned}$$

The computation of F (and similarly L) can be carried out by solving the matrix equations:

$$\begin{aligned} A_p X - X A_1 &= -B_p G \\ G &= FX \end{aligned}$$

with A_1 chosen to have the set Λ_1 as its eigenvalues and G chosen so that G, A_1 is observable. It has been proved in the literature on the above matrix equation that, under these conditions, the matrix X is "almost always" invertible so that the second equation gives a solution F.

1.4.3 Pole placement compensators

In the above approach the compensator that results has dynamic order n and allows us to assign the $2n$ eigenvalues arbitrarily. The same property of arbitrary pole placement can be retained even with compensators of lower order. The first result is that even when the compensator order is $n_c = n - m$ it is possible to assign the $2n - m$ eigenvalues arbitrarily.

The statement of the second result requires the definition of controllability and observability indices. The *controllability index* of a controllable pair A_p, B_p is defined as

$$\mu^* = min\,\{j : \;\; rank[B_p, A_p B_p, A_p^2 B_p, \cdots, A_p^j B_p] = n\}.$$

Similarly the *observability index* of an observable pair C_p, A_p is defined as

$$\nu^* = min\,\{k : \;\; rank[C_p', A_p' C_p', {A_p'}^2 C_p', \cdots, {A_p'}^k C_p'] = n\}.$$

It is known that if the compensator order is taken to be

$$n_c = min\{\mu^*, \nu^*\} \tag{1.4}$$

it is again possible to assign all $n + n_c$ eigenvalues of the closed loop system arbitrarily. This therefore is an upper bound on the order of a stabilizing compensator.

1.4.4 YJBK Parametrization

Another approach to stabilization is via the so-called YJBK parametrization of all rational, proper, stabilizing controllers. Write the proper rational plant transfer function as:

$$G(s) = N(s)D^{-1}(s) = \tilde{D}^{-1}(s)\tilde{N}(s)$$

where $N(s), D(s), \tilde{D}(s), \tilde{N}(s)$ are real, rational, stable, proper (RRSP) matrices with $N(s), D(s)$ being right coprime and $\tilde{D}(s), \tilde{N}(s)$ being left coprime over the ring of RRSP matrices. Let $\tilde{A}(s), \tilde{B}(s)$ be any RRSP matrices satisfying

$$\tilde{N}(s)A(s) + \tilde{D}(s)B(s) = I.$$

The YJBK parametrization states that *every* stabilizing controller for $G(s)$ can be generated by the formula

$$C(s) = (A(s) + D(s)X(s))(B(s) - N(s)X(s))^{-1}$$

by letting $X(s)$ range over all RRSP matrices for which $\det[B(s) - N(s)X(s)] \neq 0$. The utility of this formula is that an optimum controller may now be selected by searching over the set of stable matrices $X(s)$. It is important to point out that, even though *all* stabilizing controllers are described by the YJBK formula, it is in general not possible to use it to extract controllers of a prescribed dynamic order or even to decide about the existence of a stabilizing controller of a prescribed order.

1.5 Optimal Control

A different route to finding a stabilizing controller is to minimize some index of performance. In this case one would like to get stability as an automatic byproduct of optimality.

1.5.1 Linear Quadratic Regulator (LQR)

In the LQR problem the plant dynamics are represented by the state space model

$$\dot{x}(t) = Ax(t) + Bu(t)$$

and the objective of control is to keep x close to zero without excessive control effort. This objective is to be achieved by minimizing the quadratic cost function

$$I = \int_0^\infty (x'(t)Qx(t) + u'(t)Ru(t)dt.$$

The optimal solution is provided by the **state feedback** control

$$u(t) = Fx(t)$$

where the state feedback matrix F is calculated from the algebraic Riccati equation:

$$A'P + PA + Q - PBR^{-1}B'P = 0$$
$$F = -R^{-1}B'P.$$

The optimal state feedback control produced by the LQR problem is guaranteed to be stabilizing for any performance index of the above form, provided only that the pair (Q, A) is detectable and (A, B) is stabilizable. This means that the characteristic roots of the closed loop system which equal the eigenvalues $\sigma(A+BF)$ of $A+BF$ lie in the left half of the complex plane. Indeed, this automatic guarantee of stability is the main attraction of LQR theory, and accounts for its long standing popularity.

In implementations, the state variables, which are generally unavailable for direct measurement, would be substituted by their "estimates" generated by an observer or Kalman filter. This takes the form

$$\dot{\hat{x}}(t) = A\hat{x}(t) + Bu(t) + L(y(t) - C\hat{x}(t))$$

where $\hat{x}(t)$ is the estimate of the state $x(t)$ at time t. From the above equations it follows that

$$(\dot{x} - \dot{\hat{x}})(t) = (A - LC)(x - \hat{x})(t)$$

so that the estimation error converges to zero, regardless of initial conditions and the input $u(t)$, provided that L is chosen so that $A - LC$ has stable eigenvalues.

To close the feedback loop the optimal feedback control $u = Fx$ would be replaced by the **suboptimal** observed state feedback control $\hat{u} = F\hat{x}$. It is easily shown that the resulting closed loop system has characteristic roots which are precisely the eigenvalues of $A + BF$ and those of $A - LC$. This means that the "optimal" eigenvalues were preserved in an output feedback implementation and suggested that the design of the state estimator could be decoupled from that of the optimal controller. This and some related facts regarding the stochastic version of this problem came to be known as the **separation principle**.

1.5.2 H_∞ Optimal Control

In this approach to feedback control, one considers a system subject to controls u and disturbances v as inputs, and with outputs required to be controlled z and available for measurement y, respectively:

$$\begin{aligned} \dot{x} &= Ax + B_1 v + B_2 u \\ z &= C_1 x + D_{11} v + D_{12} u \\ y &= C_2 x + D_{21} v + D_{22} u. \end{aligned}$$

The objective of feedback control is to minimize the effect of the disturbance on the controlled output and this performance specification is quantified by the H_∞ norm of the closed loop disturbance transfer function. Let the size of a vector signal $x(t)$ be measured by the L_2-norm:

$$\|x(t)\|_2^2 = \int_0^\infty |x(t)|^2 dt \tag{1.5}$$

where

$$|x(t)|^2 = \sum_{i=1}^{i=n} x_i^2(t) \tag{1.6}$$

corresponds to the standard Euclidean norm of the vector $x(t)$ at time t. The definition of the H_∞ norm of a transfer function $H(s)$ with input v and output z is:

$$\|H(s)\|_\infty = \sup \left\{ \frac{\|z(t)\|_2}{\|v(t)\|_2} : v(t) \neq 0 \right\}. \tag{1.7}$$

The objective here is to find a dynamic controller so that the H_∞ norm of the disturbance (v) to output (z) transfer function is below a prescribed level $\gamma > 0$. Under some mild technical restrictions on the system, the solution can always be found for a large enough γ and proceeds as follows:

– Let $X \geq 0$ be a solution to the algebraic Ricatti equation

$$A^T X + XA + C_1^T C_1 + X(\gamma^{-2} B_1 B_1^T - B_2 B_2^T)X = 0 \tag{1.8}$$

with the eigenvalues of

$$A + (\gamma^{-2} B_1 B_1^T - B_2 B_2^T)X \tag{1.9}$$

in the left-half plane.

– Let $Y \geq 0$ be a solution to the algebraic Ricatti equation

$$AY + YA^t + B_1 B_1^T + Y(\gamma^{-2} C_1^T C_1 - C_2^T C_2)Y = 0 \tag{1.10}$$

with the eigenvalues of

$$A + Y(\gamma^{-2} C_1^T C_1 - C_2^T C_2) = 0 \tag{1.11}$$

in the left-half plane.

–

$$\rho(XY) < \gamma^2 \tag{1.12}$$

where $\rho(A)$ denotes the spectral radius (magnitude of the eigenvalue with largest magnitude) of the square matrix A

– A controller rendering the H_∞ norm of the transfer function between v and z less than γ is given by

$$\begin{aligned}
\dot{\hat{x}} &= A\hat{x} + B_1\hat{v} + B_2u + L(C_2\hat{x} - y) \\
u &= F\hat{x} \\
F &= -B_2^T X \\
L &= -(I - \gamma^{-2}YX)^{-1}YC_2^T \\
\hat{v} &= \gamma^{-2}B_1^T X\hat{x}.
\end{aligned}$$

We note that this controller, which is known in the literature as the central controller, is of the same form as the Kalman LQ regulator described earlier, with the exception that the input $\hat{v}$ drives the estimator. It can be shown that the input $\hat{v}$ is in fact the worst case exogenous input for the system. Note that the controller order is the same as that of the plant.

1.6 Notes and References

Integral Control has been known in the control field since the 19th century. The LQR problem was introduced by Kalman in 1960 [29]. The upper bound on the order of stabilizing compensators (1.4) was established by Brasch and Pearson [11]. The computation of state feedback using the Sylvester matrix equation was developed by Bhattacharyya and deSouza [5]. The H_∞ optimal controller presented here was derived by Doyle, Glover, Khargonekar and Francis in [17].

CHAPTER 2
SOME CURRENT TECHNIQUES FOR PID CONTROLLER DESIGN

The proportional-integral-derivative (PID) controller structure is the most widely used one in industrial applications. The controller has three tuning parameters which are often tuned by trial and error. Over the past decades, several PID tuning methods have been developed for industrial use. Most of these tuning techniques are based on simple characterizations of *stable* process dynamics, for instance, the characterization by a first order model with deadtime. There is an extensive amount of literature on PID tuning methods. For a comprehensive survey on tuning methods of PID controllers, we refer the reader to [2] and the references therein. In this chapter, we give an introductory discussion of some well-known PID tuning formulas.

2.1 The Ziegler-Nichols Step Response Method

The PID controller we are concerned with is implemented as follows:

$$C(s) = k_p + \frac{k_i}{s} + k_d s \tag{2.1}$$

where k_p is the proportional gain, k_i is the integral gain and k_d is the derivative gain.

The Ziegler-Nichols step response method [48] is an experimental open-loop tuning method and is only applicable to open-loop *stable* plants. This method first characterizes the plant by two parameters A and L obtained from the step response of the plant. A and L can be determined graphically from a measurement of the step response of the plant as illustrated in Fig. 2.1. First, the point on the step response curve with the maximum slope is determined and the tangent is drawn. The intersection of the tangent with the vertical axis gives A, while the intersection of the tangent with the horizontal axis gives L. Once A and L are determined, the PID controller parameters are then given in terms of A and L by the following formulas:

$$\begin{aligned} k_p &= \frac{1.2}{A} \\ k_i &= \frac{0.6}{AL} \end{aligned}$$

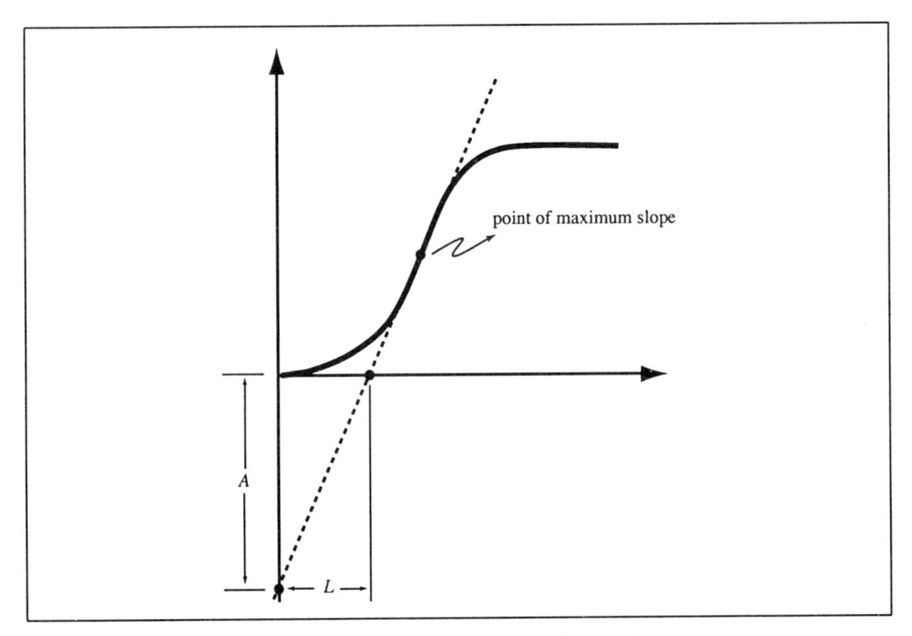

Fig. 2.1. Graphical determination of parameters A and L.

$$k_d = \frac{0.6L}{A}.$$

These formulas for the controller parameters were selected to obtain an amplitude decay ratio of 0.25, which means that the first overshoot decays to $\frac{1}{4}$th of its original value after one oscillation. Intensive experimentation showed that this criterion gives a small settling time.

2.2 The Ziegler-Nichols Frequency Response Method

The Ziegler-Nichols frequency response method is a closed-loop tuning method. This method first determines the point where the Nyquist curve of the plant $G(s)$ intersects the negative real axis. It can be obtained experimentally in the following way: Turn the integral and differential actions off and set the controller to be in the proportional mode only and close the loop as shown in Fig. 2.2. Slowly increase the proportional gain k_p until a periodic oscillation in the output is observed. This critical value of k_p is called the *ultimate gain* (k_u). The resulting period of oscillation is referred to as the *ultimate period* (T_u). Based on k_u and T_u, the Ziegler-Nichols frequency response method gives the following simple formulas for setting PID controller parameters:

$$k_p = 0.6k_u$$

$$\begin{aligned} k_i &= \frac{1.2k_u}{T_u} \\ k_d &= 0.075k_u T_u \, . \end{aligned} \tag{2.2}$$

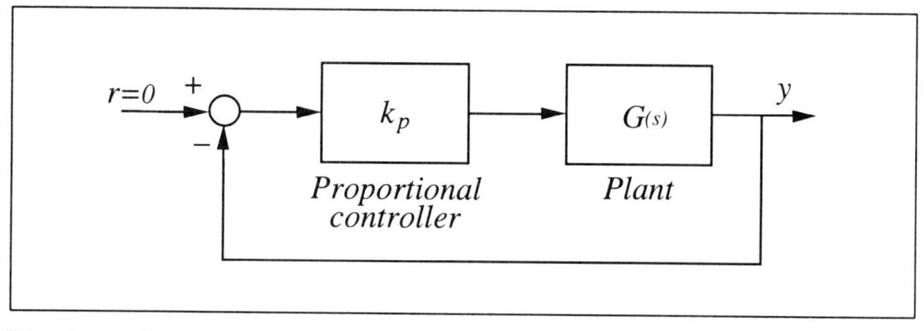

Fig. 2.2. The closed-loop system with the proportional controller.

This method can be interpreted in terms of the Nyquist plot. Using PID control it is possible to move a given point on the Nyquist curve to an arbitrary position in the complex plane. Now, the first step in the frequency response method is to determine the point $(-\frac{1}{k_u},0)$ where the Nyquist curve of the open-loop transfer function intersects the negative real axis. We will study how this point is changed by the PID controller. Using (2.2) in (2.1), the frequency response of the controller at the ultimate frequency w_u is

$$\begin{aligned} C(jw_u) &= 0.6k_u - j\left(\frac{1.2k_u}{T_u w_u}\right) + j(0.075k_u T_u w_u) \\ &= 0.6k_u(1 + j0.4671) \ \text{[since } T_u w_u = 2\pi] \, . \end{aligned}$$

From this we see that the controller gives a phase advance of 25 degrees at the ultimate frequency. The loop transfer function is then

$$G_{loop}(jw_u) = G(jw_u)C(jw_u) = -0.6(1 + j0.4671) = -0.6 - j0.28 \, .$$

Thus the point $(-\frac{1}{k_u},0)$ is moved to the point (-0.6,-0.28). The distance from this point to the critical point is almost 0.5. This means that the frequency response method gives a sensitivity greater than 2.

The procedure described above for measuring the ultimate gain and ultimate period requires that the closed-loop system has to be operated close to instability. To avoid damaging the physical system, this procedure needs to be executed carefully. Without bringing the system to the verge of instability, an alternative method was proposed in [1] using relay feedback to generate a relay oscillation for measuring the ultimate gain and ultimate period. This is

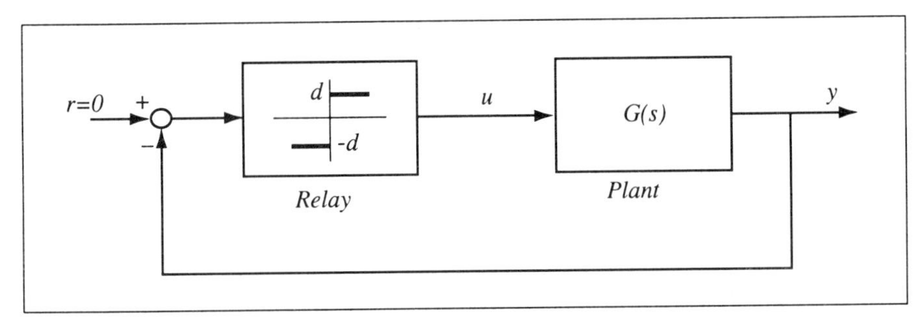

Fig. 2.3. Block diagram of relay feedback.

done by using the relay feedback configuration [2] shown in Fig. 2.3. In Fig. 2.3, the relay is adjusted to induce a self sustaining oscillation in the loop.

Now we explain why this relay feedback can be used to determine the ultimate gain and ultimate period. The relay block is a nonlinear element that can be represented by a *describing function* [31]. This describing function is obtained by applying a sinusoidal signal $asin(wt)$ at the input of the nonlinearity and calculating the ratio of the Fourier coefficient of the first harmonic at the output to a. This function can be thought of as an equivalent gain of the nonlinear system. For the case of the relay its describing function is given by

$$N(a) = \frac{4d}{a\pi}$$

where a is the amplitude of the sinusoidal output signal and d is the relay amplitude. The conditions for the presence of limit cycle oscillations can be derived by investigating the propagation of a sinusoidal signal around the loop. Since the plant $G(s)$ acts as a low pass filter, the higher harmonics produced by the nonlinear relay will be attenuated at the output of the plant. Hence, the condition for oscillation is that the fundamental sine waveform comes back with the same amplitude and phase after traversing through the loop. This means that for sustained oscillations at a frequency of ω, we must have

$$G(j\omega)N(a) = -1 \,. \tag{2.3}$$

This equation can be solved by plotting the Nyquist plot of $G(s)$ and the line $-\frac{1}{N(a)}$. As shown in Fig. 2.4, the plot of $-\frac{1}{N(a)}$ is the negative real axis, so the solution to (2.3) is given by the two conditions:

$$\begin{aligned} |G(j\omega_u)| &= \frac{a\pi}{4d} \\ &\triangleq \frac{1}{k_u} \qquad (2.4) \\ \text{and } arg\, G(j\omega_u) &= -\pi \,. \qquad (2.5) \end{aligned}$$

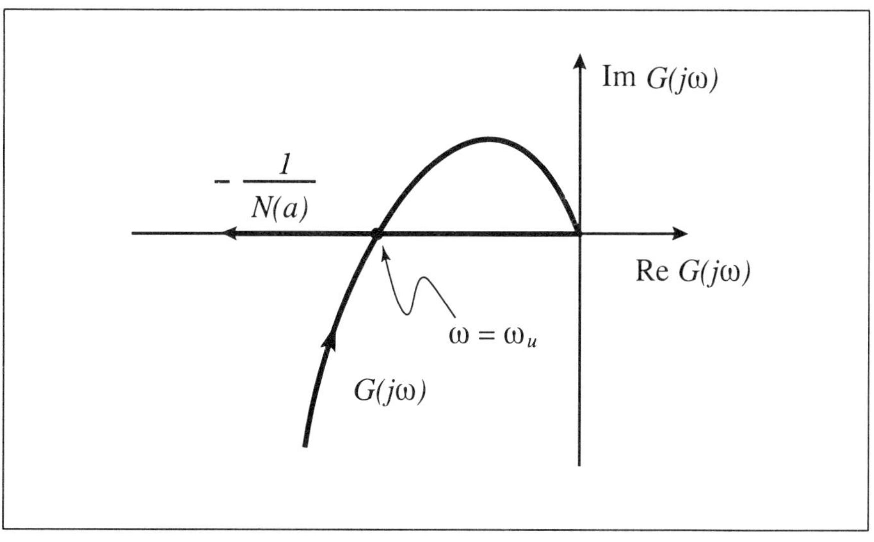

Fig. 2.4. Nyquist plots of the plant $G(j\omega)$ and the describing function $-\frac{1}{N(a)}$.

The ultimate gain and ultimate period can now be determined by measuring the amplitude and period of the oscillations. This relay feedback technique is widely used in automatic PID tuning [2, 1].

Remark 2.2.1. Both Ziegler-Nichols tuning methods require very little knowledge of the plants and also simple formulas are given for controller parameter settings. These formulas are obtained by extensive simulations of many stable and simple plants. The main design criterion of these methods is to obtain a quarter amplitude decay ratio for the load disturbance response. As pointed out in [2], little emphasis is given to measurement noise, sensitivity to process variations and setpoint response. Even though these methods provide good rejection of load disturbance, the resulting closed-loop system is poorly damped and has poor stability margins.

2.3 PID Settings using the Internal Model Controller Design Technique

The internal model controller (IMC) structure has become a popular one in process control applications [38]. This structure, in which the controller includes an explicit model of the plant, is particularly appropriate for the design and implementation of controllers for open-loop stable systems. The fact that many of the plants encountered in process control happen to be open-loop stable possibly accounts for the popularity of IMC among practicing engineers. In this section, we consider the IMC configuration for a stable plant $G(s)$ as shown in Fig.2.5. The IMC controller consists of a stable "IMC parameter" $Q(s)$ and a model of the plant $\hat{G}(s)$ which is usually referred to as the "internal model". $F(s)$ is the IMC filter [38] which is chosen to enhance robustness with respect to the modeling error and to make the overall IMC parameter $Q(s)F(s)$ proper. From Fig.2.5 the equivalent feedback controller $C(s)$ is :

$$C(s) = \frac{F(s)Q(s)}{1 - F(s)Q(s)\hat{G}(s)} .$$

The IMC design objective considered in this section is to choose $Q(s)$ which

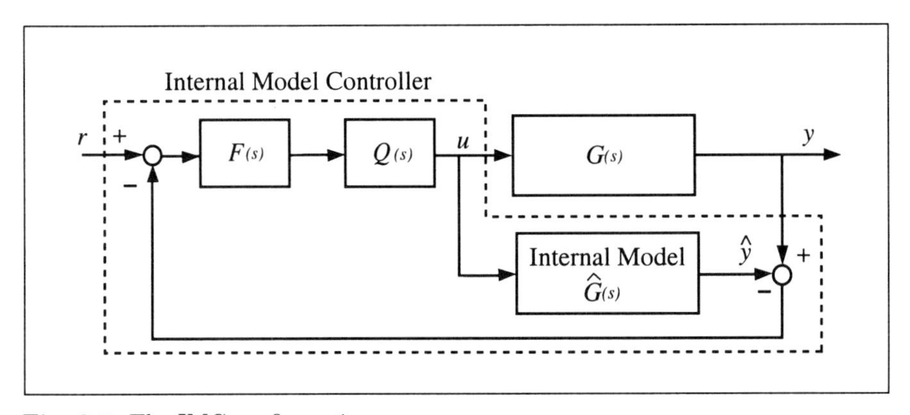

Fig. 2.5. The IMC configuration.

minimizes the L_2 norm of the tracking error $r - y$, *i.e.*, achieves an H_2-optimal control design. In general, complex models lead to complex IMC H_2-optimal controllers. However, in [38] it has been shown that, for first order plants with deadtime and a step command signal, the IMC H_2-optimal design results in a controller with a PID structure. This will be clearly borne out by the following discussion.

Assume that the plant to be controlled is a first-order model with deadtime:

$$G(s) = \frac{k}{1 + Ts} e^{-Ls} .$$

The control objective is to minimize the L_2 norm of the tracking error due to setpoint changes. Using Parseval's Theorem, this is equivalent to choosing $Q(s)$ for which $\min \|[1 - \hat{G}(s)Q(s)]R(s)\|_2$ is achieved, where $R(s) = \frac{1}{s}$ is the Laplace transform of the unit step command.

Approximating the deadtime with a first-order Padé approximation, we have

$$e^{-Ls} \cong \frac{1 - \frac{L}{2}s}{1 + \frac{L}{2}s}.$$

The resulting rational transfer function of the internal model $\hat{G}(s)$ is given by

$$\hat{G}(s) = \frac{k}{(1+Ts)} \frac{1 - \frac{L}{2}s}{1 + \frac{L}{2}s}.$$

Choosing $Q(s)$ to minimize the H_2 norm of $[1-\hat{G}(s)Q(s)]R(s)$ [38], we obtain

$$Q(s) = \frac{1+Ts}{k}.$$

Since this $Q(s)$ is improper, we choose

$$F(s) = \frac{1}{1+\lambda s}$$

where $\lambda > 0$ is a small number. The equivalent feedback controller becomes

$$\begin{aligned} C(s) &= \frac{F(s)Q(s)}{1 - F(s)Q(s)\hat{G}(s)} \\ &= \frac{(1+Ts)(1+\frac{L}{2}s)}{ks(L+\lambda+\frac{L\lambda}{2}s)} \\ &\cong \frac{(1+Ts)(1+\frac{L}{2}s)}{ks(L+\lambda)}. \end{aligned} \tag{2.6}$$

From (2.6), we can extract the following parameters for a standard PID controller:

$$\begin{aligned} k_p &= \frac{2T+L}{2k(L+\lambda)} \\ k_i &= \frac{1}{k(L+\lambda)} \\ k_d &= \frac{TL}{2k(L+\lambda)}. \end{aligned}$$

Since a first-order Padé *approximation* was used for the time delay, ensuring the robustness of the design to modelling errors is all the more important. This can be done by properly selecting the design variable λ to achieve the appropriate compromise between performance and robustness. As discussed in [38], a suitable choice for λ is $\lambda > 0.2T$ and $\lambda > 0.25L$.

Remark 2.3.1. The IMC PID design procedure minimizes the L_2 norm of the tracking error due to setpoint changes. Therefore, as expected, this design method gives good response to setpoint changes. However, for lag dominant plants the method gives poor load disturbance response because of the pole-zero cancellation inherent in the design methodology.

2.4 Dominant Pole Design: The Cohen-Coon Method

Dominant pole design attempts to position a few poles to achieve certain control performance specifications. The Cohen-Coon method [13] is a dominant pole design method. This tuning method is based on the first order plant model with deadtime:

$$G(s) = \frac{k}{1+Ts}e^{-Ls}.$$

The key feature of this tuning method is to attempt to locate three dominant poles, a pair of complex poles and one real pole, such that the amplitude decay ratio for load disturbance response is 0.25 and the integrated error $\int_0^\infty e(t)dt$ is minimized. Thus, the Cohen-Coon method gives good load disturbance rejection. Based on analytical and numerical computation, Cohen and Coon gave the following PID controller parameters in terms of k, T, and L:

$$\begin{aligned} k_p &= \frac{1.35(1-0.82b)}{a(1-b)} \\ k_i &= \frac{1.35(1-0.82b)(1-0.39b)}{aL(1-b)(2.5-2b)} \\ k_d &= \frac{1.35L(0.37-0.37b)}{a(1-b)} \end{aligned}$$

where

$$\begin{aligned} a &= \frac{kL}{T} \\ b &= \frac{L}{L+T}. \end{aligned}$$

Note that for small b, the controller parameters given by the above formulas are close to the parameters obtained by the Ziegler-Nichols step response method.

2.5 New Tuning Approaches

The tuning methods described in the previous sections are easy to use and require very little information about the plant to be controlled. However, since they do not capture all aspects of desirable PID performance, many other new approaches have been developed. These methods can be classified into three categories [20].

2.5.1 Time Domain Optimization Methods

The idea behind these methods is to choose the PID controller parameters to minimize an integral cost functional. Zhuang and Atherton [47] used an integral criterion with data from a relay experiment. The time weighted system error integral criterion was chosen as:

$$J_n(\theta) = \int_0^\infty t^n e(\theta, t)^2 dt$$

where θ is a vector containing the controller parameters and $e(\theta, t)$ represents the error signal. Experimentation showed that for $n = 1$, the controller obtained produced a step response of desirable form. This gave birth to the Integral Square Time Error (ISTE) criterion. Another contribution is due to Pessen [41] who used the Integral Absolute Error (IAE) criterion:

$$J(\theta) = \int_0^\infty |e(\theta, t)| dt \, .$$

In order to minimize the above integral cost functions, Parseval's theorem can be invoked to express the time functions in terms of their Laplace transforms. Definite integrals of the form encountered in this approach have been evaluated in terms of the coefficients of the numerator and denominator of the Laplace transforms (see [39]). Once the integration is carried out, the parameters of the PID controller are adjusted in such a way as to minimize the integral cost function. Recently Atherton and Majhi [3] proposed a modified form of the PID controller (see Fig. 2.6). In this structure an internal PD feedback is used to change the poles of the plant transfer function to more desirable locations and then a PI controller is used in the forward loop. The parameters of the controller are obtained by minimization of the ISTE criterion.

2.5.2 Frequency Domain Shaping

These methods seek a set of controller parameters that give a desired frequency response. Astrom [2] proposed the idea of using a set of rules to achieve a desired phase margin specification. In the same spirit, Ho et al [28] developed a PID self-tuning method with specifications on the gain and phase margins. Another contribution is due to Voda and Landau [45] who presented a method to shape the compensated system frequency response.

2.5.3 Optimal Control Methods

This new trend has been motivated by the desire to incorporate several control system performance objectives such as reference tracking, disturbance

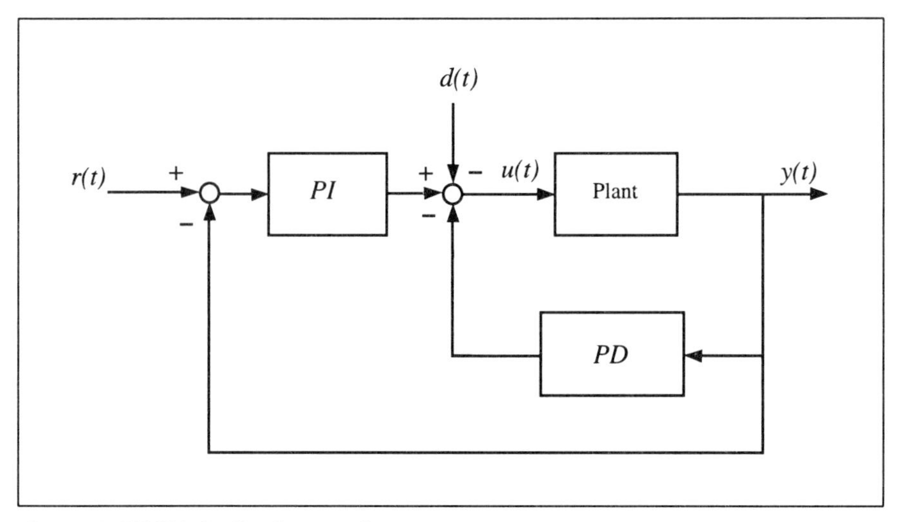

Fig. 2.6. PI-PD feedback control structure.

rejection and measurement noise rejection. Grimble and Johnson [20] incorporated all these objectives into an LQG optimal control problem. They proposed an algorithm to minimize an LQG-cost function where the controller structure is fixed to a particular PID industrial form. In a similar fashion, Panagopoulos et al [40] presented a method to design PID controllers that captures demands on load disturbance rejection, set point response, measurement noise and model uncertainty. Good load disturbance rejection was obtained by minimization of the integral control error. Good set point response was obtained by using a structure with two degrees of freedom. Measurement noise was dealt with by filtering. Robustness was achieved by requiring a maximum sensitivity of less than a specified value.

2.6 Contribution of this Book

In concluding this chapter, it is important to point out that in addition to the approaches discussed above, there are many other approaches for tuning PID controllers [2]. Despite this, for plants having order higher than two, *there is no approach that can be used to determine the set of all stabilizing PID gain values.* The principal contribution of this book to the PID literature is the development of a methodology that provides a complete answer to this longstanding open problem.

CHAPTER 3
THE HERMITE-BIEHLER THEOREM AND ITS GENERALIZATION

The classical Hermite-Biehler Theorem and our generalization of it will be described in this chapter. These results will be crucial in our characterization of stabilizing PID controllers.

3.1 Introduction

The problem of ascertaining the stability of a linear time invariant system reduces to determining conditions under which all of the roots of a given real polynomial lie in the open left-half of the complex plane. A polynomial satisfying this condition is said to be Hurwitz. This problem has intrigued researchers for more than a hundred years now and one of the earliest solutions, and the most widely known one, is the criterion of Routh-Hurwitz. There are several other equivalent conditions for ascertaining the Hurwitz stability of a given real polynomial. Of these, the classical Hermite-Biehler Theorem has recently been instrumental in studying the parametric robust stability problem, *i.e.*, the problem of guaranteeing that the roots of a given Hurwitz polynomial continue to lie in the left-half plane under real coefficient perturbations.

The Hermite-Biehler Theorem states that a given real polynomial is Hurwitz iff it satisfies a certain interlacing property. When a given real polynomial is not Hurwitz, the Hermite-Biehler Theorem, as currently known, provides absolutely no information about its root distribution. In this chapter, we present generalizations of the Hermite-Biehler Theorem to real polynomials which are not necessarily Hurwitz stable. As will be seen later on in this monograph, these generalizations are not only of mathematical interest but have practical implications in control theory.

The chapter is organized as follows. In Section 3.2, we provide a statement and proof of the Hermite-Biehler Theorem as well as some equivalent characterizations. In Section 3.3, we state the relationship between the net phase change of the "frequency response" of a real polynomial as the frequency ω varies from 0 to ∞ and the numbers of its roots in the open left-half and open right-half planes. In Section 3.4, we introduce two versions of a formula to determine the signature associated with a real polynomial. The next three sections are devoted to developing generalizations of the Hermite-Biehler Theorem under progressively less restrictive conditions. First, in Section 3.5.1, we derive generalizations of the Hermite-Biehler Theorem applicable to the case where the test polynomial has no zeros on the imaginary axis. Thereafter, in Section 3.5.2, we show that the identical Theorem statement can accomodate the presence of imaginary axis zeros, provided they are not at the origin. Finally, in Section 3.5.3, we show that the presence of one or more zeros at the origin can also be handled. In Section 3.6, as an application of the Generalized Hermite-Biehler Theorems, we present an elementary derivation of the Routh-Hurwitz criterion.

3.2 The Hermite-Biehler Theorem for Hurwitz Polynomials

In this section, we first state the Hermite-Biehler Theorem which provides necessary and sufficient conditions for the Hurwitz stability of a given real polynomial. To do so, we establish some notation.

Definition 3.2.1. *Let $\delta(s) = \delta_0 + \delta_1 s + \cdots + \delta_n s^n$ be a given real polynomial of degree n. Write*

$$\delta(s) = \delta_e(s^2) + s\delta_o(s^2)$$

where $\delta_e(s^2)$, $s\delta_o(s^2)$ are the components of $\delta(s)$ made up of even and odd powers of s respectively. For every frequency $\omega \in \mathcal{R}$, denote

$$\delta(j\omega) = p(\omega) + jq(\omega)$$

where $p(\omega) = \delta_e(-\omega^2)$, $q(\omega) = \omega\delta_o(-\omega^2)$. Let $\omega_{e_1}, \omega_{e_2}, \cdots$ denote the non-negative real zeros of $\delta_e(-\omega^2)$ and let $\omega_{o_1}, \omega_{o_2}, \cdots$ denote the non-negative real zeros of $\delta_o(-\omega^2)$, both arranged in ascending order of magnitude.

Theorem 3.2.1. *(Hermite-Biehler Theorem): Let $\delta(s) = \delta_0 + \delta_1 s + \cdots + \delta_n s^n$ be a given real polynomial of degree n. Then $\delta(s)$ is Hurwitz stable if and only if all the zeros of $\delta_e(-\omega^2)$, $\delta_o(-\omega^2)$ are real and distinct, δ_n and δ_{n-1} are of the same sign, and the non-negative real zeros satisfy the following interlacing property*

$$0 < \omega_{e_1} < \omega_{o_1} < \omega_{e_2} < \omega_{o_2} < \cdots \tag{3.1}$$

In this chapter, our objective is to obtain generalizations of the above theorem for real polynomials that are not necessarily Hurwitz. To clearly understand what it is that we are trying to generalize, we provide below some alternative characterizations and interpretations of the Hermite-Biehler Theorem. To do so, we first introduce the standard signum function sgn : $\mathcal{R} \rightarrow \{-1, 0, 1\}$ defined by

$$sgn[x] = \begin{cases} -1 \text{ if } x < 0 \\ 0 \text{ if } x = 0 \\ 1 \text{ if } x > 0 \end{cases}$$

Lemma 3.2.1. *Let $\delta(s) = \delta_0 + \delta_1 s + \cdots + \delta_n s^n$ be a given real polynomial of degree n. Then the following conditions are equivalent:*

(i) $\delta(s)$ is Hurwitz stable.
(ii) δ_n and δ_{n-1} are of the same sign and

$$n = \begin{cases} sgn[\delta_0] \cdot \{sgn[p(0)] - 2sgn[p(\omega_{o_1})] + 2sgn[p(\omega_{o_2})] + \cdots \\ +(-1)^{m-1} \cdot 2sgn[p(\omega_{o_{m-1}})] + (-1)^m \cdot sgn[p(\infty)]\}, \\ \quad \text{for } n = 2m \\ \\ sgn[\delta_0] \cdot \{sgn[p(0)] - 2sgn[p(\omega_{o_1})] + 2sgn[p(\omega_{o_2})] + \cdots \\ +(-1)^{m-1} \cdot 2sgn[p(\omega_{o_{m-1}})] + (-1)^m \cdot 2sgn[p(\omega_{o_m})]\}, \\ \quad \text{for } n = 2m+1 \end{cases} \tag{3.2}$$

(iii) δ_n and δ_{n-1} are of the same sign and

$$n = \begin{cases} sgn[\delta_0] \cdot \{2sgn[q(\omega_{e_1})] - 2sgn[q(\omega_{e_2})] + 2sgn[q(\omega_{e_3})] + \cdots \\ +(-1)^{m-2} \cdot 2sgn[q(\omega_{e_{m-1}})] + (-1)^{m-1} \cdot 2sgn[q(\omega_{e_m})]\}, \\ \quad \text{for } n = 2m \\ \\ sgn[\delta_0] \cdot \{2sgn[q(\omega_{e_1})] - 2sgn[q(\omega_{e_2})] + 2sgn[q(\omega_{e_3})] + \cdots \\ +(-1)^{m-1} \cdot 2sgn[q(\omega_{e_m})] + (-1)^m \cdot sgn[q(\infty)]\}, \\ \quad \text{for } n = 2m+1 \end{cases} \tag{3.3}$$

Proof. (1) $(i) \Leftrightarrow (ii)$

We first show that $(i) \Rightarrow (ii)$
Now a Hurwitz polynomial $\delta(s)$ has the *monotonic phase property*, *i.e.*, the phase of $\delta(j\omega)$ increases monotonically as ω increases from $-\infty$ to ∞. Using this property, we can show that the plot of $\delta(j\omega) = p(\omega) + jq(\omega)$ must move strictly counterclockwise and go through n quadrants in turn as ω increases from 0 to ∞ [7]. For Hurwitz $\delta(s)$, the admissible plots of $\delta(j\omega)$ are illustrated in Fig. 3.1. From Fig. 3.1, it is clear that

$$\begin{cases} \text{for } n = 2m \\ \text{sgn}[\delta_0] \cdot \text{sgn}[p(0)] > 0 \\ -\text{sgn}[\delta_0] \cdot \text{sgn}[p(\omega_{o_1})] > 0 \\ \quad \vdots \\ (-1)^{m-1}\text{sgn}[\delta_0] \cdot \text{sgn}[p(\omega_{o_{m-1}})] > 0 \\ (-1)^m \text{sgn}[\delta_0] \cdot \text{sgn}[p(\infty)] > 0 \end{cases} \tag{3.4}$$

and

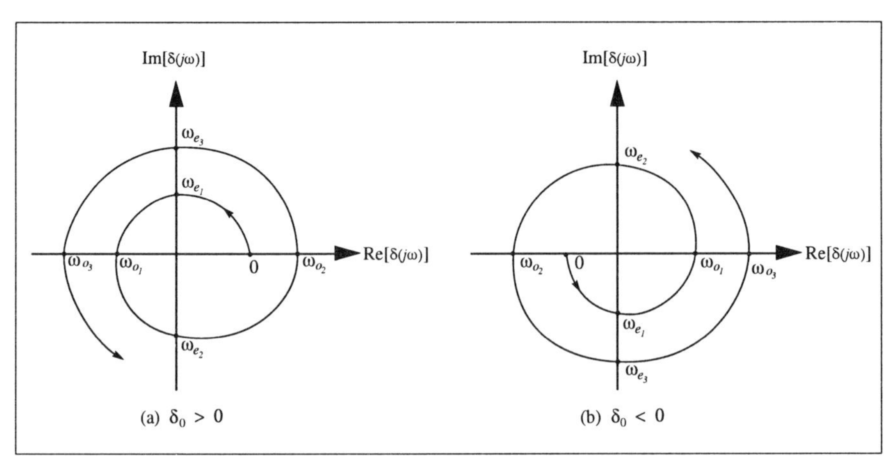

Fig. 3.1. Monotonic phase increase property for Hurwitz polynomials $\delta(s)$.

$$\begin{cases} \text{for } n = 2m+1 \\ \text{sgn}[\delta_0] \cdot \text{sgn}[p(0)] > 0 \\ -\text{sgn}[\delta_0] \cdot \text{sgn}[p(\omega_{o_1})] > 0 \\ \quad \vdots \\ (-1)^{m-1}\text{sgn}[\delta_0] \cdot \text{sgn}[p(\omega_{o_{m-1}})] > 0 \\ (-1)^{m}\text{sgn}[\delta_0] \cdot \text{sgn}[p(\omega_{o_m})] > 0 \end{cases} \tag{3.5}$$

From (3.4) and (3.5), it follows that (3.2) holds.

(ii) $\Rightarrow$ *(i)*
Let $\omega_{o_0} = 0$ and for $n = 2m$, define $\omega_{o_m} = \infty$. Equation 3.2 holds if and only if $[p(\omega_{o_{l-1}})]$ and $[p(\omega_{o_l})]$ are of opposite signs for $l = 1, 2, \cdots, m$. By the continuity of $p(\omega)$, there exists at least one $\omega_e \in \mathcal{R}$, $\omega_{o_{l-1}} < \omega_e < \omega_{o_l}$ such that $p(\omega_e) = 0$. Moreover, since the maximum possible number of non-negative real roots of $p(\cdot)$ is m, it follows that there exists one and only one $\omega_e \in (\omega_{o_{l-1}}, \omega_{o_l})$ such that $p(\omega_e) = 0$, thereby leading us to the interlacing property (3.1).

(2) *(i)* $\Leftrightarrow$ *(iii)*
The proof of (2) follows along the same lines as that of (1). ♣

The interlacing property in Theorem 3.2.1 gives a graphical interpretation of the Hermite-Biehler Theorem while Lemma 3.2.1 gives an equivalent analytical characterization. Note that from Lemma 3.2.1 if $\delta(s)$ is Hurwitz stable then all the zeros of $p(\omega)$ and $q(\omega)$ must be real and distinct, otherwise (3.2) and (3.3) will fail to hold. Furthermore, the signs of $p(\omega)$ at the successive zero crossings of $q(\omega)$ must alternate. This is also true for the signs of $q(\omega)$ at the successive zero crossings of $p(\omega)$.

We now present an example to illustrate the application of Theorem 3.2.1 and Lemma 3.2.1 to verify the interlacing property.

Example 3.2.1. Consider the real polynomial

$$\delta(s) = s^7 + 5s^6 + 14s^5 + 25s^4 + 31s^3 + 26s^2 + 14s + 4$$

Then

$$\delta(j\omega) = p(\omega) + jq(\omega)$$

where

$$\begin{aligned} p(\omega) &= -5\omega^6 + 25\omega^4 - 26\omega^2 + 4 \\ q(\omega) &= \omega(-\omega^6 + 14\omega^4 - 31\omega^2 + 14) \end{aligned}$$

The plots of $p(\omega)$ and $q(\omega)$ are shown in Fig. 3.2. They show that the polynomial $\delta(s)$ satisfies the interlacing property.

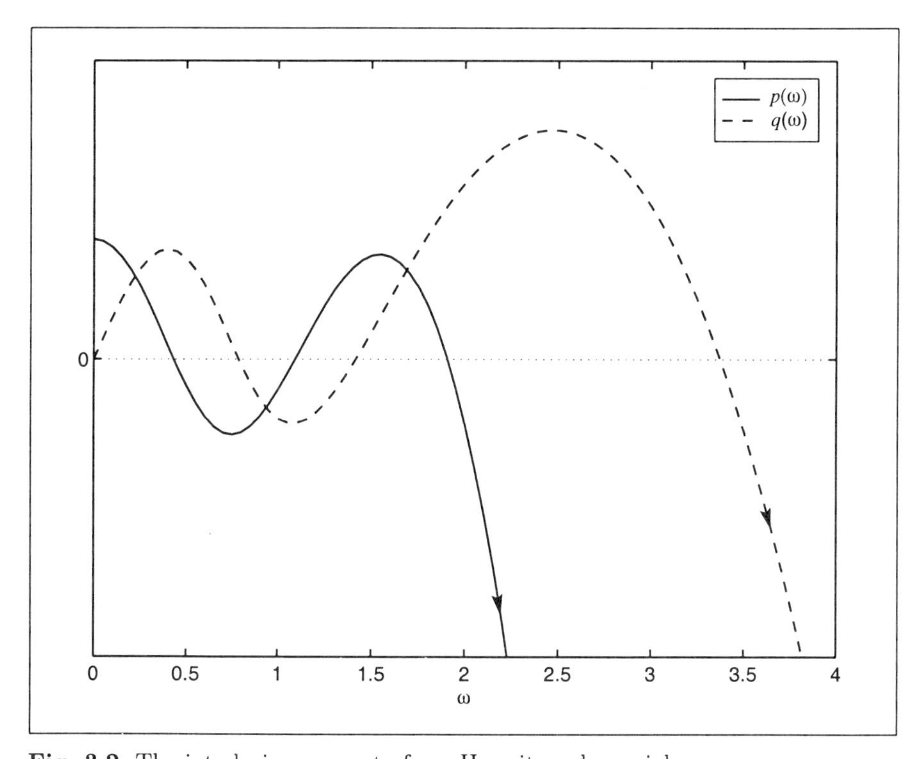

Fig. 3.2. The interlacing property for a Hurwitz polynomial.

Also

$$\omega_{e_1} = 0.43106,\ \omega_{e_2} = 1.08950,\ \omega_{e3} = 1.90452$$
$$\omega_{o_1} = 0.78411,\ \omega_{o_2} = 1.41421,\ \omega_{o_3} = 3.37419$$

and

$$\mathrm{sgn}[p(0)] = 1,\ \mathrm{sgn}[p(\omega_{o_1})] = -1,\ \mathrm{sgn}[p(\omega_{o_2})] = 1,\ \mathrm{sgn}[p(\omega_{o_3})] = -1.$$

Now $\delta(s)$ is of degree $n = 7$ which is odd and

$$\mathrm{sgn}[\delta_0] \cdot [\mathrm{sgn}[p(0)] - 2\mathrm{sgn}[p(\omega_{o_1})] + 2\mathrm{sgn}[p(\omega_{o_2})] - 2\mathrm{sgn}[p(\omega_{o_3})]] = 7$$

which shows that (3.2) holds.

Also, we have

$$\mathrm{sgn}[q(\omega_{e_1})] = 1,\ \mathrm{sgn}[q(\omega_{e_2})] = -1,\ \mathrm{sgn}[q(\omega_{e_3})] = 1,\ \mathrm{sgn}[q(\infty)] = -1$$

so that

$$\mathrm{sgn}[\delta_0] \cdot [2\mathrm{sgn}[q(\omega_{e_1})] - 2\mathrm{sgn}[q(\omega_{e_2})] + 2\mathrm{sgn}[q(\omega_{e_3})] - \mathrm{sgn}[q(\infty)]] = 7.$$

Once again, this checks with (3.3).

To verify that $\delta(s)$ is indeed a Hurwitz polynomial, we solve for the roots of $\delta(s)$:

$$\begin{array}{cc} -0.5 \pm 1.3229j & -0.5 \pm 0.8660j \\ -1 \pm j & -1 \end{array}$$

We see that all the roots of $\delta(s)$ are in the left-half-plane so that $\delta(s)$ is Hurwitz.

Now consider $\delta(j\omega) = p(\omega) + jq(\omega)$ where $p(\omega)$ and $q(\omega)$ are as illustrated in Fig. 3.3. From Fig. 3.3, we know that the polynomial $\delta(s)$ is not a Hurwitz polynomial because it fails to satisfy the interlacing property since, for instance, ω_{e_1}, ω_{e_2}, ω_{e_3} are three successive roots of $p(\omega)$ between 0 and ω_{o_1}; similarly, ω_{e_4}, ω_{e_5} are two successive roots of $p(\omega)$ between ω_{o_1} and ω_{o_2}; and ω_{o_2}, ω_{o_3} are two successive roots of $q(\omega)$ between ω_{e_5} and ω_{e_6}. However, it is logical to ask: does Fig. 3.3 provide us with any more information about $\delta(s)$, beyond whether or not it is Hurwitz? As it turns out it is possible to know the number of right-half plane roots of $\delta(s)$ from the above graph. This motivates us to derive generalized versions of the Hermite-Biehler Theorem for not necessarily Hurwitz polynomials. Such a generalization is also needed for solving the fixed order stabilization problem, as will be clearly borne out by the following discussion.

Motivation for generalizing the Hermite-Bieler Theorem

Consider the constant gain stabilization problem shown in Fig. 3.4. Here r is the command signal, y is the output, $G(s) = \frac{N(s)}{D(s)}$ is the plant to be controlled, $N(s)$ and $D(s)$ are coprime polynomials, and $C(s) = k$, *i.e.*, the

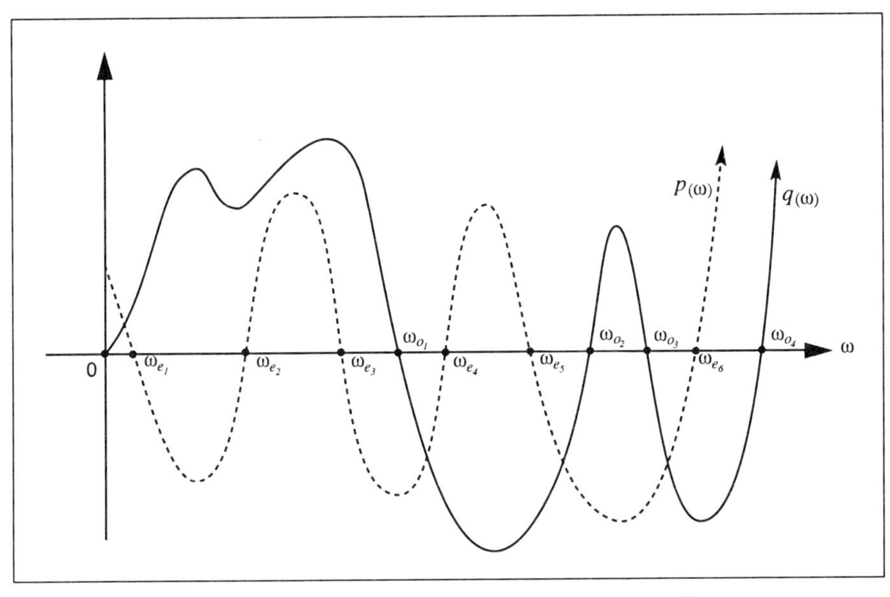

Fig. 3.3. Interlacing property fails for Non-Hurwitz polynomials.

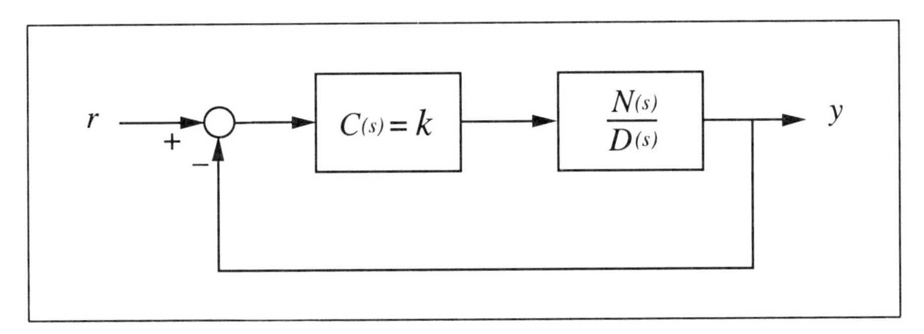

Fig. 3.4. Feedback control system.

controller is of zeroth order. Then the closed loop characteristic polynomial is

$$\delta(s,k) = D(s) + kN(s),$$

where k is a scalar. Our objective is to analytically determine k, if any, for which $\delta(s,k)$ is Hurwitz. Now let $D(s)$, $N(s)$ have the following even-odd decompositions:

$$\begin{aligned} D(s) &= D_e(s^2) + sD_o(s^2) \\ N(s) &= N_e(s^2) + sN_o(s^2) \end{aligned}$$

Then

$$\delta(s,k) = [D_e(s^2) + kN_e(s^2)] + s[D_o(s^2) + kN_o(s^2)]$$

Substituting $s = j\omega$, we have

$$\delta(j\omega,k) = [D_e(-\omega^2) + kN_e(-\omega^2)] + j\omega[D_o(-\omega^2) + kN_o(-\omega^2)]$$

Denote

$$\begin{aligned} \bar{p}(\omega,k) &= D_e(-\omega^2) + kN_e(-\omega^2) \\ \bar{q}(\omega,k) &= \omega[D_o(-\omega^2) + kN_o(-\omega^2)]. \end{aligned}$$

Then

$$\delta(j\omega,k) = \bar{p}(\omega,k) + j\bar{q}(\omega,k).$$

Now, if we were to use Lemma 3.2.1 to determine the values of k for which $\delta(s,k)$ is Hurwitz, then we would have to first solve for the frequencies ω as a function of k at which $\bar{p}(\omega,k) = 0$ or $\bar{q}(\omega,k) = 0$ and determine the values of k for which condition (iii) or (ii) respectively of Lemma 3.2.1 holds. The fact that both $\bar{p}(\omega,k)$ and $\bar{q}(\omega,k)$ depend on k makes this quite a formidable problem to solve, even when $\bar{p}(\omega,k)$, $\bar{q}(\omega,k)$ are of low order. This difficulty can be overcome by recasting the problem with $\bar{p}(\omega,k)$ or $\bar{q}(\omega,k)$ independent of k. This can be done as follows.

Let

$$\begin{aligned} N^*(s) &= N(-s) \\ &= N_e(s^2) - sN_o(s^2) \end{aligned}$$

Then

$$\begin{aligned} \delta(s,k)N^*(s) &= [D_e(s^2)N_e(s^2) - s^2D_o(s^2)N_o(s^2) + kN_e^2(s^2) - ks^2N_o^2(s^2)] \\ &\quad + s[N_e(s^2)D_o(s^2) - D_e(s^2)N_o(s^2)]. \end{aligned}$$

Substituting $s = j\omega$, we obtain

$$\delta(j\omega,k)N^*(j\omega) = p(\omega,k) + jq(\omega) \tag{3.6}$$

where

$$\begin{aligned} p(\omega, k) &= D_e(-\omega^2)N_e(-\omega^2) + \omega^2 D_o(-\omega^2)N_o(-\omega^2) + kN_e^2(-\omega^2) \\ &\quad + k\omega^2 N_o^2(-\omega^2) \qquad (3.7) \\ q(\omega) &= \omega[N_e(-\omega^2)D_o(-\omega^2) - D_e(-\omega^2)N_o(-\omega^2)] \qquad (3.8) \end{aligned}$$

Suppose $N^*(s)$ is a Hurwitz polynomial. Then $\delta(s,k)$ is Hurwitz if and only if $\delta(s,k)N^*(s)$ is Hurwitz. From (3.8), it is clear that $q(\omega)$ is independent of k. Now let $\omega_{o_0}, \omega_{o_1}, \cdots, \omega_{o_l}$ denote the non-negative real zeros of $q(\omega)$. Suppose the degree of $\delta(s,k)N^*(s)$ is n. Since

$$\delta(j\omega, k)N^*(j\omega) = p(\omega, k) + jq(\omega),$$

condition (ii) of Lemma 3.2.1 can be used to determine the values of k for which $\delta(s,k)N^*(s)$ is Hurwitz. However, in general, $N^*(s)$ is *not* Hurwitz to start with and the above approach cannot be used since Lemma 3.2.1 is not applicable to non-Hurwitz polynomials. This motivates us to derive an appropriate generalization of Lemma 3.2.1 for not necessarily Hurwitz polynomials. This is carried out in the next five sections of this chapter.

3.3 Root Distribution and Net Accumulated Phase

In this section we develop, as a preliminary step to the Generalized Hermite-Biehler Theorems, a fundamental relationship between the net accumulated phase of the frequency response of a real polynomial and the difference between the numbers of roots of the polynomial in the open left-half and open right-half planes. To this end, let $\mathcal{C}$ denote the complex plane, $\mathcal{C}^-$ the open left-half plane and $\mathcal{C}^+$ the open right-half plane.

In the beginning, we focus on polynomials without zeros on the imaginary axis. Consider a real polynomial $\delta(s)$ of degree n:

$$\begin{aligned} \delta(s) &= \delta_0 + \delta_1 s + \delta_2 s^2 + \ldots + \delta_n s^n,\ \delta_i \in \mathcal{R},\ i = 0, 1, \cdots, n,\ \delta_n \neq 0 \\ &\quad \text{such that } \delta(j\omega) \neq 0,\ \forall \omega \in (-\infty,\ \infty) \end{aligned}$$

Let $p(\omega)$ and $q(\omega)$ be two functions defined pointwise by $p(\omega) = Re[\delta(j\omega)]$, $q(\omega) = Im[\delta(j\omega)]$.
With this definition, we have

$$\delta(j\omega) = p(\omega) + jq(\omega) \ \ \forall \omega$$

Furthermore $\theta(\omega) \stackrel{\Delta}{=} \angle\delta(j\omega) = \arctan\left[\frac{q(\omega)}{p(\omega)}\right]$. Let $\Delta_0^\infty\theta$ denote the net change in the argument $\theta(\omega)$ as ω increases from 0 to ∞ and let $l(\delta)$ and $r(\delta)$ denote the numbers of roots of $\delta(s)$ in $\mathcal{C}^-$ and $\mathcal{C}^+$ respectively. Then we can state the following Lemma [19]:

Lemma 3.3.1. *Let $\delta(s)$ be a real polynomial with no imaginary axis roots. Then*

$$\Delta_0^\infty \theta = \frac{\pi}{2}(l(\delta) - r(\delta))$$

Proof. Each C^- root contributes $+\frac{\pi}{2}$ and each C^+ root contributes $-\frac{\pi}{2}$ to the net change in argument.

3.4 Imaginary and Real Signatures Associated with a Real Polynomial

In this section, we define, mainly for notational convenience, the imaginary and real signatures associated with a real polynomial. These definitions are useful because they facilitate an elegant statement of the generalizations of the Hermite-Biehler Theorem.

Definition 3.4.1. *Let $\delta(s)$ be any given real polynomial of degree n with k denoting the multiplicity of a root at the origin. Define*

$$p_f(\omega) := \frac{p(\omega)}{(1+\omega^2)^{\frac{n}{2}}}, \; q_f(\omega) := \frac{q(\omega)}{(1+\omega^2)^{\frac{n}{2}}}.$$

Let $0 = \omega_0 < \omega_1 < \omega_2 < \cdots < \omega_{m-1}$ be the real, non-negative, distinct finite zeros of $q_f(\omega)$ with odd multiplicities. Also define $\omega_m = \infty$. Then the imaginary signature *$\sigma_i(\delta)$ of $\delta(s)$ is defined by*

$$\sigma_i(\delta) := \begin{cases} \{sgn[p_f^{(k)}(\omega_0)] - 2sgn[p_f(\omega_1)] + 2sgn[p_f(\omega_2)] + \cdots + (-1)^{m-1} \\ \cdot 2sgn[p_f(\omega_{m-1})] + (-1)^m sgn[p_f(\omega_m)]\} \cdot (-1)^{m-1} sgn[q(\infty)] \\ \qquad \text{if } n \text{ is even} \\ \\ \{sgn[p_f^{(k)}(\omega_0)] - 2sgn[p_f(\omega_1)] + 2sgn[p_f(\omega_2)] + \cdots + (-1)^{m-1} \\ \cdot 2sgn[p_f(\omega_{m-1})]\} \cdot (-1)^{m-1} sgn[q(\infty)] \\ \qquad \text{if } n \text{ is odd} \end{cases} \tag{3.9}$$

where $p_f^{(k)}(\omega_0) := \frac{d^k}{d\omega^k}[p_f(\omega)]|_{\omega=\omega_0}$.

Definition 3.4.2. *Let $\delta(s)$ be any given real polynomial of degree n with k denoting the multiplicity of a root at the origin. Let $p_f(\omega)$, $q_f(\omega)$ be as in the last definition, and let $0 < \omega_1 < \omega_2 < \cdots < \omega_{m-1}$ be the real, non-negative, distinct finite zeros of $p_f(\omega)$ with odd multiplicities. Also define $\omega_0 = 0$, $\omega_m = \infty$. Then the* real signature *$\sigma_r(\delta)$ of $\delta(s)$ is defined by*

$$\sigma_r(\delta) := \begin{cases} \{sgn[q^{(k)}(\omega_0)] - 2sgn[q_f(\omega_1)] + 2sgn[q_f(\omega_2)] + \cdots \\ +(-1)^{m-1}2 \cdot sgn[q_f(\omega_{m-1})]\} \cdot (-1)^m sgn[p(\infty)] \\ \qquad \textit{if } n \textit{ is even} \\ \\ \{sgn[q^{(k)}(\omega_0)] - 2sgn[q_f(\omega_1)] + 2sgn[q_f(\omega_2)] + \cdots \\ +(-1)^{m-1}2 \cdot sgn[q_f(\omega_{m-1})] + (-1)^m sgn[q_f(\omega_m)]\} \cdot (-1)^m \\ \cdot sgn[p(\infty)] \qquad \textit{if } n \textit{ is odd} \end{cases} \tag{3.10}$$

where $q_f^{(k)}(\omega_0) := \frac{d^k}{d\omega^k}[q_f(\omega)]|_{\omega=\omega_0}$.

3.5 Generalizations of the Hermite-Biehler Theorem

3.5.1 No Imaginary Axis Roots

In this subsection, we focus on real polynomials with no imaginary axis roots and derive two generalizations of the Hermite-Biehler Theorem by first developing a procedure for systematically determining the net accumulated phase change of the "frequency response" of a polynomial. We first recall that at any given frequency ω, the phase angle of $\delta(j\omega)$ is given by

$$\theta(\omega) = \tan^{-1}\frac{q(\omega)}{p(\omega)}.$$

Hence the rate of change of phase with respect to frequency at any given frequency ω is given by

$$\begin{aligned} \frac{d\theta(\omega)}{d\omega} &= \frac{1}{1+\frac{q^2(\omega)}{p^2(\omega)}}\frac{\dot{q}(\omega)p(\omega)-\dot{p}(\omega)q(\omega)}{p^2(\omega)} \\ &= \frac{\dot{q}(\omega)p(\omega)-\dot{p}(\omega)q(\omega)}{p^2(\omega)+q^2(\omega)} \end{aligned} \tag{3.11}$$

If $p(\omega)$ and $q(\omega)$ are known for all ω, we can integrate (3.11) to obtain the net phase accumulation. However, to calculate the net accumulation of phase over all frequencies it is *not necessary* to know the precise rate of change of phase at each and every frequency. This is because, we know that over the frequencies where the polar plot of $\delta(j\omega)$ makes a transition from the real axis to the imaginary axis, or vice versa, there can be at most a net phase change of $\pm\frac{\pi}{2}$ radians. The actual sign of the phase change can be determined by examining Equation 3.11 at the real or imaginary axis crossings of the $\delta(j\omega)$ plot. Since at a real or imaginary axis crossing, one of the two terms in the numerator of (3.11) vanishes and the denominator is always positive, the actual determination of the sign of the phase change becomes even simpler.

Now, given any polynomial $\delta(s)$ of degree greater than or equal to one, either the real part or the imaginary part or both of $\delta(j\omega)$ become infinitely large as $\omega \to \pm\infty$. However, if we wish to count the total phase accumulation in integral multiples of axis crossings, it is imperative that the frequency response plot used approach either the real or imaginary axis as $\omega \to \pm\infty$. To accomplish this, one can normalize the plot of $\delta(j\omega)$ by scaling it with $\frac{1}{f(\omega)}$ where $f(\omega) = (1+\omega^2)^{\frac{n}{2}}$. Since $f(\omega)$ does not have any real roots, this scaling will ensure that the normalized frequency response plot

$$\delta_f(j\omega) = p_f(\omega) + jq_f(\omega) \text{ where } p_f(\omega) := \frac{p(\omega)}{(1+\omega^2)^{\frac{n}{2}}},\ q_f(\omega) := \frac{q(\omega)}{(1+\omega^2)^{\frac{n}{2}}}$$

actually intersects either the real axis or the imaginary axis at a finite point at $\omega = \pm\infty$, while at the same time leaving unchanged the finite frequencies at which $\delta(j\omega)$ intersects the real and imaginary axes. The subsequent development in this chapter makes use of the normalized frequency response plot for determining the net accumulated phase change as we move from $\omega = 0$ to $\omega = +\infty$. Here it should be pointed out that the normalized frequency response plot being used in this chapter is similar to the well known Mikhailov plot for Hurwitz polynomials.

Let $p(\omega)$, $q(\omega)$, $p_f(\omega)$, $q_f(\omega)$ be as already defined and let

$$0 = \omega_0 < \omega_1 < \omega_2 < \cdots < \omega_{m-1}$$

be the real, non-negative distinct finite zeros of $q_f(\omega)$ with odd multiplicities. The function $q_f(\omega)$ does not change sign while passing through a real zero of even multiplicity; hence such zeros can be skipped while counting the net phase accumulation. Also define $\omega_m = +\infty$.

Then we can make the following simple observations:

1. If ω_i, ω_{i+1} are both zeros of $q_f(\omega)$ then

$$\Delta_{\omega_i}^{\omega_{i+1}}\theta = \frac{\pi}{2}\left[\text{sgn}[p_f(\omega_i)] - \text{sgn}[p_f(\omega_{i+1})]\right] \cdot \text{sgn}[q_f(\omega_i^+)] \quad (3.12)$$

2. If ω_i is a zero of $q_f(\omega)$ while ω_{i+1} is not a zero of $q_f(\omega)$, a situation possible only when $\omega_{i+1} = \infty$ is a zero of $p_f(\omega)$ and n is odd, then

$$\Delta_{\omega_i}^{\omega_{i+1}}\theta = \frac{\pi}{2}\text{sgn}[p_f(\omega_i)] \cdot \text{sgn}[q_f(\omega_i^+)] \quad (3.13)$$

3.

$$\text{sgn}[q_f(\omega_{i+1}^+)] = -\text{sgn}[q_f(\omega_i^+)],\ i = 0, 1, 2, \cdots, m-2 \quad (3.14)$$

Equation 3.12 above is obvious while Equation 3.14 simply states that $q_f(\omega)$ changes sign when it passes through a zero of odd multiplicity. Equation 3.13, on the other hand, follows directly from Equation 3.11.

Using (3.14) repeatedly, we obtain

$$\text{sgn}[q_f(\omega_i^+)] = (-1)^{m-1-i} \cdot \text{sgn}[q_f(\omega_{m-1}^+)], \quad i = 0,\ 1, \cdots,\ m-1 \quad (3.15)$$

Substituting (3.15) into (3.12), we see that if ω_i, ω_{i+1} are both zeros of $q_f(\omega)$ then

$$\begin{aligned} \Delta_{\omega_i}^{\omega_{i+1}}\theta &= \frac{\pi}{2}\left[\operatorname{sgn}[p_f(\omega_i)] - \operatorname{sgn}[p_f(\omega_{i+1})]\right] \\ &\quad \cdot(-1)^{m-1-i} \cdot \operatorname{sgn}[q_f(\omega_{m-1}^+)] \end{aligned} \tag{3.16}$$

The above observations enable us to state and prove the following Theorem concerning $l(\delta) - r(\delta)$.

Theorem 3.5.1. *Let $\delta(s)$ be a given real polynomial of degree n with no roots on the $j\omega$ axis,* i.e., *the normalized plot $\delta_f(j\omega)$ does not pass through the origin. Then*

$$l(\delta) - r(\delta) = \sigma_i(\delta) \tag{3.17}$$

Proof. We note that under the conditions of this Theorem, $k = 0$ in Definition 3.4.1 so that $p_f^{(k)}(\omega_0) = p_f(\omega_0)$. Now first let us suppose that n is even. Then $\omega_m = \infty$ is a zero of $q_f(\omega)$. By repeatedly using (3.16) to determine $\Delta_0^\infty\theta$, applying Lemma 3.3.1, and then using the fact that $\operatorname{sgn}[q_f(\omega_{m-1}^+)] = \operatorname{sgn}[q(\infty)]$, it follows that $l(\delta) - r(\delta)$ is equal to the first expression in (3.9). Hence (3.17) holds for n even.

Next let us consider the case in which n is odd. Then $\omega_m = \infty$ is not a zero of $q_f(\omega)$. Hence,

$$\begin{aligned} \Delta_0^\infty\theta &= \sum_{i=0}^{m-2} \Delta_{\omega_i}^{\omega_{i+1}}\theta + \Delta_{\omega_{m-1}}^{\infty}\theta \\ &= \sum_{i=0}^{m-2} \frac{\pi}{2}\left[\operatorname{sgn}[p_f(\omega_i)] - \operatorname{sgn}[p_f(\omega_{i+1})]\right] \cdot (-1)^{m-1-i}\operatorname{sgn}[q_f(\omega_{m-1}^+)] \\ &\quad + \frac{\pi}{2}\operatorname{sgn}[p_f(\omega_{m-1})] \cdot \operatorname{sgn}[q_f(\omega_{m-1}^+)] \\ &\quad \text{(using (3.16) and (3.13))} \end{aligned} \tag{3.18}$$

Applying Lemma 3.3.1, and then using the fact that $\operatorname{sgn}[q_f(\omega_{m-1}^+)] = \operatorname{sgn}[q(\infty)]$, it follows that $l(\delta) - r(\delta)$ is equal to the second expression in (3.9). Hence (3.17) also holds for n odd. ♣

We now state the result analogous to Theorem 3.5.1 where $l(\delta) - r(\delta)$ of a real polynomial $\delta(s)$ is to be determined using the values of the frequencies where $\delta_f(j\omega)$ crosses the imaginary axis. The proof is omitted since it follows along the same lines as that of Theorem 3.5.1.

Theorem 3.5.2. *Let $\delta(s)$ be a given real polynomial of degree n with no roots on the $j\omega$ axis,* i.e., *the normalized plot $\delta_f(j\omega)$ does not pass through the origin. Then*

$$l(\delta) - r(\delta) = \sigma_r(\delta) \tag{3.19}$$

Remark 3.5.1. Theorems 3.5.1 and 3.5.2 generalize Lemma 3.2.1, parts (ii) and (iii) to the case of not necessarily Hurwitz polynomials. It is precisely in this sense that Theorems 3.5.1 and 3.5.2 are generalizations of the Hermite-Biehler Theorem.

3.5.2 No Roots at the Origin

In this subsection, we extend Theorems 3.5.1 and 3.5.2 so that $\delta(s)$ is now allowed to have *non-zero* imaginary axis roots. Theorems 3.5.3 and 3.5.4 below show that the statements of Theorems 3.5.1 and 3.5.2 continue to hold for this case. We will present a detailed proof of only Theorem 3.5.3; the proof of Theorem 3.5.4 follows along the same lines and is therefore omitted.

Theorem 3.5.3. *Let $\delta(s)$ be a given real polynomial of degree n with no roots at the origin. Then*

$$l(\delta) - r(\delta) = \sigma_i(\delta) \tag{3.20}$$

Proof. Now, $\delta(s)$ can be factored as

$$\delta(s) = \delta_o^*(s)\delta_e^*(s)\delta^{'}(s)$$

where $\delta_o^*(s)$ contains all the $j\omega$ axis roots of $\delta(s)$ with odd multiplicities, $\delta_e^*(s)$ contains all the $j\omega$ axis roots of $\delta(s)$ with even multiplicities, while $\delta^{'}(s)$ has no $j\omega$ axis roots. $\delta_o^*(s)$ and $\delta_e^*(s)$ must necessarily be of the form

$$\delta_o^*(s) = \prod_{i_o=1,2,3,\cdots} (s^2+\alpha_{i_o}^2)^{n_{i_o}}, \ \alpha_{i_o}>0, \ n_{i_o}\geq 0, \ n_{i_o} \text{ is odd,}$$

and

$$\alpha_1 < \alpha_2 < \cdots;$$

$$\delta_e^*(s) = \prod_{i_e=1,2,3,\cdots} (s^2+\beta_{i_e}^2)^{n_{i_e}}, \ \beta_{i_e}>0, \ n_{i_e}\geq 0, \ n_{i_e} \text{ is even.}$$

The proof is carried out in two steps. First, we show that multiplying $\delta^{'}(s)$ by $\delta_e^*(s)$ has no effect on (3.17). Thereafter, we use an inductive argument to show that multiplying $\delta_e^*(s)\delta^{'}(s)$ by $\delta_o^*(s)$ also does not affect (3.17).
<u>Step I:</u> Define[1]

$$\begin{aligned}\delta_0(s) &= \delta_e^*(s)\delta^{'}(s)\\ &= \prod_{i_e}(s^2+\beta_{i_e}^2)^{n_{i_e}}\delta^{'}(s).\end{aligned} \tag{3.21}$$

[1] Note that in this proof $\delta_0(s), \delta_1(s), \cdots, \delta_k(s)$, etc. represent particular polynomials which should not be confused with the coefficients of $\delta(s)$ defined earlier.

We want to show that $\delta_0(s)$ satisfies (3.20).

Define

$$\begin{aligned}\delta'(j\omega) &:= p'(\omega)+jq'(\omega)\\ \delta_0(j\omega) &:= p_0(\omega)+jq_0(\omega)\end{aligned}$$

so that $p'(\omega)$, $p_0(\omega)$, $q'(\omega)$, $q_0(\omega)$ are related by

$$p_0(\omega) = \prod_{i_e}(-\omega^2+\beta_{i_e}^2)^{n_{i_e}}p'(\omega) \tag{3.22}$$

$$q_0(\omega) = \prod_{i_e}(-\omega^2+\beta_{i_e}^2)^{n_{i_e}}q'(\omega). \tag{3.23}$$

Let $0=\omega_0 < \omega_1 < \omega_2 < \cdots < \omega_{m-1}$ be the real, non-negative, distinct finite zeros of $q_f'(\omega)$ with odd multiplicities. Also define $\omega_m=\infty$. First let us assume that $\delta'(s)$ is of even degree. Then, from Theorem 3.5.1, we have

$$\begin{aligned}l(\delta')-r(\delta') &= \sigma_i(\delta')\\ &= \{\text{sgn}[p_f'(\omega_0)]-2\text{sgn}[p_f'(\omega_1)]+2\text{sgn}[p_f'(\omega_2)]+\cdots\\ &\quad +(-1)^{m-1}2\text{sgn}[p_f'(\omega_{m-1})]+(-1)^m\text{sgn}[p_f'(\omega_m)]\}\\ &\quad \cdot(-1)^{m-1}\text{sgn}[q'(\infty)].\end{aligned}$$

From (3.23), it follows that ω_i, $i=0,1,\ldots,m-1$ are also the real, non-negative, distinct finite zeros of $q_{0_f}(\omega)$ with odd multiplicities. Furthermore, from (3.22) and (3.23), we have

$$\begin{aligned}\text{sgn}[p_f'(\omega_i)] &= \text{sgn}[p_{0_f}(\omega_i)],\ i=0,1,\ldots,m\\ \text{sgn}[q'(\infty)] &= \text{sgn}[q_0(\infty)].\end{aligned}$$

Since

$$l(\delta_0)-r(\delta_0)=l(\delta')-r(\delta')$$

it follows that (3.20) is true for $\delta_0(s)$ of even degree. The fact that (3.20) is also true for $\delta_0(s)$ of odd degree, can be verified by proceeding along exactly the same lines.

Step II Proof by Induction: Let the induction index j be equal to 1 and consider

$$\begin{aligned}\delta_1(s) &= (s^2+\alpha_1^2)^{n_1}\prod_{i_e}(s^2+\beta_{i_e}^2)^{n_{i_e}}\delta'(s)\\ &= (s^2+\alpha_1^2)^{n_1}\delta_0(s).\end{aligned} \tag{3.24}$$

Define

$$\delta_1(j\omega) = p_1(\omega)+jq_1(\omega)$$

so that $p_1(\omega)$, $p_0(\omega)$, $q_1(\omega)$, $q_0(\omega)$ are related by

$$\begin{aligned} p_1(\omega) &= (-\omega^2+\alpha_1^2)^{n_1} p_0(\omega) && (3.25)\\ q_1(\omega) &= (-\omega^2+\alpha_1^2)^{n_1} q_0(\omega). && (3.26) \end{aligned}$$

Let $0=\omega_0 < \omega_1 < \omega_2 < \cdots < \omega_{m-1}$ be the real, non-negative, distinct finite zeros of $q_{0_f}(\omega)$ with odd multiplicities. Also define $\omega_m=\infty$. First let us assume that $\delta_0(s)$ has even degree. Then, from Step I

$$\begin{aligned} l(\delta_0)-r(\delta_0) &= \sigma_i(\delta_0)\\ &= \{\operatorname{sgn}[p_{0_f}(\omega_0)] - 2\operatorname{sgn}[p_{0_f}(\omega_1)] + 2\operatorname{sgn}[p_{0_f}(\omega_2)]\\ &\quad +\cdots+(-1)^{m-1}2\operatorname{sgn}[p_{0_f}(\omega_{m-1})]\\ &\quad +(-1)^m\operatorname{sgn}[p_{0_f}(\omega_m)]\}\\ &\quad \cdot(-1)^{m-1}\operatorname{sgn}[q_0(\infty)]. \qquad (3.27) \end{aligned}$$

From (3.26), it follows that ω_i, $i=0, 1, \ldots, m-1$; α_1 are the real, non-negative, distinct finite zeros of $q_{1_f}(\omega)$ with odd multiplicities. Let us assume that $\omega_l < \alpha_1 < \omega_{l+1}$. Then, from (3.25) and (3.26), we have

$$\left.\begin{aligned} \operatorname{sgn}[p_{0_f}(\omega_i)] &= \operatorname{sgn}[p_{1_f}(\omega_i)],\ i=0, 1, \ldots, l\\ \operatorname{sgn}[p_{0_f}(\omega_i)] &= -\operatorname{sgn}[p_{1_f}(\omega_i)],\ i=l+1, l+2, \ldots, m\\ \operatorname{sgn}[p_{1_f}(\alpha_1)] &= 0\\ \operatorname{sgn}[q_0(\infty)] &= -\operatorname{sgn}[q_1(\infty)]. \end{aligned}\right\} \qquad (3.28)$$

Since $l(\delta_1)-r(\delta_1)=l(\delta_0)-r(\delta_0)$, using (3.27), (3.28), we obtain

$$\begin{aligned} l(\delta_1)-r(\delta_1) &= \{\operatorname{sgn}[p_{1_f}(\omega_0)] - 2\operatorname{sgn}[p_{1_f}(\omega_1)] + 2\operatorname{sgn}[p_{1_f}(\omega_2)]\\ &\quad +\cdots+(-1)^l 2\operatorname{sgn}[p_{1_f}(\omega_l)]\\ &\quad +(-1)^{l+1}2\operatorname{sgn}[p_{1_f}(\alpha_1)] + (-1)^{l+2}2\operatorname{sgn}[p_{1_f}(\omega_{l+1})]\\ &\quad +\cdots+(-1)^m 2\operatorname{sgn}[p_{1_f}(\omega_{m-1})]\\ &\quad +(-1)^{m+1}\operatorname{sgn}[p_{1_f}(\omega_m)]\}\cdot(-1)^m\operatorname{sgn}[q_1(\infty)]\\ &= \sigma_i(\delta_1) \end{aligned}$$

which shows that (3.20) is true for $\delta_1(s)$ of even degree. The fact that (3.20) holds for $\delta_1(s)$ of odd degree, can be verified likewise. This completes the first step of the induction argument.

Now let $j=k$ and consider

$$\delta_k(s) = \prod_{i_o=1}^{k}(s^2+\alpha_{i_o}^2)^{n_{i_o}} \prod_{i_e}(s^2+\beta_{i_e}^2)^{n_{i_e}}\delta'(s). \qquad (3.29)$$

Assume that (3.20) is true for $\delta_k(s)$ (inductive assumption). Then

$$
\begin{aligned}
\delta_{k+1}(s) &= \prod_{i_o=1}^{k+1}(s^2+\alpha_{i_o}^2)^{n_{i_o}}\prod_{i_e}(s^2+\beta_{i_e}^2)^{n_{i_e}}\delta'(s) \\
&= (s^2+\alpha_{k+1}^2)^{n_{k+1}}\delta_k(s). \qquad (3.30)
\end{aligned}
$$

Now, define

$$
\begin{aligned}
\delta_k(j\omega) &= p_k(\omega)+jq_k(\omega) \\
\delta_{k+1}(j\omega) &= p_{k+1}(\omega)+jq_{k+1}(\omega)
\end{aligned}
$$

so that $p_{k+1}(\omega)$, $p_k(\omega)$, $q_{k+1}(\omega)$, $q_k(\omega)$ are related by

$$
\begin{aligned}
p_{k+1}(\omega) &= (-\omega^2+\alpha_{k+1}^2)^{n_{k+1}}p_k(\omega) \qquad (3.31)\\
q_{k+1}(\omega) &= (-\omega^2+\alpha_{k+1}^2)^{n_{k+1}}q_k(\omega). \qquad (3.32)
\end{aligned}
$$

Let $0=\omega_0 < \omega_1 < \omega_2 < \cdots < \omega_{m-1}$ be the real, non-negative, distinct finite zeros of $q_{k_f}(\omega)$ with odd multiplicities. Also define $\omega_m=\infty$. First let us assume that $\delta_k(s)$ is of even degree. Then from the inductive assumption, we have

$$
\begin{aligned}
l(\delta_k)-r(\delta_k) &= \sigma_i(\delta_k) \\
&= \{\text{sgn}[p_{k_f}(\omega_0)]-2\text{sgn}[p_{k_f}(\omega_1)]+2\text{sgn}[p_{k_f}(\omega_2)] \\
&\quad +\cdots+(-1)^{m-1}2\text{sgn}[p_{k_f}(\omega_{m-1})]+(-1)^m \\
&\quad \text{sgn}[p_{k_f}(\omega_m)]\}\cdot(-1)^{m-1}\text{sgn}[q_k(\infty)]. \qquad (3.33)
\end{aligned}
$$

Now from (3.32), it follows that ω_i, $i=0, 1, \ldots, m-1$; α_{k+1} are the real, non-negative, distinct finite zeros of $q_{k+1_f}(\omega)$ with odd multiplicities. Let us assume that $\omega_l < \alpha_{k+1} < \omega_{l+1}$. Then from (3.31) and (3.32), we have

$$
\left.
\begin{aligned}
\text{sgn}[p_{k_f}(\omega_i)] &= \text{sgn}[p_{k+1_f}(\omega_i)],\ i=0, 1, \ldots, l \\
\text{sgn}[p_{k_f}(\omega_i)] &= -\text{sgn}[p_{k+1_f}(\omega_i)],\ i=l+1, \ldots, m \\
\text{sgn}[p_{k+1_f}(\alpha_{k+1})] &= 0 \\
\text{sgn}[q_k(\infty)] &= -\text{sgn}[q_{k+1}(\infty)].
\end{aligned}
\right\} \qquad (3.34)
$$

Since $l(\delta_{k+1})-r(\delta_{k+1})=l(\delta_k)-r(\delta_k)$, using (3.33), (3.34), we obtain

$$
\begin{aligned}
l(\delta_{k+1})-r(\delta_{k+1}) &= \{\text{sgn}[p_{k+1_f}(\omega_0)]-2\text{sgn}[p_{k+1_f}(\omega_1)] \\
&\quad +2\text{sgn}[p_{k+1_f}(\omega_2)]+\cdots \\
&\quad +(-1)^l2\text{sgn}[p_{k+1_f}(\omega_l)] \\
&\quad +(-1)^{l+1}\cdot 2\text{sgn}[p_{k+1_f}(\alpha_{k+1})] \\
&\quad +(-1)^{l+2}2\text{sgn}[p_{k+1_f}(\omega_{l+1})]+\cdots+ \\
&\quad (-1)^m2\text{sgn}[p_{k+1_f}(\omega_{m-1})]+(-1)^{m+1} \\
&\quad \cdot\text{sgn}[p_{k+1_f}(\omega_m)]\}\cdot(-1)^m\text{sgn}[q_{k+1}(\infty)] \\
&= \sigma_i(\delta_{k+1}). \qquad (3.35)
\end{aligned}
$$

which shows that (3.20) is true for $\delta_{k+1}(s)$ of even degree. The fact that (3.20) is true for $\delta_{k+1}(s)$ of odd degree, can be similarly verified. This completes the induction argument and hence the proof. ♣

Theorem 3.5.4. *Let $\delta(s)$ be a given real polynomial of degree n with no roots at the origin. Then*

$$l(\delta) - r(\delta) = \sigma_r(\delta). \tag{3.36}$$

3.5.3 No Restriction on Root Locations

Theorems 3.5.3 and 3.5.4 presented in the last subsection require that the polynomial $\delta(s)$ have no roots at the origin. In this subsection, we present two theorems showing that such restrictions can be removed. A proof is presented only for the first theorem; the proof of the second one is similar and, is therefore, omitted.

Theorem 3.5.5. *Let $\delta(s)$ be a given real polynomial of degree n. Then*

$$l(\delta) - r(\delta) = \sigma_i(\delta). \tag{3.37}$$

Proof. If $\delta(s)$ has no roots at the origin, then (3.37) follows from Theorem 3.5.3. So, let us assume that $\delta(s)$ has a root of multiplicity k at the origin. Then, we can write

$$\delta(s) = s^k \delta^{'}(s)$$

where $\delta^{'}(s)$ is a real polynomial of degree $n^{'}$ with no roots at the origin. Define

$$\begin{aligned} \delta^{'}(j\omega) &:= p^{'}(\omega) + jq^{'}(\omega) \\ \delta(j\omega) &:= p(\omega) + jq(\omega). \end{aligned}$$

The proof can be completed by considering four different cases, namely $k = 4l$, $k = 4l + 1$, $k = 4l + 2$ and $k = 4l + 3$. These four cases correspond to the four different ways in which multiplication by $(j\omega)^k$ affects the real and imaginary parts of $\delta^{'}(j\omega)$. Due to the fact that each of these cases is handled by proceeding along similar lines, we do not treat all of the cases here. Instead, we focus on a representative case, say $k = 4l + 1$, and provide a detailed treatment for it.

Now, for $k = 4l + 1$, we have

$$\begin{aligned} \delta(j\omega) &= p(\omega) + jq(\omega) \\ &= -\omega^{4l+1} q^{'}(\omega) + j\omega^{4l+1} p^{'}(\omega). \end{aligned}$$

First let us assume that n' is even. Then, from Theorem 3.5.4, we have

$$\begin{aligned} l(\delta') - r(\delta') &= \sigma_r(\delta') \\ &= -\{2\mathrm{sgn}[q_f'(\omega_1)] - 2\mathrm{sgn}[q_f'(\omega_2)] + \cdots \\ &\quad +(-1)^{m-2}2\mathrm{sgn}[q_f'(\omega_{m-1})]\} \cdot (-1)^m \mathrm{sgn}[p'(\infty)] \quad (3.38) \end{aligned}$$

where $0 < \omega_1 < \omega_2 < \cdots < \omega_{m-1}$ are the real, non-negative, distinct finite zeros of $p_f'(\omega)$ with odd multiplicities.

Define $w_0 := 0$. Since

$$p(\omega) = -\omega^{4l+1} q'(\omega),$$

$$p^{(4l+1)}(\omega_0) = -(4l+1)!\, q'(\omega_0) = 0,$$

$$q(\omega) = \omega^{4l+1} p'(\omega),$$

and

$$p_f^{(4l+1)}(\omega_0) = p^{(4l+1)}(\omega_0),$$

we have

$$\mathrm{sgn}[p_f^{(4l+1)}(\omega_0)] = 0 \quad (3.39)$$

$$\mathrm{sgn}[q_f'(\omega_i)] = -\mathrm{sgn}[p_f(\omega_i)],\ i = 1, 2, \ldots, m \quad (3.40)$$

$$\mathrm{sgn}[p'(\infty)] = \mathrm{sgn}[q(\infty)]. \quad (3.41)$$

Since n' is even and $k = 4l+1$, it follows that n is odd. Moreover, since

$$l(\delta) - r(\delta) = l(\delta') - r(\delta'),$$

using (3.38), (3.39), (3.40), (3.41), we have

$$\begin{aligned} l(\delta) - r(\delta) &= \{\mathrm{sgn}[p_f^{(k)}(\omega_0)] - 2\mathrm{sgn}[p_f(\omega_1)] + 2\mathrm{sgn}[p_f(\omega_2)] + \cdots \\ &\quad +(-1)^{m-1}2\mathrm{sgn}[p_f(\omega_{m-1})]\} \cdot (-1)^{m-1}\mathrm{sgn}[q(\infty)] \\ &= \sigma_i(\delta) \quad (3.42) \end{aligned}$$

which shows that (3.37) holds for $\delta(s)$ of odd degree. The fact that (3.37) also holds for $\delta(s)$ of even degree or equivalently n' odd, can be verified by proceeding along exactly the same lines. ♣

Theorem 3.5.6. *Let $\delta(s)$ be a given real polynomial of degree n. Then*

$$l(\delta) - r(\delta) = \sigma_r(\delta). \quad (3.43)$$

Remark 3.5.2. In view of the theorems presented so far, it is clear that for any real polynomial $\delta(s)$, the value of the real signature $\sigma_r(\delta)$ is equal to the value of the imaginary signature $\sigma_i(\delta)$ and each is in fact equal to $l(\delta) - r(\delta)$. Thus either of them could be referred to as the signature $\sigma(\delta)$ of $\delta(s)$ with the subscript 'r' or 'i' indicating the formula that is being used in a particular situation to compute the value.

Remark 3.5.3. From the proof of Theorem 3.5.5, it is clear that if $k = 1$, *i.e.*, the root of $\delta(s)$ at the origin is simple, then the even part of $\delta(s)$ has a double root at $s = 0$ so that $p(\omega_0) = p_f(\omega_0) = 0$ and $p_f^{(1)}(\omega_0) = 0$. Thus for $k = 0$ and $k = 1$, $p_f^{(k)}(\omega_0)$ in the expression for $\sigma_i(\delta)$ can be replaced by 0. These two special cases $k = 0$ and $k = 1$ are of importance since they arise repeatedly in the stabilization problems to be studied in the next chapter. Accordingly, whenever the imaginary signature $\sigma_i(\delta)$ appears in the next chapter, we will tacitly assume that the term $p_f^{(k)}(\omega_0)$ has been replaced by $p_f(\omega_0)$.

We conclude this section by presenting the following example to verify Theorem 3.5.5.

Example 3.5.1. Consider the real polynomial

$$\delta(s) = s^3(s^2+1)^2(s^2+5)(s-3)(s^2+s+1).$$

Substituting $s = j\omega$, we have

$$\delta(j\omega) = p(\omega) + jq(\omega)$$

where

$$p(\omega) = \omega^{12} - 5\omega^{10} - 3\omega^8 + 17\omega^6 - 10\omega^4$$

and

$$q(\omega) = 2\omega^{11} - 17\omega^9 + 43\omega^7 - 43\omega^5 + 15\omega^3.$$

The real, positive finite zeros of $q_f(\omega)$ with odd multiplicities are $\omega_1 = 1.22474$ and $\omega_2 = \sqrt{5}$. Also define $\omega_0 = 0$ and $\omega_3 = \infty$. Hence,

$$\mathrm{sgn}[p^{(3)}(\omega_0)] = 0, \mathrm{sgn}[p_f(\omega_1)] = -1, \mathrm{sgn}[p_f(\omega_2)] = 0 \text{ and } \mathrm{sgn}[p_f(\omega_3)] = 1.$$

Since $\delta(s)$ is of even degree and with a root at the origin of multiplicity 3, from formula (3.9), it follows that

$$\begin{aligned}\sigma_i(\delta) &= \{\mathrm{sgn}[p^{(3)}(\omega_0)] - 2\mathrm{sgn}[p_f(\omega_1)] + 2\mathrm{sgn}[p_f(\omega_2)] - \mathrm{sgn}[p_f(\omega_3)]\} \\ &\quad \cdot(-1)^2\mathrm{sgn}[q(\infty)] \\ &= 0 + 2 + 0 - 1 = 1.\end{aligned}$$

This agrees with the value for $l(\delta) - r(\delta)$ obtained from visual inspection of the factored form of $\delta(s)$, so that Theorem 3.5.5 is verified.

3.6 An Elementary Derivation of the Routh-Hurwitz Criterion

In this section, we use Theorems 3.5.1 and 3.5.2 to give, as an application, a simple proof of the Routh-Hurwitz criterion. First we consider a real polynomial $\delta(s)$ of degree n:

$$\delta(s) = \delta_0 + \delta_1 s + \delta_2 s^2 + \ldots + \delta_n s^n, \ \delta_n \neq 0.$$

and denote, as usual,

$$\delta(s) = \delta^{even}(s) + \delta^{odd}(s).$$

To avoid singularities of the "first type" and the "second type" [19] in Routh's Algorithm, we make the following assumptions:

(1) $\delta_{n-1} \neq 0$.
(2) $\delta^{even}(s)$ and $\delta^{odd}(s)$ are coprime.

To derive Routh algorithm we now introduce the parametrized family of polynomials $\delta_\lambda(s), \lambda \in [0,1]$ defined below:
If n is even, then

$$\delta - \lambda(s) = \left[\delta^{even}(s) - \lambda \frac{\delta_n}{\delta_{n-1}} \cdot s \cdot \delta^{odd}(s)\right] + \delta^{odd}(s). \tag{3.44}$$

If n is odd, then

$$\delta_\lambda(s) = \left[\delta^{odd}(s) - \lambda \frac{\delta_n}{\delta_{n-1}} \cdot s \cdot \delta^{even}(s)\right] + \delta^{even}(s) \tag{3.45}$$

Note that $\delta_0(s) = \delta(s)$, $\delta_\lambda(s)$ has degree n for $\lambda \in [0,1)$ and $\delta_1(s)$ is a polynomial of degree $n-1$. It is easily seen from the coprimeness assumption above that no roots cross the imaginary axis along this line segment of polynomials, that is $\delta_\lambda(j\omega) \neq 0, \omega \in [0,\infty], \lambda \in [0,1]$. Thus along this line segment the *numbers* of LHP and RHP roots are preserved except that a real root is lost at the endpoint $\lambda = 1$. Let $\bar{\delta}(s)$ denote the polynomial $\delta_1(s)$ of degree $n-1$:
If n is even

$$\bar{\delta}(s) = \left[\delta^{even}(s) - \frac{\delta_n}{\delta_{n-1}} \cdot s \cdot \delta^{odd}(s)\right] + \delta^{odd}(s). \tag{3.46}$$

If n is odd

$$\bar{\delta}(s) = \left[\delta^{odd}(s) - \frac{\delta_n}{\delta_{n-1}} \cdot s \cdot \delta^{even}(s)\right] + \delta^{even}(s) \tag{3.47}$$

The following Theorem relates $l(\delta) - r(\delta)$ to $l(\bar{\delta}) - r(\bar{\delta})$.

Theorem 3.6.1. *Let $\delta(s)$, $\bar{\delta}(s)$ be as already defined. Then*

$$(l(\delta) - r(\delta)) - (l(\bar{\delta}) - r(\bar{\delta})) = \begin{cases} 1 & \text{if } \delta_n\delta_{n-1} > 0 \\ -1 & \text{if } \delta_n\delta_{n-1} < 0 \end{cases}$$

Proof. Suppose

$$\delta(j\omega) = p(\omega) + jq(\omega) \tag{3.48}$$

First let us consider the case when n is even. Then from (3.46),

$$\bar{\delta}(j\omega) = \left[p(\omega) + \frac{\delta_n}{\delta_{n-1}}\omega q(\omega)\right] + jq(\omega) \tag{3.49}$$

From (3.48), (3.49) it follows that the finite zeros of $q_f(\omega)$ are the same for both $\delta(j\omega)$ and $\bar{\delta}(j\omega)$. Moreover, at these frequencies both $\delta(j\omega)$ and $\bar{\delta}(j\omega)$ have the same real part so that $\text{sgn}[p_f(\omega)]$ is also identical for both these polynomials at these frequencies. Thus, using (3.9), we obtain

$$\sigma_i(\delta) - \sigma_i(\bar{\delta}) = -\text{sgn}[p_f(\infty)] \cdot \text{sgn}[q(\infty)]$$

Now for large positive ω,

$$\begin{aligned} p(\omega) &\simeq (-1)^{\frac{n}{2}}\delta_n\omega^n \\ q(\omega) &\simeq (-1)^{\frac{n-2}{2}}\delta_{n-1}\omega^{n-1} \end{aligned}$$

so that

$$\text{sgn}[p_f(\infty)] \cdot \text{sgn}[q(\infty)] = -\text{sgn}[\delta_n\delta_{n-1}]$$

Thus

$$\sigma_i(\delta) - \sigma_i(\bar{\delta}) = \begin{cases} 1 & \text{if } \delta_n\delta_{n-1} > 0 \\ -1 & \text{if } \delta_n\delta_{n-1} < 0 \end{cases} \tag{3.50}$$

We now consider the case that n is odd. Then from (3.47)

$$\bar{\delta}(j\omega) = p(\omega) + j\left[q(\omega) - \frac{\delta_n}{\delta_{n-1}}\omega p(\omega)\right] \tag{3.51}$$

From (3.48), (3.51) it follows that the finite zeros of $p_f(\omega)$ are the same for both $\delta(j\omega)$ and $\bar{\delta}(j\omega)$. Moreover at these frequencies, both $\delta(j\omega)$ and $\bar{\delta}(j\omega)$ have the same imaginary part so that $\text{sgn}[q_f(\omega)]$ is also identical for both these polynomials at these frequencies. Thus, using (3.10) we obtain

$$\begin{aligned} \sigma_r(\delta) - \sigma_r(\bar{\delta}) &= (-1)^m(-1)^m\text{sgn}[q_f(\infty)] \cdot \text{sgn}[p(\infty)] \\ &= \text{sgn}[p(\infty)] \cdot \text{sgn}[q_f(\infty)] \end{aligned}$$

Now for large positive ω,

$$\begin{aligned} p(\omega) &\simeq (-1)^{\frac{n-1}{2}}\delta_{n-1}\omega^{n-1} \\ q(\omega) &\simeq (-1)^{\frac{n-1}{2}}\delta_n\omega^n \end{aligned}$$

so that

$$\operatorname{sgn}[p(\infty)] \cdot \operatorname{sgn}[q_f(\infty)] = \operatorname{sgn}[\delta_n \delta_{n-1}]$$

Thus

$$\sigma_r(\delta) - \sigma_r(\bar{\delta}) = \begin{cases} 1 & \text{if } \delta_n \delta_{n-1} > 0 \\ -1 & \text{if } \delta_n \delta_{n-1} < 0 \end{cases} \tag{3.52}$$

Combining (3.50), (3.52) and using Theorems 3.5.1 and 3.5.2, the desired result follows. ♣

Using Theorem 3.6.1, we obtain the following Corollary.

Corollary 3.6.1. *Let $\delta(s)$ be a given real polynomial and let $\bar{\delta}(s)$ be defined by (3.46) or (3.47) as appropriate. Then*

$$\left.\begin{array}{ll} l(\bar{\delta}) = l(\delta) - 1,\ r(\bar{\delta}) = r(\delta) & \textit{if } \delta_n \delta_{n-1} > 0 \\ l(\bar{\delta}) = l(\delta),\ r(\bar{\delta}) = r(\delta) - 1 & \textit{if } \delta_n \delta_{n-1} < 0 \end{array}\right\} \tag{3.53}$$

Proof. Theorem 3.6.1 implies that

$$l(\delta) - r(\delta) - l(\bar{\delta}) + r(\bar{\delta}) = \begin{cases} 1 & \text{if } \delta_n \delta_{n-1} > 0 \\ -1 & \text{if } \delta_n \delta_{n-1} < 0 \end{cases} \tag{3.54}$$

But, since $\bar{\delta}$ has degree one less than δ,

$$(l(\delta) + r(\delta)) - (l(\bar{\delta}) + r(\bar{\delta})) = 1 \tag{3.55}$$

Adding (3.54), (3.55) we obtain

$$l(\delta) - l(\bar{\delta}) = \begin{cases} 1 & \text{if } \delta_n \delta_{n-1} > 0 \\ 0 & \text{if } \delta_n \delta_{n-1} < 0 \end{cases} \tag{3.56}$$

Again, subtracting (3.54) from (3.55), we obtain

$$r(\delta) - r(\bar{\delta}) = \begin{cases} 0 & \text{if } \delta_n \delta_{n-1} > 0 \\ 1 & \text{if } \delta_n \delta_{n-1} < 0 \end{cases} \tag{3.57}$$

The desired result now follows from (3.56) and (3.57). ♣

Now given a real polynomial $\delta(s)$, Routh's algorithm is equivalent to reducing the degree of $\delta(s)$ by one at each step using (3.46) and (3.47) alternately. Thus Corollary 3.6.1 leads us to the immediate conclusion that $\delta(s)$ will be Hurwitz iff the leading coefficients of all the polynomials that result from alternately applying (3.46) and (3.47) to $\delta(s)$ are of the same sign. Furthermore, it is also clear that the number of open right half plane roots of $\delta(s)$ is equal to the number of sign changes in the leading coefficients of the successive polynomials. This is exactly the Routh-Hurwitz criterion.

3.6.1 Singular Cases

The derivation of the Routh-Hurwitz criterion presented above dealt with only the so called "regular" case, *i.e.*, the case in which the degree of $\delta(s)$ can be successively reduced by one at a time by the alternate application of (3.46) and (3.47) until we finally end up with a zeroth order polynomial. This process would, however, terminate prematurely if, while trying to apply (3.46) or (3.47), we encounter $\delta_{n-1} = 0$. Then we have what are called "singular" cases and this subsection is devoted to their treatment.

Starting with a given real polynomial $\delta_0(s)$ of degree n:

$$\delta_0(s) = \delta_0^0 + \delta_1^0 s + \delta_2^0 s^2 + \ldots + \delta_n^0 s^n,$$

suppose using (3.46) and (3.47) alternately, we obtain a sequence of polynomials $\{\delta_0(s), \delta_1(s), \delta_2(s), \cdots, \delta_m(s)\}$, where the leading coefficient of each $\delta_i(s), i = 0, 1, 2, \cdots, m$ is non zero. Let

$$\delta_m(s) = \delta_0^m + \delta_1^m s + \delta_2^m s^2 + \ldots + \delta_{n-m-1}^m s^{n-m-1} + \delta_{n-m}^m s^{n-m}$$

where $\delta_{n-m}^m \neq 0$. Now, if $\delta_{n-m-1}^m = 0$, then it is clear that Routh's algorithm stops because to proceed with Routh's algorithm using (3.46) or (3.47), we would need to divide by δ_{n-m-1}^m which is now equal to zero. To handle such singularities we consider the three distinct possibilities that can occur:

Case (I): $\delta_{n-m-1}^m = 0$ but there exists at least one k, $k = 3, 5, 7, 9, \cdots$ such that $\delta_{n-m-k}^m \neq 0$, *i.e.*, the first element in any one row of the Routh table vanishes but there is at least one non-zero element in that row.

If we know before hand that $\delta_0(s)$ has no imaginary axis roots, then we can proceed as follows. Replace $\delta_{n-m-1}^m = 0$ with a "small" nonzero number ϵ of arbitrary sign and then continue to proceed with Routh's algorithm. If a similar singularity is encountered later, introduce another parameter to replace the offending zero element, and so on.

By replacing $\delta_{n-m-1}^m = 0$ with ϵ, we in fact modify the original polynomial $\delta_0(s)$. From (3.46) and (3.47), for $\delta_{n-m-1}^m = \epsilon$, we can work backwards to obtain a modified polynomial $\delta_0(s, \epsilon)$, where the coefficients of $\delta_0(s, \epsilon)$ are rational functions of ϵ. Since $\delta_0(s)$ has no roots on the imaginary axis, it follows by continuity that for $|\epsilon|$ small enough, $l(\delta_0(s)) = l(\delta_0(s, \epsilon))$ and $r(\delta_0(s)) = r(\delta_0(s, \epsilon))$. This is the reason why this modification can be used to handle a singularity of this type and still provide a count of the number of open right-half plane roots.

Case (II): $\delta_{n-m-k}^m = 0$ for $k = 1, 3, 5, 7, \cdots$, *i.e.*, all the elements in one row of the Routh table vanish.

For this case, since $\delta^m_{n-m-k} = 0$ for $k = 1, 3, 5, 7, \cdots$, it follows that $\delta_0(s)$ must have one or more pairs of complex conjugate roots symmetrically distributed about the origin of the complex plane. This includes the case of purely imaginary roots as well as the case of purely real roots having opposite signs.

To take care of this kind of singularity, one can simply replace $\delta_0(s)$ with $\delta_0(s - \epsilon)$, where ϵ is a sufficiently "small" positive number, and then proceed with Routh's algorithm. The net result is that the number of *closed* right half plane roots of $\delta_0(s)$ equals the number of sign changes in the leading coefficients of the successive polynomials.

Case (III): Case(I) and **Case(II)** occur at different stages in the same problem when proceeding with Routh's algorithm.

Once again, we can replace $\delta_0(s)$ with $\delta_0(s-\epsilon)$, where ϵ is a sufficiently "small" positive number and then proceed with Routh's algorithm. Alternatively, we can factor out the imaginary axis roots as in [19] and then apply Routh's algorithm to the new polynomial.

Remark 3.6.1. The derivation of the Routh-Hurwitz criterion in [19] is carried out using the Cauchy Index which disregards the imaginary axis roots. Consequently, in [19], even in singular cases, it is possible to obtain a count of the number of *open* right half plane roots by appropriately modifying Routh's algorithm. The modifications proposed here, however, allow us to count the number of *closed* right half plane roots when the original polynomial has roots on the imaginary axis.

3.7 Notes and References

The Hermite-Biehler Theorem and its classical proof can be found in [19]. An alternative proof using the Boundary Crossing Theorem was developed in [12]. References [32, 6] contain instances of the application of the Hermite-Biehler Theorem to the parametric robust stability problem. The generalizations of the Hermite-Biehler Theorem and the elementary derivation of the Routh-Hurwitz criterion are due to Ho, Datta and Bhattacharyya [26, 27]. An appropriate reference for the Mikhailov plot is [43].

CHAPTER 4

STABILIZATION OF LINEAR TIME-INVARIANT PLANTS USING PID CONTROLLERS

In this chapter we utilize the Generalized Hermite-Biehler Theorem to give a solution to the problem of feedback stabilization of a given linear time-invariant (LTI) plant by a PID controller. The solution so obtained gives a constructive condition for existence and also characterizes the entire family of stabilizing controllers in terms of a linear programming (LP) problem. Some applications of this characterization are also discussed.

4.1 Introduction

Modern optimal control techniques such as H_2, H_∞ [17] and L_1 Optimal [14] control are incapable of accomodating constraints on the controller order or structure into their design methods, and consequently cannot be used for designing optimal or robust PID controllers. Since PID controllers are so widely used in industrial applications, it is important that such a design methodology be developed.

Motivated by this fact, in this chapter, we first provide a new and complete analytical solution to the problem of stabilizing a given plant using a constant gain, *i.e.*, a zeroth order controller. This solution which is presented in Section 4.2 is based on the Generalized Hermite-Biehler Theorem developed in Chapter 3. In Sections 4.3, 4.4 and 4.5 we derive a computational characterization of all stabilizing PI and PID controllers. The characterization for PI controllers is in a quasi-closed form while that for PID controllers involves the solution of a linear programming (LP) problem. These characterizations are analogous to the YJBK parametrization [46] of all stabilizing controllers with the difference that the YJBK parametrization cannot accommodate constraints on the controller order or structure whereas the characterization here works with these constraints from the very outset. Some possible applications of the LP characterization of all stabilizing PID gain values is discussed in Section 4.6. In addition, the quasi-closed form and LP characterization of all stabilizing PI and PID controllers opens up the possibility of optimizing various performance criteria when the controller structure is constrained to be either of the PI or PID type. Such optimization problems will be discussed in the next chapter.

4.2 A Characterization of All Stabilizing Feedback Gains

In this section, we make use of Theorem 3.5.5 to provide a complete *analytical* solution to the constant gain stabilization problem. To this end, consider the feedback system shown in Fig. 4.1. Here r is the command signal, y is the

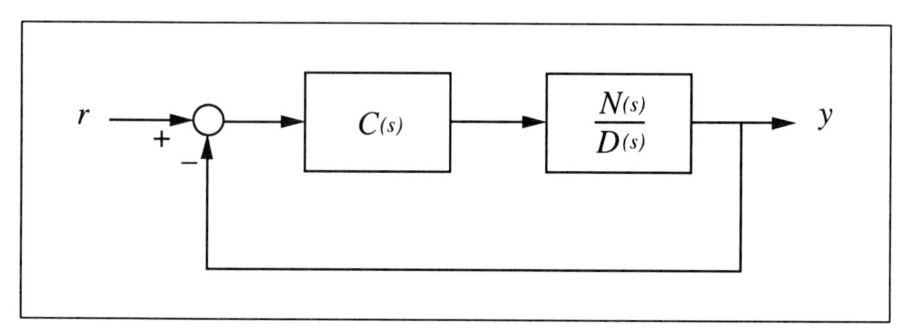

Fig. 4.1. Feedback control system.

output,

$$G(s) = \frac{N(s)}{D(s)}$$

is the plant to be controlled, $N(s)$ and $D(s)$ are coprime polynomials, and $C(s)$ is the controller to be designed. In the case of constant gain stabilization,

$$C(s) = k$$

so that the closed loop characteristic polynomial $\delta(s,k)$ is given by

$$\delta(s,k) = D(s) + kN(s). \tag{4.1}$$

Our objective is to determine those values of k, if any, for which the closed loop system is stable, *i.e.*, $\delta(s,k)$ is Hurwitz.

There are several classical approaches for solving this problem: the root locus technique, the Nyquist stability criterion, and the Routh-Hurwitz criterion. Of these approaches, the root locus technique and the Nyquist stability criterion solve this problem by plotting the root loci of $\delta(s,k)$ and the Nyquist plot of $G(s) = \frac{N(s)}{D(s)}$, respectively. Both of these methods are graphical in nature and fail to provide us with an analytical characterization of all stabilizing k's. The Routh-Hurwitz criterion, on the other hand, does provide us with an analytical solution. However, the set of stabilizing k's must be determined by solving a set of polynomial inequalities, a task which is not straight forward especially for higher order plants.

Let us now see how the results of the last chapter can be used to determine the values of k for which $\delta(s,k)$ in (4.1) is Hurwitz stable. As discussed in

Section 3.2, both the even as well as the odd parts of $\delta(s,k)$ depend on k and this creates difficulties when trying to use Lemma 3.2.1 to ensure the Hurwitz stability of $\delta(s,k)$. Consequently, starting from $\delta(s,k)$, we will now construct a polynomial for which only the even part depends on k, and to which Theorem 3.5.5 is applicable.

To this end, consider (4.1) with the even-odd decompositions:

$$\begin{aligned} N(s) &= N_e(s^2) + sN_o(s^2) \\ D(s) &= D_e(s^2) + sD_o(s^2). \end{aligned}$$

Suppose that the degree of $D(s)$ is n while the degree of $N(s)$ is m and $m \leq n$. Define

$$N^*(s) := N(-s) = N_e(s^2) - sN_o(s^2).$$

Multiplying $\delta(s,k)$ by $N^*(s)$ and examining the resulting polynomial, we obtain

$$\begin{aligned} l(\delta(s,k)N^*(s)) - r(\delta(s,k)N^*(s)) &= l(\delta(s,k)) - r(\delta(s,k)) \\ &\quad + l(N^*(s)) - r(N^*(s)) \\ &= l(\delta(s,k)) - r(\delta(s,k)) \\ &\quad + l(N(-s)) - r(N(-s)) \\ &= l(\delta(s,k)) - r(\delta(s,k)) \\ &\quad - (l(N(s)) - r(N(s))). \end{aligned}$$

Now, $\delta(s,k)$ of degree n is Hurwitz if and only if $l(\delta(s,k)) = n$ and $r(\delta(s,k)) = 0$. Furthermore, from Theorem 3.5.5

$$\sigma_i(\delta(s,k)N^*(s)) = l(\delta(s,k)N^*(s)) - r(\delta(s,k)N^*(s)).$$

Therefore we have the following.

Lemma 4.2.1. *$\delta(s,k)$ is Hurwitz if and only if*

$$\sigma_i(\delta(s,k)N^*(s)) = n - (l(N(s)) - r(N(s))). \tag{4.2}$$

Our task now is to determine those values of k, if any, for which (4.2) holds. It is straight forward to verify that

$$\delta(s,k)N^*(s) = h_1(s^2) + kh_2(s^2) + sg_1(s^2)$$

where

$$\begin{aligned} h_1(s^2) &= D_e(s^2)N_e(s^2) - s^2D_o(s^2)N_o(s^2) \\ h_2(s^2) &= N_e(s^2)N_e(s^2) - s^2N_o(s^2)N_o(s^2) \\ g_1(s^2) &= N_e(s^2)D_o(s^2) - D_e(s^2)N_o(s^2). \end{aligned}$$

Substituting $s = j\omega$, we obtain

$$\delta(j\omega, k)N^*(j\omega) = p(\omega, k) + jq(\omega) \tag{4.3}$$

where

$$\begin{aligned} p(\omega, k) &= p_1(\omega) + kp_2(\omega) &(4.4)\\ p_1(\omega) &= [D_e(-\omega^2)N_e(-\omega^2) + \omega^2 D_o(-\omega^2)N_o(-\omega^2)] &(4.5)\\ p_2(\omega) &= [N_e(-\omega^2)N_e(-\omega^2) + \omega^2 N_o(-\omega^2)N_o(-\omega^2)] &(4.6)\\ q(\omega) &= \omega[N_e(-\omega^2)D_o(-\omega^2) - D_e(-\omega^2)N_o(-\omega^2)]. &(4.7) \end{aligned}$$

Also, define

$$\begin{aligned} p_f(\omega, k) &= \frac{p(\omega, k)}{(1+\omega^2)^{\frac{m+n}{2}}} \\ q_f(\omega) &= \frac{q(\omega)}{(1+\omega^2)^{\frac{m+n}{2}}}. \end{aligned}$$

Since $l(N(s)) - r(N(s))$ is known and fixed, the stabilizing values of k can be determined from (4.2).

The formal statement of our main result on constant gain stabilization involves certain *strings* of the real numbers 0, 1 and -1. These strings are used to essentially capture all the different possibilities for the sign of $p_f(\omega, k)$ at the real zeros of $q_f(\omega)$ (with odd multiplicities). Among these, we are interested in only those possibilities, which when substituted into the expression for $\sigma_i(\delta(s, k)N^*(s))$, calculated from formula (3.9), yields a value for which (4.2) holds.

For clarity of presentation, we first introduce these strings and motivate some definitions before formally stating the result.

Definition 4.2.1. *Let the integers m, n and the function $q_f(\omega)$ be as already defined. Let $0 = \omega_0 < \omega_1 < \omega_2 < \cdots < \omega_{l-1}$ be the real, non-negative, distinct finite zeros of $q_f(\omega)$ with odd multiplicities*[1]*. Define a sequence of numbers $i_0, i_i, i_2, \cdots, i_l$ as follows:*
(i) If $N^*(j\omega_t) = 0$ for some $t = 1, 2, \cdots, l-1$, then define

$$i_t = 0;$$

(ii) If $N^*(s)$ has a zero of multiplicity k_n at the origin, then define

$$i_0 = sgn[p_{1_f}^{(k_n)}(0)]$$

where

$$p_{1_f}(\omega) := \frac{p_1(\omega)}{(1+\omega^2)^{\frac{(m+n)}{2}}};$$

[1] Note that these zeros are independent of k.

(iii) For all other $t = 0, 1, 2, \cdots, l$,

$$i_t \in \{-1, 1\}.$$

With $i_0, i_1, \cdots$ *defined in this way, we define the set* A *as*

$$A := \begin{cases} \{\{i_0, i_1, \cdots, i_l\}\} & \text{if } n+m \text{ is even} \\ \{\{i_0, i_1, \cdots, i_{l-1}\}\} & \text{if } n+m \text{ is odd.} \end{cases}$$

In other words A *is the set of all possible strings of 1's, 0's and -1's, whose length is* l *or* $l+1$ *depending on the value of* $n+m$*, and subject to the restrictions outlined in (i), (ii) and (iii).*

Next we introduce the set $A(\gamma)$ of strings in A with a prescribed "imaginary signature" γ. To do so, we first need to define the "imaginary signature" $\gamma(\mathcal{I})$ associated with any element $\mathcal{I} \in A$. This definition is motivated by Theorem 4.2.1 to follow.

Definition 4.2.2. *Let the integers* m*,* n *and the functions* $q(\omega)$*,* $q_f(\omega)$ *be as already defined. Let* $0 = \omega_0 < \omega_1 < \omega_2 < \cdots < \omega_{l-1}$ *be the real, non-negative, distinct finite zeros of* $q_f(\omega)$ *with odd multiplicities. Also define* $\omega_l = \infty$*. For each string* $\mathcal{I} = \{i_0, i_1, \cdots\}$ *in* A*, let* $\gamma(\mathcal{I})$ *denote the "imaginary signature" associated with the string* $\mathcal{I}$ *defined by*

$$\gamma(\mathcal{I}) : = \begin{cases} \{i_0 - 2i_1 + 2i_2 + \cdots + (-1)^{l-1} 2i_{l-1} \\ +(-1)^l i_l\} \cdot (-1)^{l-1} sgn[q(\infty)] \\ \qquad \text{for } m+n \text{ even} \\ \\ \{i_0 - 2i_1 + 2i_2 + \cdots + (-1)^{l-1} 2i_{l-1}\} \\ \cdot(-1)^{l-1} sgn[q(\infty)] \\ \qquad \text{for } m+n \text{ odd} \end{cases} \tag{4.8}$$

Remark 4.2.1. Note that if we make the identification $i_0 = sgn[p_f^{(k_n)}(0,k)]$, $i_t = sgn[p_f(\omega_t, k)]$ fot $t \neq 0$, then the imaginary signature of $\delta(s,k)N^*(s)$ as determined from (3.9) is the same as the quantity $\gamma(\mathcal{I})$ defined above. Hence, referring to $\gamma(\mathcal{I})$ as the "imaginary signature" of $\mathcal{I}$ is appropriate terminology.

Definition 4.2.3. *The set of strings in* A *with a prescribed imaginary signature* $\gamma = \psi$ *is denoted by* $A(\psi)$*. We also define the set of feasible strings for the constant gain stabilization problem as*

$$F^* = A(n - (l(N(s)) - r(N(s)))).$$

The following example illustrates these definitions.

Example 4.2.1. From (4.3), we have

$$\delta(j\omega, k)N^*(j\omega) = p(\omega, k) + jq(\omega)$$

where

$$p(\omega, k) = p_1(\omega) + kp_2(\omega).$$

Now suppose, for example, that $\delta(s, k)$ is of degree $n = 6$, the degree of $N^*(s)$ is 4, and $l(N(s)) - r(N(s)) = 2$. Let $q(\omega)$ have 3 real, non-negative, distinct finite zeros ω_0, ω_1, ω_2 with odd multiplicities. Also define $\omega_3 = \infty$ and let $N^*(j\omega_i) \neq 0$ for $i = 0, 1, 2$. Suppose $q(\omega)$ is as shown in Fig. 4.2.

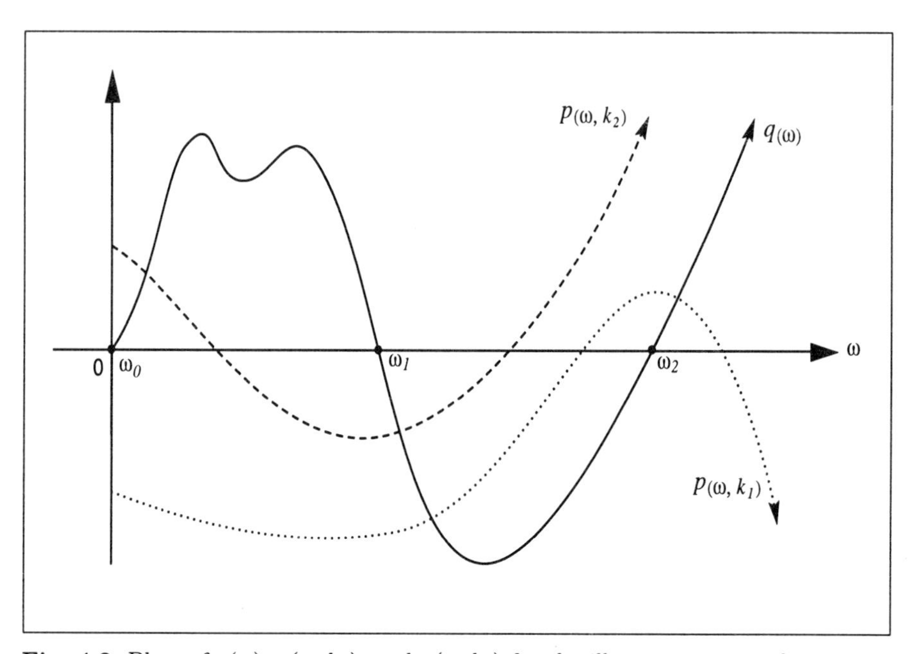

Fig. 4.2. Plots of $q(\omega)$, $p(\omega, k_1)$, and $p(\omega, k_2)$ for the illustrative example

The set A of all the possible strings $\{i_0, i_1, i_2, i_3\}$ is as follows:

$$A = \left\{ \begin{array}{ll} \{-1,-1,-1,-1\} & \{1,-1,-1,-1\} \\ \{-1,-1,-1,1\} & \{1,-1,-1,1\} \\ \{-1,-1,1,-1\} & \{1,-1,1,-1\} \\ \{-1,-1,1,1\} & \{1,-1,1,1\} \\ \{-1,1,-1,-1\} & \{1,1,-1,-1\} \\ \{-1,1,-1,1\} & \{1,1,-1,1\} \\ \{-1,1,1,-1\} & \{1,1,1,-1\} \\ \{-1,1,1,1\} & \{1,1,1,1\} \end{array} \right\}$$

From Lemma 4.2.1, we have $\delta(s, k)$ is Hurwitz if and only if

$$\sigma_i(\delta(s,k)N^*(s)) \quad = \quad n - (l(N(s)) - r(N(s))) = 4.$$

Since

$$(-1)^{(l-1)}\text{sgn}[q(\infty)] = 1,$$

it follows using Definition 4.2.3 that every string $\mathcal{I} = \{i_0, i_1, i_2, i_3\} \in F^*$ must satisfy

$$i_0 - 2i_1 + 2i_2 - i_3 = 4 \tag{4.9}$$

so that

$$F^* = \{\{-1,-1,1,-1\},\{1,-1,1,1\}\}.$$

Therefore, the constant gain stabilization problem now reduces to the problem of determining the values of k, if any, such that $\text{sgn}[p_f(\omega_j,k)] = i_j$, $j = 0,\ 1,\ 2,\ 3$ and $\{i_0,i_1,i_2,i_3\} \in F^*$. For instance, for $k = k_1$, $p(\omega,k_1)$ shown in Fig. 4.2 has

$$\begin{aligned}\{\text{sgn}[p_f(\omega_0,k_1)],\text{sgn}[p_f(\omega_1,k_1)],\text{sgn}[p_f(\omega_2,k_1)],\text{sgn}[p_f(\omega_3,k_1)]\}\\ = \{-1,-1,1,-1\}.\end{aligned}$$

On the other hand, for $k = k_2$, $p(\omega,k_2)$ shown in Fig. 4.2 has

$$\begin{aligned}\{\text{sgn}[p_f(\omega_0,k_2)],\text{sgn}[p_f(\omega_1,k_2)],\text{sgn}[p_f(\omega_2,k_2)],\text{sgn}[p_f(\omega_3,k_2)]\}\\ = \{1,-1,1,1\}.\end{aligned}$$

Thus for both $k = k_1$ and $k = k_2$, $\delta(s,k)$ is Hurwitz.

Having illustrated the above definitions we are now ready to state the main result of this section.

Theorem 4.2.1. *(Constant Gain Stabilization) The constant gain feedback stabilization problem is solvable for a given plant with transfer function $G(s)$ if and only if the following conditions hold:*
(i) F^* is not empty where F^* is as already defined, i.e., *at least one feasible string exists*
and
(ii) There exists a string $\mathcal{I} = \{i_0, i_1, \cdots\} \in F^*$ such that

$$\max_{i_t\in\mathcal{I},\, i_t\,>\,0}\left[-\frac{1}{G(j\omega_t)}\right] < \min_{i_t\in\mathcal{I},\, i_t\,<\,0}\left[-\frac{1}{G(j\omega_t)}\right]$$

where ω_0, ω_1, $\omega_2, \cdots$ are as already defined. Furthermore, if the above condition is satisfied by the feasible strings $\mathcal{I}_1, \mathcal{I}_2, \cdots, \mathcal{I}_s \in F^*$, then the set of all stabilizing gains is given by $K = \cup_{r=1}^{s} K_r$ where

$$K_r = \left(\max_{i_t\in\mathcal{I}_r,\, i_t\,>\,0}\left[-\frac{1}{G(j\omega_t)}\right],\ \min_{i_t\in\mathcal{I}_r,\, i_t\,<\,0}\left[-\frac{1}{G(j\omega_t)}\right]\right),\quad r = 1,2,\cdots,s.$$

Proof. Now from (4.2), we know that $\delta(s,k)$ is Hurwitz if and only if

$$\sigma_i(\delta(s,k)N^*(s)) = n - (l(N(s)) - r(N(s))).$$

Thus $\delta(s,k)$ is Hurwitz if and only if

$$\mathcal{I} \in A(n - (l(N(s)) - r(N(s))))$$

where

$$\mathcal{I} = \{i_0, i_1, \cdots\},$$
$$i_0 = sgn[p_f^{(k_n)}(\omega_0, k)],$$
$$i_j = sgn[p_f(\omega_j, k)],$$

and

$$j = 1, 2, \cdots, l$$

or

$$j = 1, 2, \cdots, l-1$$

accordingly as $m+n$ is even or odd. Let us now consider two different cases:
Case 1 $N^*(s)$ does not have any zeros on the imaginary axis: In this case, for all stabilizing values of the gain k, $\delta(s,k)N^*(s)$ will also not have any zeros on the $j\omega$-axis so that $i_j \in \{-1, 1\}$ for $j = 0, 1, 2, \cdots, l$. Next we consider the two different possibilities:
(a) $i_j > 0$: If $i_j > 0$, then the stability requirement is

$$p_1(\omega_j) + kp_2(\omega_j) > 0.$$

From (4.6), we note that

$$p_2(\omega) = |N(j\omega)|^2.$$

Since $N^*(s)$ does not have any zeros on the $j\omega$-axis, it follows that $p_2(\omega_j) > 0$. Hence

$$k > -\frac{p_1(\omega_j)}{p_2(\omega_j)}. \tag{4.10}$$

(b) $i_j < 0$: If $i_j < 0$, then the stability requirement is

$$p_1(\omega_j) + kp_2(\omega_j) < 0.$$

Once again, since $p_2(\omega_j) > 0$, it follows that

$$k < -\frac{p_1(\omega_j)}{p_2(\omega_j)}. \tag{4.11}$$

Case 2 $N^*(s)$ has one or more zeros on the $j\omega$-axis including a zero of multiplicity k_n at the origin: In this case, for all stabilizing values of the gain k, $\delta(s,k)N^*(s)$ will also have the same set of $j\omega$-axis zeros. Furthermore, it is clear that these zero locations will be a subset of $\{\omega_0, \omega_1, \cdots, \omega_{l-1}\}$. Since

the location of these zeros depends on $N^*(s)$ and is independent of the gain k, it is reasonable to expect that such a zero, at ω_m say, will not impose any additional constraint on k. Instead, it will only mandate that $i_m \in \mathcal{I}$ be constrained to a particular value. We next proceed to establish rigorously these facts. We consider two possibilities:
(a) $m \neq 0$: Here $N^*(s)$ has a zero at $j\omega_m$ where $\omega_m \neq 0$. This implies that

$$N_e(-\omega_m^2) = N_o(-\omega_m^2) = 0$$

so that from (4.5), (4.6) we obtain

$$p_1(\omega_m) = 0$$

and

$$p_2(\omega_m) = 0.$$

Thus from (4.4), it follows that

$$p(\omega_m, k) = 0.$$

Thus $i_m = 0$ independent of k and this constraint on $\mathcal{I}$ was already incorporated into the definition of A.
(b) $m = 0$: Here $N^*(s)$ has a zero at the origin of multiplicity k_n. Since

$$N^*(j\omega) = N_e(-\omega^2) - j\omega N_o(-\omega^2)$$

it follows that $N_e(-\omega^2)$ and $\omega N_o(-\omega^2)$ must each have zeros at the origin of multiplicity at least k_n. Thus from (4.6), we see that $p_2(\omega)$ will have a zero at the origin of multiplicity $2k_n$ so that for $k_n \geq 1$,

$$p_{2_f}^{(k_n)}(0) = 0.$$

Since

$$p_f^{(k_n)}(0, k) = p_{1_f}^{(k_n)}(0) + k p_{2_f}^{(k_n)}(0)$$

it follows that for $k_n \geq 1$

$$p_f^{(k_n)}(0, k) = p_{1_f}^{(k_n)}(0)$$

independent of k. Thus, although no constraints on k appear, we must have

$$i_0 = sgn[p_{1_f}^{(k_n)}(0)].$$

Once again, we note that this condition has been explicitly incorporated into the definition of A.

Of the two cases discussed above, only Case 1 imposes constraints on k as given by (4.10) and (4.11). This leads us to the conclusion that each $i_j > 0$ in the string $\mathcal{I} \in A(n - (l(N(s)) - r(N(s))))$ contributes a lower bound on k while each $i_j < 0$ contributes an upper bound on k. Thus, if the string

$\mathcal{I} \in A(n-(l(N(s))-r(N(s))))$ is to correspond to a stabilizing k then we must have

$$\max_{i_t \in \mathcal{I}, i_t > 0} \left[-\frac{p_1(\omega_t)}{p_2(\omega_t)}\right] < \min_{i_t \in \mathcal{I}, i_t < 0} \left[-\frac{p_1(\omega_t)}{p_2(\omega_t)}\right]. \tag{4.12}$$

Now

$$\begin{aligned} G(s) &= \frac{N(s)}{D(s)} \\ &= \frac{N_e(s^2)+sN_o(s^2)}{D_e(s^2)+sD_o(s^2)} \end{aligned}$$

so that

$$\begin{aligned} \frac{1}{G(j\omega)} &= \frac{D_e(-\omega^2)+j\omega D_o(-\omega^2)}{N_e(-\omega^2)+j\omega N_o(-\omega^2)} \\ &= \frac{[D_e(-\omega^2)+j\omega D_o(-\omega^2)][N_e(-\omega^2)-j\omega N_o(-\omega^2)]}{[N_e(-\omega^2)+j\omega N_o(-\omega^2)][N_e(-\omega^2)-j\omega N_o(-\omega^2)]} \\ &= \frac{[D_e(-\omega^2)N_e(-\omega^2)+\omega^2 D_o(-\omega^2)N_o(-\omega^2)]}{[N_e(-\omega^2)N_e(-\omega^2)+\omega^2 N_o(-\omega^2)N_o(-\omega^2)]} \\ &\quad +j\frac{\omega[N_e(-\omega^2)D_o(-\omega^2)-D_e(-\omega^2)N_o(-\omega^2)]}{[N_e(-\omega^2)N_e(-\omega^2)+\omega^2 N_o(-\omega^2)N_o(-\omega^2)]} \\ &= \frac{p_1(\omega)+jq(\omega)}{p_2(\omega)}. \end{aligned} \tag{4.13}$$

Since $q(\omega_t)=0$, it follows that

$$-\frac{p_1(\omega_t)}{p_2(\omega_t)} = -\frac{1}{G(j\omega_t)}.$$

Thus, from (4.12), we must have

$$\max_{i_t \in \mathcal{I}, i_t > 0} \left[-\frac{1}{G(j\omega_t)}\right] < \min_{i_t \in \mathcal{I}, i_t < 0} \left[-\frac{1}{G(j\omega_t)}\right]$$

which is condition (ii) in the Theorem statement. This completes the proof of the necessary and sufficient conditions for the existence of a stabilizing k. The set of all stabilizing k's is now determined by taking the union of all k's that are obtained from all the feasible strings which satisfy (ii). ♣

Remark 4.2.2. It is appropriate to point out here that Theorem 4.2.1 parts (i) and (ii) do provide a characterization of all plants that are stabilizable by a constant gain. Also note that a necessary condition for F^* to be nonempty is that for $m+n$ even,

$$l \geq \frac{|n-(l(N(s))-r(N(s)))|}{2}$$

and for $m+n$ odd,

$$l \geq \frac{|n-(l(N(s))-r(N(s)))|+1}{2}.$$

We now present examples to illustrate the detailed calculations involved when using Theorem 4.2.1 to analytically determine the set of all stabilizing feedback gains.

Example 4.2.2. Consider the constant gain stabilization problem with

$$\begin{aligned} D(s) &= s^5+11s^4+22s^3+60s^2+47s+25 \\ N(s) &= s^4+6s^3+12s^2+54s+16. \end{aligned}$$

The closed loop characteristic polynomial is

$$\delta(s,k)=D(s)+kN(s)$$

Here $N_e(s^2)=s^4+12s^2+16$ and $N_o(s^2)=6s^2+54$ so that

$$\begin{aligned} N^*(s) &= N(-s) \\ &= N_e(s^2)-sN_o(s^2). \end{aligned}$$

Therefore

$$\begin{aligned} \delta(s,k)N^*(s) &= (5s^8+6s^6-549s^4-1278s^2+400) \\ &\quad +k(s^8-12s^6-472s^4-2532s^2+256) \\ &\quad +s(s^8-32s^6-627s^4-2474s^2-598) \end{aligned}$$

so that

$$\delta(j\omega,k)N^*(j\omega) = p_1(\omega)+kp_2(\omega)+jq(\omega)$$

with

$$\begin{aligned} p_1(\omega) &:= 5\omega^8-6\omega^6-549\omega^4+1278\omega^2+400 \\ p_2(\omega) &:= \omega^8+12\omega^6-472\omega^4+2532\omega^2+256 \\ q(\omega) &:= \omega(\omega^8+32\omega^6-627\omega^4+2474\omega^2-598). \end{aligned}$$

The real, non-negative, distinct finite zeros of $q_f(\omega)$ with odd multiplicities are

$$\omega_0=0,\ \omega_1=0.50834,\ \omega_2=2.41735,\ \omega_3=2.91515.$$

Since $n+m=9$, which is odd, and $N^*(s)$ has no roots on the $j\omega$ axis, from Definition 4.2.1, the set A becomes

$$A = \left\{ \begin{array}{ll} \{-1,-1,-1,-1\} & \{1,-1,-1,-1\} \\ \{-1,-1,-1,1\} & \{1,-1,-1,1\} \\ \{-1,-1,1,-1\} & \{1,-1,1,-1\} \\ \{-1,-1,1,1\} & \{1,-1,1,1\} \\ \{-1,1,-1,-1\} & \{1,1,-1,-1\} \\ \{-1,1,-1,1\} & \{1,1,-1,1\} \\ \{-1,1,1,-1\} & \{1,1,1,-1\} \\ \{-1,1,1,1\} & \{1,1,1,1\} \end{array} \right\}$$

Since $l(N(s)) - r(N(s)) = 4$ and $(-1)^{l-1} sgn[q(\infty)] = -1$, it follows using Definition 4.2.3 that every string $\mathcal{I} = \{i_0,\ i_1,\ i_2,\ i_3\} \in F^*$ must satisfy

$$-(i_0 - 2i_1 + 2i_2 - 2i_3) = 1.$$

Hence $F^* = \{I_1, I_2, I_3\}$ where

$$\begin{aligned} \mathcal{I}_1 &= \{1,-1,-1,1\} \\ \mathcal{I}_2 &= \{1,1,1,1\} \\ \mathcal{I}_3 &= \{1,1,-1,-1\}. \end{aligned}$$

Furthermore,

$$\begin{aligned} -\frac{1}{G(j\omega_0)} &= -1.56250, \\ -\frac{1}{G(j\omega_1)} &= -0.78898, \\ -\frac{1}{G(j\omega_2)} &= 2.50345, \\ -\frac{1}{G(j\omega_3)} &= 22.49390. \end{aligned}$$

Hence from Theorem 4.2.1, we have

$$\left\{ \begin{array}{l} K_1 = \emptyset \text{ for } \mathcal{I}_1 \\ K_2 = (22.49390,\ \infty) \text{ for } \mathcal{I}_2 \\ K_3 = (-0.78898,\ 2.50345) \text{ for } \mathcal{I}_3. \end{array} \right.$$

Therefore $\delta(s,k)$ is Hurwitz for $k \in (-0.78898,\ 2.50345) \cup (22.49390,\ \infty)$.

Example 4.2.3. Consider the constant gain stabilization problem with

$$\begin{aligned} D(s) &= s^5 - 2s^4 + 3s^3 + 7s^2 + 10s + 7 \\ N(s) &= s^5 + 2s^4 + 3s^2 + 4s + 1. \end{aligned}$$

The closed loop characteristic polynomial is

$$\delta(s,k) = D(s) + kN(s).$$

Here $N_e(s^2) = 2s^4 + 3s^2 + 1$ and $N_o(s^2) = s^4 + 4$ so that

$$\begin{aligned} N^*(s) &= N(-s) \\ &= N_e(s^2) - sN_o(s^2). \end{aligned}$$

Therefore

$$\begin{aligned} \delta(s,k)N^*(s) &= (-s^{10} - 7s^8 - 6s^6 + 21s^4 - 12s^2 + 7) \\ &\quad +k(-s^{10} + 4s^8 + 4s^6 + 13s^4 - 10s^2 + 1) \\ &\quad +s(4s^8 + 2s^6 + 31s^4 + 5s^2 - 18) \end{aligned}$$

so that

$$\delta(j\omega,k)N^*(j\omega) = p_1(\omega) + kp_2(\omega) + jq(\omega)$$

with

$$\begin{aligned} p_1(\omega) &:= \omega^{10} - 7\omega^8 + 6\omega^6 + 21\omega^4 + 12\omega^2 + 7 \\ p_2(\omega) &:= \omega^{10} + 4\omega^8 - 4\omega^6 + 13\omega^4 + 10\omega^2 + 1 \\ q(\omega) &:= \omega(4\omega^8 - 2\omega^6 + 31\omega^4 - 5\omega^2 - 18). \end{aligned}$$

The real, non-negative, distinct finite zeros of $q_f(\omega)$ with odd multiplicities are

$$\omega_0 = 0,\ \omega_1 = 0.91146.$$

Since $n+m = 10$, which is even, and $N^*(s)$ has no roots on the $j\omega$-axis, from Definition 4.2.1, the set A becomes

$$A = \left\{ \begin{array}{ll} \{-1,-1,-1\} & \{1,-1,-1\} \\ \{-1,-1,1\} & \{1,-1,1\} \\ \{-1,1,-1\} & \{1,1,-1\} \\ \{-1,1,1\} & \{1,1,1\}. \end{array} \right\}$$

Since $l(N(s)) - r(N(s)) = 1$ and $(-1)^{l-1}sgn[q(\infty)] = -1$, it follows using Definition 4.2.3 that every string $\mathcal{I} = \{i_0,\ i_1,\ i_2\} \in F^*$ must satisfy

$$-(i_0 - 2i_1 + i_2) = 4.$$

Hence, $F^* = \{\mathcal{I}_1\}$ where

$$\mathcal{I}_1 = \{-1,1,-1\}.$$

Furthermore,

$$\begin{aligned} -\frac{1}{G(j\omega_0)} &= -7.0, \\ -\frac{1}{G(j\omega_1)} &= -1.74789, \\ -\frac{1}{G(j\omega_2)} &= -1.0. \end{aligned}$$

Hence from Theorem 4.2.1 we have

$$K_1 = \emptyset \text{ for } \mathcal{I}_1.$$

Therefore, in this example $G(s)$ is not stabilizable by a constant gain.

4.3 A Characterization of All Stabilizing PI Controllers

In this section, we show how the results developed in Section 4.2 for the constant gain stabilization problem can be extended to solve the problem of PI stabilization. Once again, we consider the feedback control system shown in Fig. 4.1. Since we now have a PI controller, this time $C(s)$ is given by

$$C(s) = k_p + \frac{k_i}{s} = \frac{k_i + k_p s}{s}.$$

The closed loop characteristic polynomial is

$$\delta(s, k_p, k_i) = sD(s) + (k_i + k_p s)N(s).$$

Let n be the degree of $\delta(s, k_p, k_i)$. The problem of stabilization using a PI controller, the first step in PI design, is to determine the values of k_p and k_i for which the closed loop characteristic polynomial $\delta(s, k_p, k_i)$ is Hurwitz.

Clearly, k_p and k_i both affect the even and odd parts of $\delta(s, k_p, k_i)$. Motivated by the approach used in Section 4.2, we now proceed to construct a new polynomial whose even part depends on k_i and odd part depends on k_p. Consider the even-odd decompositions

$$\begin{aligned} N(s) &= N_e(s^2) + sN_o(s^2) \\ D(s) &= D_e(s^2) + sD_o(s^2). \end{aligned}$$

Define

$$N^*(s) = N(-s) = N_e(s^2) - sN_o(s^2).$$

Let m be the degree of $N(s)$. Now, multiplying $\delta(s, k_p, k_i)$ by $N^*(s)$ and examining the resulting polynomial, we obtain

$$\begin{aligned} l(\delta(s, k_p, k_i)N^*(s)) - r(\delta(s, k_p, k_i)N^*(s)) &= l(\delta(s, k_p, k_i)) - r(\delta(s, k_p, k_i)) \\ &\quad -(l(N(s)) - r(N(s))). \end{aligned}$$

Now, $\delta(s, k_p, k_i)$ of degree n is Hurwitz if and only if $l(\delta(s, k_p, k_i)) = n$ and $r(\delta(s, k_p, k_i)) = 0$. Therefore, in view of Theorem 3.5.5, we have the following.

Lemma 4.3.1. *$\delta(s, k_p, k_i)$ is Hurwitz if and only if*

$$\sigma_i(\delta(s, k_p, k_i)N^*(s)) = n - (l(N(s)) - r(N(s))) \tag{4.14}$$

Our task now is to determine those values of k_p, k_i for which (4.14) holds. It can be verified that

$$\begin{aligned}\delta(s,k_p,k_i)N^*(s) &= [s^2(N_e(s^2)D_o(s^2)-D_e(s^2)N_o(s^2))\\ &\quad +k_i(N_e(s^2)N_e(s^2)-s^2N_o(s^2)N_o(s^2))]\\ &\quad +s[D_e(s^2)N_e(s^2)-s^2D_o(s^2)N_o(s^2)\\ &\quad +k_p(N_e(s^2)N_e(s^2)-s^2N_o(s^2)N_o(s^2))] \qquad (4.15)\end{aligned}$$

Substituting $s=j\omega$, we obtain

$$\delta(j\omega,k_p,k_i)N^*(j\omega) = p(\omega,\ k_i)+jq(\omega,\ k_p)$$

where

$$\begin{aligned}p(\omega,\ k_i) &= p_1(\omega)+k_ip_2(\omega)\\ q(\omega,\ k_p) &= q_1(\omega)+k_pq_2(\omega)\\ p_1(\omega) &= -\omega^2(N_e(-\omega^2)D_o(-\omega^2)-D_e(-\omega^2)N_o(-\omega^2))\\ p_2(\omega) &= N_e(-\omega^2)N_e(-\omega^2)+\omega^2N_o(-\omega^2)N_o(-\omega^2))\\ q_1(\omega) &= \omega(D_e(-\omega^2)N_e(-\omega^2)+\omega^2D_o(-\omega^2)N_o(-\omega^2))\\ q_2(\omega) &= \omega(N_e(-\omega^2)N_e(-\omega^2)+\omega^2N_o(-\omega^2)N_o(-\omega^2)).\end{aligned}$$

Also, define

$$\begin{aligned}p_f(\omega,k_i) &= \frac{p(\omega,k_i)}{(1+\omega^2)^{\frac{m+n}{2}}}\\ q_f(\omega,k_p) &= \frac{q(\omega,k_p)}{(1+\omega^2)^{\frac{m+n}{2}}}.\end{aligned}$$

We first note that k_i, k_p appear affinely in $p(\omega,k_i)$, $q(\omega,k_p)$ respectively. Furthermore, for every fixed k_p, the zeros of $q(\omega,k_p)$ do not depend on k_i, and so the results of Section 4.2 are applicable verbatim. Thus, by sweeping over all real k_p and solving a constant gain stabilization problem at each stage, we can determine the set of all stabilizing (k_p,k_i) values for the given plant.

The range of k_p values over which the sweeping needs to be carried out can be considerably reduced in many cases. We recall from Section 4.2, Remark 4.2.2 that for a fixed k_p, a necessary condition for the existence of a stabilizing k_i value is that $q(\omega,k_p)$ have at least

$$\frac{|n-(l(N(s))-r(N(s)))|}{2}$$

or

$$\frac{|n-(l(N(s))-r(N(s)))|+1}{2}$$

real, non-negative, distinct finite zeros of odd multiplicities, accordingly as $m+n$ is even or odd. Such a necessary condition can be checked by proceeding as follows.

We first recall that

$$q(\omega,\ k_p) = \omega[U(\omega) + k_p V(\omega)] \tag{4.16}$$

where

$$\begin{aligned} U(\omega) &= D_e(-\omega^2)N_e(-\omega^2) + \omega^2 D_o(-\omega^2)N_o(-\omega^2) \\ V(\omega) &= N_e(-\omega^2)N_e(-\omega^2) + \omega^2 N_o(-\omega^2)N_o(-\omega^2). \end{aligned}$$

From (4.16), we see that $q(\omega,\ k_p)$ has at least one real non-negative root at the origin. Now applying the root locus ideas discussed in Appendix A, we can determine the real root distributions of $q(\omega,\ k_p)$ corresponding to different ranges of k_p. Then, using the fact that

$$n - (l(N(s)) - r(N(s)))$$

is known, one can identify the ranges of k_p for which $q(\omega, k_p)$ does not satisfy the necessary condition stated above. Such k_p ranges do not need to be swept over and can, therefore, be safely discarded.

We now present a simple example to illustrate the detailed calculations involved in determining the stabilizing (k_p, k_i) values for a given plant. This example also shows how the root locus ideas in Appendix A can be used to narrow down the sweeping range for k_p.

Example 4.3.1. Consider the problem of choosing stabilizing PI gains for the plant $G(s) = \frac{N(s)}{D(s)}$ where

$$\begin{aligned} D(s) &= s^5 + 3s^4 + 29s^3 + 15s^2 - 3s + 60 \\ N(s) &= s^3 + 6s^2 - 2s + 1. \end{aligned}$$

The closed loop characteristic polynomial is

$$\delta(s, k_p, k_i) = sD(s) + k_i N(s) + k_p s N(s).$$

We consider the even-odd decompositions for the polynomials $N(s)$ and $D(s)$:

$$\begin{aligned} D(s) &= D_e(s^2) + sD_o(s^2) \\ N(s) &= N_e(s^2) + sN_o(s^2) \end{aligned}$$

where

$$\begin{aligned} D_e(s^2) &= 3s^4 + 15s^2 + 60 \\ D_o(s^2) &= s^4 + 29s^2 - 3 \\ N_e(s^2) &= 6s^2 + 1 \\ N_o(s^2) &= s^2 - 2. \end{aligned}$$

Now

$$\begin{aligned} N^*(s) &= N(-s) \\ &= N_e(s^2) - sN_o(s^2) \\ &= (6s^2+1) - s(s^2-2). \end{aligned}$$

Therefore, from (4.15) we obtain

$$\begin{aligned} \delta(s,k_p,k_i)N^*(s) &= [s^2(3s^6+166s^4-19s^2+117)+k_i(-s^6+40s^4 \\ &\quad +8s^2+1)] + s[(-s^8-9s^6+154s^4+369s^2+60) \\ &\quad +k_p(-s^6+40s^4+8s^2+1)] \end{aligned}$$

so that

$$\delta(j\omega,k_p,k_i)N^*(j\omega) = [p_1(\omega)+k_ip_2(\omega)] + j[q_1(\omega)+k_pq_2(\omega)]$$

with

$$\begin{aligned} p_1(\omega) &= 3\omega^8 - 166\omega^6 - 19\omega^4 - 117\omega^2 \\ p_2(\omega) &= \omega^6 + 40\omega^4 - 8\omega^2 + 1 \\ q_1(\omega) &= -\omega^9 + 9\omega^7 + 154\omega^5 - 369\omega^3 + 60\omega \\ q_2(\omega) &= \omega^7 + 40\omega^5 - 8\omega^3 + \omega. \end{aligned}$$

Now we use the root locus results from Appendix A to specify the range of k_p values over which the sweeping should be carried out. To this end, let

$$\begin{aligned} U(\omega) &= -\omega^8 + 9\omega^6 + 154\omega^4 - 369\omega^2 + 60 \\ V(\omega) &= \omega^6 + 40\omega^4 - 8\omega^2 + 1 \end{aligned}$$

so that

$$q(\omega,\ k_p) = \omega[U(\omega) + k_pV(\omega)].$$

Now

$$\begin{aligned} &\frac{U(\omega)\frac{dV(\omega)}{d\omega} - V(\omega)\frac{dU(\omega)}{d\omega}}{U^2(\omega)} \\ &= [2\omega^{13} + 160\omega^{11} - 460\omega^9 - 1180\omega^7 - 26750\omega^5 + 8984\omega^3 - 222\omega]/ \\ &\quad [(-\omega^8 + 9\omega^6 + 154\omega^4 - 369\omega^2 + 60)^2] \end{aligned}$$

Then the distinct, finite k_i producing either real breakaway points or a root at the origin are

$$k_1 = -61.67086,\ k_2 = -60,\ k_3 = -2.54119,\ k_4 = 16.44309$$

and the corresponding zeros ω_i are

$$\omega_1 = \pm 0.16390,\ \omega_2 = 0,\ \omega_3 = \pm 2.60928,\ \omega_4 = \pm 0.55140$$

The real root distributions of $U(\omega) + k_p V(\omega) = 0$ with respect to the origin, corresponding to the different ranges of k_p, are given below:

$$
\begin{array}{rcl}
k_p \in (-\infty,\ -61.67086) & : & \text{no real roots} \\
k_p \in (-61.67086,\ -60) & : & \text{2 positive simple real roots} \\
 & & \text{2 negative simple real roots} \\
k_p \in (-60,\ -2.54119) & : & \text{1 positive simple real root} \\
 & & \text{1 negative simple real root} \\
k_p \in (-2.54119,\ 16.44309) & : & \text{3 positive simple real roots} \\
 & & \text{3 negative simple real roots} \\
k_p \in (16.44309,\ \infty) & : & \text{1 positive simple real root} \\
 & & \text{1 negative simple real root}
\end{array}
$$

Now, for this example, $m + n$ is odd and

$$n - (l(N(s)) - r(N(s))) = 6 - (1 - 2) = 7.$$

Hence, for a given fixed k_p, a necessary condition for the existence of a stabilizing k_i value is that $q(\omega, k_p)$ must have at least four distinct, real non-negative roots of odd multiplicities. The root distributions presented above show that this is possible only for

$$k_p \in (-2.54119, 16.44309).$$

For each k_p in this range we can use the constant gain stabilization result of Section 4.2 to determine the exact ranges of stabilizing k_i. By sweeping over all

$$k_p \in (-2.54119,\ 16.44309)$$

and using the constant gain stabilization result at each stage, we obtained the stabilizing region sketched in Fig. 4.3.

We now present another PI stabilization example. In this case, the set of all stabilizing gain values is the union of two disconnected sets.

Example 4.3.2. Consider the problem of choosing stabilizing PI gains for the plant $G(s) = \frac{N(s)}{D(s)}$ where

$$
\begin{aligned}
D(s) &= s^5 + 2s^4 + 23s^3 + 44s^2 + 97s + 98 \\
N(s) &= s^4 + 4s^3 + 23s^2 + 46s - 12.
\end{aligned}
$$

The closed loop characteristic polynomial is

$$\delta(s, k_p, k_i) = sD(s) + k_i N(s) + k_p s N(s).$$

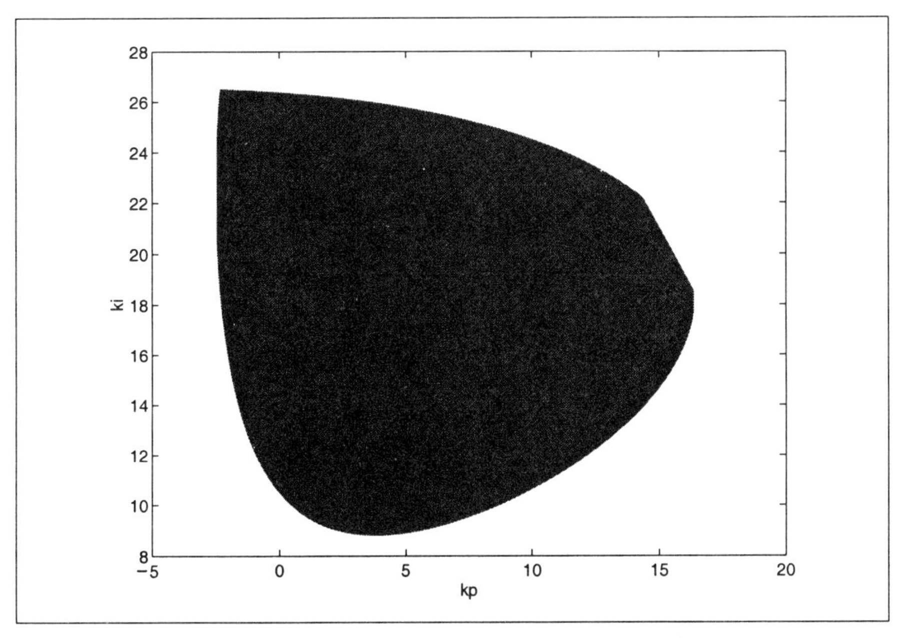

Fig. 4.3. The stabilizing set of $(k_p,\ k_i)$ values (Example 4.3.1).

We consider the even-odd decompositions for the polynomials $N(s)$ and $D(s)$, *i.e.*,

$$\begin{aligned} D(s) &= D_e(s^2) + sD_o(s^2) \\ N(s) &= N_e(s^2) + sN_o(s^2) \end{aligned}$$

where

$$\begin{aligned} D_e(s^2) &= 2s^4 + 44s^2 + 98 \\ D_o(s^2) &= s^4 + 23s^2 + 97 \\ N_e(s^2) &= s^4 + 23s^2 - 12 \\ N_o(s^2) &= 4s^2 + 46. \end{aligned}$$

Now

$$\begin{aligned} N^*(s) &= N(-s) \\ &= N_e(s^2) - sN_o(s^2) \\ &= (s^4 + 23s^2 - 12) - s(4s^2 + 46) \end{aligned}$$

Therefore, from (4.15) we obtain

$$\begin{aligned} \delta(s, k_p, k_i)N^*(s) = & \; [s^2(s^8 + 38s^6 + 346s^4 - 461s^2 - 5672) \\ & + k_i(s^8 + 30s^6 + 137s^4 - 2668s^2 + 144)] \\ & + s[(-2s^8 - 48s^6 - 360s^4 - 2736s^2 - 1176) \\ & + k_p(s^8 + 30s^6 + 137s^4 - 2668s^2 + 144)] \end{aligned}$$

so that

$$\delta(j\omega, k_p, k_i)N^*(j\omega) = [p_1(\omega) + k_i p_2(\omega)] + j[q_1(\omega) + k_p q_2(\omega)]$$

with

$$\begin{aligned} p_1(\omega) &= -\omega^{10} + 38\omega^8 - 346\omega^6 - 461\omega^4 + 5672\omega^2 \\ p_2(\omega) &= \omega^8 - 30\omega^6 + 137\omega^4 + 2668\omega^2 + 144 \\ q_1(\omega) &= -2\omega^9 + 48\omega^7 - 360\omega^5 + 2736\omega^3 - 1176\omega \\ q_2(\omega) &= \omega^9 - 30\omega^7 + 137\omega^5 + 2668\omega^3 + 144\omega. \end{aligned}$$

Using the root locus ideas presented in Appendix A, the range of k_p values over which the sweeping needs to be carried out was narrowed down to

$$k_p \in (-1.06347,\ 8.16667).$$

For each k_p in this range we can use the constant gain stabilization result of Section 4.2 to determine the exact ranges of stabilizing k_i. By sweeping over all

$$k_p \in (-1.06347,\ 8.16667)$$

and using the constant gain stabilization result at each stage, we obtained two disconnected stabilizing regions sketched in Fig. 4.4.

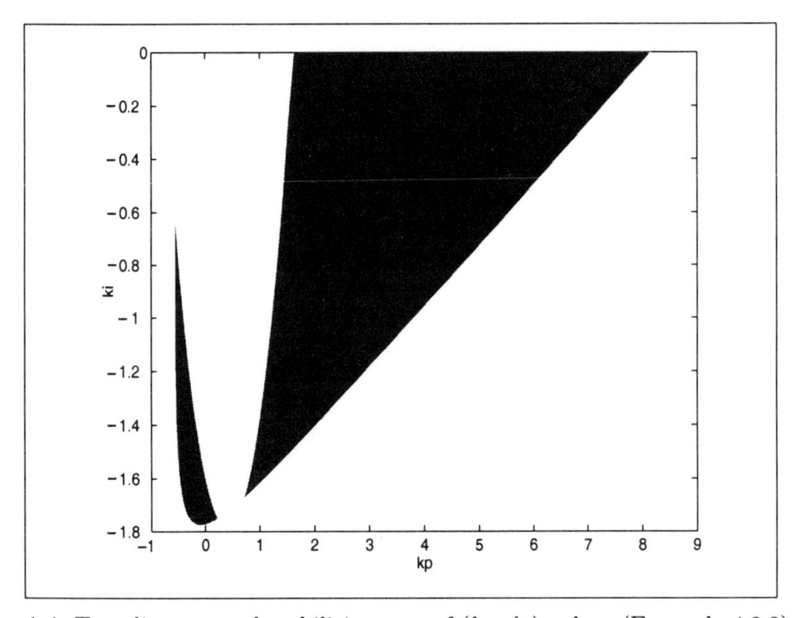

Fig. 4.4. Two disconnected stabilizing sets of $(k_p,\ k_i)$ values (Example 4.3.2).

4.4 A Characterization of All Stabilizing PID Controllers

In this section we show that the results developed in Section 4.2 for solving the constant gain stabilization problem can be extended to yield an elegant procedure for determining stabilizing PID gains k_p, k_i and k_d for a given plant. To this end, we consider the feedback control system of Fig. 4.1 where $C(s)$ is now chosen as:

$$C(s) = k_p + \frac{k_i}{s} + k_d s = \frac{k_i + k_p s + k_d s^2}{s}.$$

The closed loop characteristic polynomial becomes

$$\delta(s, k_p, k_i, k_d) = sD(s) + (k_i + k_d s^2)N(s) + k_p s N(s). \tag{4.17}$$

The problem of stabilization using a PID controller is to determine the values of k_p, k_i and k_d for which the closed loop characteristic polynomial $\delta(s, k_p, k_i, k_d)$ is Hurwitz. As shown by the following example, this can be quite a complicated task when using some of the classical stability tests such as the Routh-Hurwitz criterion.

Example 4.4.1. Consider the problem of choosing stabilizing PID gains for the plant $G(s) = \frac{N(s)}{D(s)}$ where

$$\begin{aligned} D(s) &= s^5 + 8s^4 + 32s^3 + 46s^2 + 46s + 17 \\ N(s) &= s^3 - 4s^2 + s + 2. \end{aligned}$$

The closed loop characteristic polynomial is

$$\begin{aligned} \delta(s, k_p, k_i, k_d) &= sD(s) + (k_i + k_p s + k_d s^2)N(s) \\ &= s^6 + (k_d + 8)s^5 + (k_p - 4k_d + 32)s^4 \\ &\quad + (k_i - 4k_p + k_d + 46)s^3 + (-4k_i + k_p + 2k_d + 46)s^2 \\ &\quad + (k_i + 2k_p + 17)s + 2k_i \end{aligned}$$

Using the Routh-Hurwitz criterion to determine the stabilizing values for k_p, k_i and k_d, one can determine that the following inequalities must hold:

$$
\left\{
\begin{array}{l}
k_d + 8 > 0 \\[2ex]
k_p k_d - 4k_d^2 - k_i + 12k_p - k_d + 210 > 0 \\[2ex]
k_i k_p k_d - 4k_p^2 k_d + 16k_p k_d^2 - 6k_d^3 - k_i^2 + 16k_i k_p + 63k_i k_d - 48k_p^2 \\
+48k_p k_d - 263k_d^2 + 428k_i - 336k_p - 683k_d + 6852 > 0 \\[2ex]
-4k_i^2 k_p k_d + 16k_i k_p^2 k_d - 52k_i k_p k_d^2 - 6k_p^3 k_d + 24k_p^2 k_d^2 - 6k_p k_d^3 - 12k_d^4 \\
+4k_i^3 - 64k_i^2 k_p - 264k_i^2 k_d + 198k_i k_p^2 - 9k_i k_p k_d + 1238k_i k_d^2 - 72k_p^3 \\
-213k_p^2 k_d + 957k_p k_d^2 - 1074k_d^3 - 1775k_i^2 + 2127k_i k_p + 7688k_i k_d \\
-3924k_p^2 + 3027k_p k_d - 11322k_d^2 - 10746k_i - 31338k_p - 1836k_d \\
+206919 > 0 \\[2ex]
-6k_i^3 k_p k_d + 24k_i^2 k_p^2 k_d - 84k_i^2 k_p k_d^2 - 6k_i k_p^3 k_d + 60k_i k_p^2 k_d^2 \\
-102k_i k_p k_d^3 - 12k_p^4 k_d + 48k_p^3 k_d^2 - 12k_p^2 k_d^3 - 24k_p k_d^4 + 6k_i^4 \\
-96k_i^3 k_p - 390k_i^3 k_d + 294k_i^2 k_p^2 - 285k_i^2 k_p k_d + 1476k_i^2 k_d^2 \\
-60k_i k_p^3 + 969k_i k_p^2 k_d - 1221k_i k_p k_d^2 - 132k_i k_d^3 - 144k_p^4 - 528k_p^3 k_d \\
+2322k_p^2 k_d^2 - 2250k_p k_d^3 - 204k_d^4 - 2487k_i^3 + 273k_i^2 k_p - 2484k_i^2 k_d \\
+5808k_i k_p^2 + 10530k_i k_p k_d + 34164k_i k_d^2 - 9072k_p^3 \\
+2433k_p^2 k_d - 6375k_p k_d^2 - 18258k_d^3 - 92961k_i^2 + 79041k_i k_p \\
+184860k_i k_d - 129384k_p^2 + 47787k_p k_d - 192474k_d^2 \\
-549027k_i - 118908k_p - 31212k_d + 3517623 > 0 \\[2ex]
2k_i > 0
\end{array}
\right.
\tag{4.18}
$$

Clearly, the above inequalities are highly nonlinear and there is no straight forward method for obtaining a solution.

Remark 4.4.1. The constant gain stabilization problem considered in Section 4.2 can be solved by several classical approaches such as the root locus technique, the Nyquist stability criterion, and the Routh-Hurwitz criterion. However, the same is not true for PID stabilization where none of these classical techniques are of much help. For the Routh-Hurwitz criterion, this is already clear from the above example.

Now let us consider an approach similar to that of Section 4.2 for determining the values of $(k_p,\ k_i,\ k_d)$ for which $\delta(s, k_p, k_i, k_d)$ is Hurwitz. Recall that

$$\delta(s, k_p, k_i, k_d) = sD(s) + (k_i + k_d s^2)N(s) + k_p s N(s).$$

Clearly, all three parameters k_p, k_i and k_d affect both the even and odd parts of $\delta(s, k_p, k_i, k_d)$. We now proceed as in Section 4.2 to construct a new polynomial whose even part depends on (k_i, k_d) and whose odd part depends on k_p.

Consider the even-odd decompositions

$$\begin{aligned} N(s) &= N_e(s^2) + sN_o(s^2) \\ D(s) &= D_e(s^2) + sD_o(s^2). \end{aligned}$$

Define

$$N^*(s) = N(-s) = N_e(s^2) - sN_o(s^2).$$

Also let n, m be the degrees of $\delta(s, k_p, k_i, k_d)$ and $N(s)$ respectively. Now, multiplying $\delta(s, k_p, k_i, k_d)$ by $N^*(s)$ and examining the resulting polynomial, we obtain

$$\begin{aligned} & l(\delta(s, k_p, k_i, k_d)N^*(s)) - r(\delta(s, k_p, k_i, k_d)N^*(s)) \\ = \ & (l(\delta(s, k_p, k_i, k_d)) - r(\delta(s, k_p, k_i, k_d))) - (l(N(s)) - r(N(s))). \end{aligned}$$

Now, $\delta(s, k_p, k_i, k_d)$ of degree n is Hurwitz if and only if

$$l(\delta(s, k_p, k_i, k_d)) = n$$

and

$$r(\delta(s, k_p, k_i, k_d)) = 0.$$

Therefore, in view of Theorem 3.5.5, we have the following.

Lemma 4.4.1. *$\delta(s, k_p, k_i, k_d)$ is Hurwitz if and only if*

$$\sigma_i(\delta(s, k_p, k_i, k_d)N^*(s)) = n - (l(N(s)) - r(N(s))) \tag{4.19}$$

Our task now is to determine those values of k_p, k_i, k_d for which (4.19) holds. It is straight forward to verify that

$$\begin{aligned} \delta(s, k_p, k_i, k_d)N^*(s) = \ & [s^2(N_e(s^2)D_o(s^2) - D_e(s^2)N_o(s^2)) \\ & +(k_i + k_d s^2)(N_e(s^2)N_e(s^2) - s^2N_o(s^2)N_o(s^2))] \\ & +s[D_e(s^2)N_e(s^2) - s^2D_o(s^2)N_o(s^2) \\ & +k_p(N_e(s^2)N_e(s^2) - s^2N_o(s^2)N_o(s^2))]. \end{aligned} \tag{4.20}$$

Substituting $s = j\omega$, we obtain

$$\delta(j\omega, k_p, k_i, k_d)N^*(j\omega) = p(\omega, k_i, k_d) + jq(\omega, k_p)$$

where

$$\begin{aligned} p(\omega, k_i, k_d) &= p_1(\omega) + (k_i - k_d\omega^2)p_2(\omega) \\ q(\omega, k_p) &= q_1(\omega) + k_p q_2(\omega) \\ p_1(\omega) &= -\omega^2(N_e(-\omega^2)D_o(-\omega^2) - D_e(-\omega^2)N_o(-\omega^2)) \\ p_2(\omega) &= N_e(-\omega^2)N_e(-\omega^2) + \omega^2N_o(-\omega^2)N_o(-\omega^2)) \\ q_1(\omega) &= \omega(D_e(-\omega^2)N_e(-\omega^2) + \omega^2D_o(-\omega^2)N_o(-\omega^2)) \\ q_2(\omega) &= \omega(N_e(-\omega^2)N_e(-\omega^2) + \omega^2N_o(-\omega^2)N_o(-\omega^2)) \end{aligned}$$

Also, define

$$p_f(\omega, k_i, k_d) = \frac{p(\omega, k_i, k_d)}{(1+\omega^2)^{\frac{m+n}{2}}}$$
$$q_f(\omega, k_p) = \frac{q(\omega, k_p)}{(1+\omega^2)^{\frac{m+n}{2}}}.$$

We first note that k_i, k_d appear affinely in $p(.,.,.)$ while k_p appears affinely in $q(.,.)$. Furthermore, for every fixed k_p, the zeros of $q(\omega, k_p)$ will not depend on k_i or k_d and so we can use the approach of Section 4.2 to determine stabilizing values for k_i and k_d. Since there are two variables here, we are no longer able to obtain a closed form solution. Instead, a linear programming problem has to be solved for each fixed k_p. As k_p is varied, we will have a one parameter family of linear programming problems to solve. Before formally stating our main result on PID stabilization, we first introduce some definitions. These definitions are essentially the same as those introduced in Section 4.2 with the only difference that the present definitions *are conditioned on* k_p *being held at some fixed value.*

Definition 4.4.1. *Let* m, n, $q_f(\omega, k_p)$ *be as already defined. For a given fixed* k_p, *let* $0 = \omega_0 < \omega_1 < \omega_2 < \cdots < \omega_{l-1}$ *be the real, non-negative, distinct finite zeros of* $q_f(\omega, k_p)$ *with odd multiplicities*[2]. *Define a sequence of numbers* $i_0, i_i, i_2, \cdots, i_l$ *as follows:*
(i) If $N^*(j\omega_t) = 0$ *for some* $t = 1, 2, \cdots, l-1$, *then define*

$$i_t = 0;$$

(ii) If $N^*(s)$ *has a zero of multiplicity* k_n *at the origin, then define*

$$i_0 = sgn[p_{1_f}^{(k_n)}(0)]$$

where

$$p_{1_f}(\omega) := \frac{p_1(\omega)}{(1+\omega^2)^{\frac{(m+n)}{2}}};$$

(iii) For all other $t = 0, 1, 2, \cdots, l$,

$$i_t \in \{-1, 1\}.$$

With $i_0, i_1, \cdots$ *defined in this way, we define the set* A_{k_p} *as*

$$A_{k_p} := \begin{cases} \{\{i_0, i_1, \cdots, i_l\}\} & \text{if } n+m \text{ is even} \\ \{\{i_0, i_1, \cdots, i_{l-1}\}\} & \text{if } n+m \text{ is odd.} \end{cases}$$

In other words A_{k_p} *is the set of all possible strings of 1's, 0's and -1's, whose length is* l *or* $l+1$ *depending on the value of* $n+m$, *and subject to the restrictions outlined in (i), (ii) and (iii).*

[2] Note that these zeros are independent of k_i or k_d.

Next we introduce the set $A_{k_p}(\gamma)$ of strings in A_{k_p} with a prescribed "imaginary signature" γ. To do so, we first need to define the "imaginary signature" $\gamma(\mathcal{I})$ associated with any element $\mathcal{I} \in A_{k_p}$. This definition is motivated by Theorem 4.4.1 to follow.

Definition 4.4.2. *Let m, n, $q(\omega, k_p)$, $q_f(\omega, k_p)$ be as already defined. For a given fixed k_p, let $0 = \omega_0 < \omega_1 < \omega_2 < \cdots < \omega_{l-1}$ be the real, non-negative, distinct finite zeros of $q_f(\omega, k_p)$ with odd multiplicities. Also define $\omega_l = \infty$. For each string $\mathcal{I} = \{i_0, i_1, \cdots\}$ in A_{k_p}, let $\gamma(\mathcal{I})$ denote the "imaginary signature" associated with the string $\mathcal{I}$ defined by*

$$\gamma(\mathcal{I}) := \begin{cases} \{i_0 - 2i_1 + 2i_2 + \cdots + (-1)^{l-1}2i_{l-1} \\ +(-1)^l i_l\} \cdot (-1)^{l-1} sgn[q(\infty, k_p)] \\ \qquad \text{for } m+n \text{ even} \\ \\ \{i_0 - 2i_1 + 2i_2 + \cdots + (-1)^{l-1}2i_{l-1}\} \\ \cdot(-1)^{l-1} sgn[q(\infty, k_p)] \\ \qquad \text{for } m+n \text{ odd} \end{cases} \tag{4.21}$$

Note that referring to $\gamma(\mathcal{I})$ as the "imaginary signature" of $\mathcal{I}$ can be justified as in Section 4.2.

Definition 4.4.3. *The set of strings in A_{k_p} with a prescribed imaginary signature $\gamma = \psi$ is denoted by $A_{k_p}(\psi)$. For a given fixed k_p, we also define the set of feasible strings for the PID stabilization problem as*

$$F^*_{k_p} = A_{k_p}(n - (l(N(s)) - r(N(s))))$$

We are now ready to state the main result of this section.

Theorem 4.4.1. *(Main Result on PID Stabilization) The PID stabilization problem, with a fixed k_p, is solvable for a given plant with transfer function $G(s)$ if and only if the following conditions hold:*
(i) $F^*_{k_p}$ is not empty where $F^*_{k_p}$ is as already defined, i.e., *at least one feasible string exists*
and
(ii) There exists a string $\mathcal{I} = \{i_0, i_1, \cdots\} \in F^*_{k_p}$ and values of k_i and k_d such that $\forall\, t = 0, 1, 2, \cdots$ for which $N^*(j\omega_t) \neq 0$

$$p(\omega_t, k_i, k_d)i_t > 0. \tag{4.22}$$

where $p(\omega, k_i, k_d)$ is as already defined. Furthermore, if there exist values of k_i and k_d such that the above condition is satisfied for the feasible strings $\mathcal{I}_1, \mathcal{I}_2, \cdots, \mathcal{I}_s \in F^*_{k_p}$, then the set of stabilizing (k_i, k_d) values corresponding to the fixed k_p is the union of the (k_i, k_d) values satisfying (4.22) for $\mathcal{I}_1, \mathcal{I}_2, \cdots, \mathcal{I}_s$.

Proof. From (4.19), we know that $\delta(s, k_p, k_i, k_d)$ is Hurwitz if and only if

$$\sigma_i(\delta(s, k_p, k_i, k_d)N^*(s)) = n - (l(N(s)) - r(N(s))).$$

Thus, for a fixed k_p, it follows that $\delta(s, k_p, k_i, k_d)$ is Hurwitz if and only if there exists $\mathcal{I} \in F^*_{k_p}$ where $\mathcal{I} = \{i_0, i_1, \cdots\}$, $i_0 = sgn[p_f^{(k_n)}(0, k_i, k_d)]$, $i_j = sgn[p_f(\omega_j, k_i, k_d)]$, and $j = 1, 2, \cdots, l$ or $j = 1, 2, \cdots, l-1$ accordingly as $m + n$ is even or odd. As in the proof of Theorem 4.2.1, it can be easily shown that if $N^*(j\omega_t) = 0$ for some $t = 0, 1, 2, \cdots$, then the value of the corresponding entry $i_t \in \mathcal{I}$ is pre-determined and is independent of k_i and k_d. The definition of A already accounts for such special cases. Focussing now on the other case, *i.e.*, $N^*(j\omega_t) \neq 0$, leads us to (4.22). This completes the proof of (i) and (ii). The characterization of all stabilizing (k_i, k_d) values, corresponding to the fixed k_p, now follows as an immediate consequence. ♣

Remark 4.4.2. It should be noted that since the constraint set is linear, the admissible set for (4.22) is either a convex polygon or an intersection of half planes, which is again a *convex set.* Therefore, for each fixed k_p, the region in the $(k_i,\ k_d)$ plane, if any, for which $\delta(s, k_p, k_i, k_d)$ is Hurwitz is a union of convex sets.

Remark 4.4.3. As in the case of PI stabilization, the range of k_p values over which the sweeping has to be carried out can be *a priori* narrowed down by using the root locus ideas presented in Appendix A.

By using Theorem 4.4.1 and sweeping over $k_p \in [k_{p_{min}},\ k_{p_{max}}]$, we can determine the entire set of values of $(k_p,\ k_i,\ k_d)$ for which $\delta(s, k_p, k_i, k_d)$ is Hurwitz. We now present an example to illustrate the details of the calculations.

Example 4.4.2. Consider the problem of choosing stabilizing PID gains for the same plant considered in Example 4.4.1, *i.e.*,

$$\begin{aligned} D(s) &= s^5 + 8s^4 + 32s^3 + 46s^2 + 46s + 17 \\ N(s) &= s^3 - 4s^2 + s + 2 \end{aligned}$$

and

$$C(s) = k_p + \frac{k_i}{s} + k_d s = \frac{k_i + k_p s + k_d s^2}{s}.$$

The closed loop characteristic polynomial is

$$\delta(s, k_p, k_i, k_d) = sD(s) + (k_i + k_d s^2)N(s) + k_p s N(s).$$

Write

$$\begin{aligned} D(s) &= D_e(s^2) + sD_o(s^2) \\ N(s) &= N_e(s^2) + sN_o(s^2) \end{aligned}$$

where

$$\begin{aligned} D_e(s^2) &= 8s^4 + 46s^2 + 17 \\ D_o(s^2) &= s^4 + 32s^2 + 46 \\ N_e(s^2) &= -4s^2 + 2 \\ N_o(s^2) &= s^2 + 1. \end{aligned}$$

Now

$$\begin{aligned} N^*(s) &= N(-s) \\ &= N_e(s^2) - sN_o(s^2) \\ &= (-4s^2 + 2) - s(s^2 + 1). \end{aligned}$$

Therefore, from (4.20) we obtain

$$\begin{aligned} \delta(s, k_p, k_i, k_d)N^*(s) &= [s^2(-12s^6 - 180s^4 - 183s^2 + 75) + (k_i + k_d s^2) \\ &\quad (-s^6 + 14s^4 - 17s^2 + 4)] + s[(-s^8 - 65s^6 \\ &\quad -246s^4 - 22s^2 + 34) + k_p(-s^6 + 14s^4 - 17s^2 + 4)] \end{aligned}$$

so that

$$\delta(j\omega, k_p, k_i, k_d)N^*(j\omega) = [p_1(\omega) + (k_i - k_d\omega^2)p_2(\omega)] + j[q_1(\omega) + k_p q_2(\omega)]$$

where

$$\begin{aligned} p_1(\omega) &= -12\omega^8 + 180\omega^6 - 183\omega^4 - 75\omega^2 \\ p_2(\omega) &= \omega^6 + 14\omega^4 + 17\omega^2 + 4 \\ q_1(\omega) &= -\omega^9 + 65\omega^7 - 246\omega^5 + 22\omega^3 + 34\omega \\ q_2(\omega) &= \omega^7 + 14\omega^5 + 17\omega^3 + 4\omega. \end{aligned}$$

Now for a fixed k_p, for instance $k_p = 1$,we have

$$\begin{aligned} q(\omega, 1) &= q_1(\omega) + q_2(\omega) \\ &= -\omega^9 + 66\omega^7 - 232\omega^5 + 39\omega^3 + 38\omega \end{aligned}$$

Then the real, non-negative, distinct finite zeros of $q_f(\omega, 1)$ with odd multiplicities are

$$\omega_0 = 0,\ \omega_1 = 0.74230,\ \omega_2 = 1.86590,\ \omega_3 = 7.89211.$$

Since $n+m = 9$ which is odd, and $N^*(s)$ has no $j\omega$-axis roots, from Definition 4.4.1, the set A_1 becomes

$$A_1 = \left\{ \begin{array}{ll} \{-1,-1,-1,-1\} & \{1,-1,-1,-1\} \\ \{-1,-1,-1,1\} & \{1,-1,-1,1\} \\ \{-1,-1,1,-1\} & \{1,-1,1,-1\} \\ \{-1,-1,1,1\} & \{1,-1,1,1\} \\ \{-1,1,-1,-1\} & \{1,1,-1,-1\} \\ \{-1,1,-1,1\} & \{1,1,-1,1\} \\ \{-1,1,1,-1\} & \{1,1,1,-1\} \\ \{-1,1,1,1\} & \{1,1,1,1\}. \end{array} \right\}$$

Since

$$l(N(s)) - r(N(s)) = -1$$

and

$$(-1)^{l-1} sgn[q(\infty, 1)] = 1,$$

it follows using Definition 4.4.3 that every string

$$\mathcal{I} = \{i_0, i_1, i_2, i_3\} \in F_1^*$$

must satisfy

$$i_0 - 2i_1 + 2i_2 - 2i_3 = 7.$$

Hence

$$F_1^* = \{1,\ -1,\ 1,\ -1\}.$$

Thus, it follows from Theorem 4.4.1 that the stabilizing (k_i, k_d) values corresponding to $k_p = 1$ must satisfy the string of inequalities:

$$\left\{ \begin{array}{l} p_1(\omega_0) + (k_i - k_d\omega_0^2)p_2(\omega_0) > 0 \\ p_1(\omega_1) + (k_i - k_d\omega_1^2)p_2(\omega_1) < 0 \\ p_1(\omega_2) + (k_i - k_d\omega_2^2)p_2(\omega_2) > 0 \\ p_1(\omega_3) + (k_i - k_d\omega_3^2)p_2(\omega_3) < 0 \end{array} \right.$$

Substituting for ω_0, ω_1, ω_2 and ω_3 in the above expressions, we obtain

$$\left\{ \begin{array}{l} k_i > 0 \\ k_i - 0.55101k_d < 3.81670 \\ k_i - 3.48158k_d > -12.19183 \\ k_i - 62.28540k_d < 464.03862 \end{array} \right. \tag{4.23}$$

The admissible set of values of $(k_i,\ k_d)$ for which (4.23) holds can be solved by linear programming and is shown in Fig. 4.5. Similarly, for k_p fixed at 5 say, we have

$$\begin{aligned} q(\omega, 5) &= q_1(\omega) + 5q_2(\omega) \\ &= -\omega^9 + 70\omega^7 - 176\omega^5 + 107\omega^3 + 54\omega \end{aligned}$$

Then the real, non-negative, distinct finite zeros of $q_f(\omega, 5)$ with odd multiplicities are

$$\omega_0 = 0,\ \omega_1 = 8.21054.$$

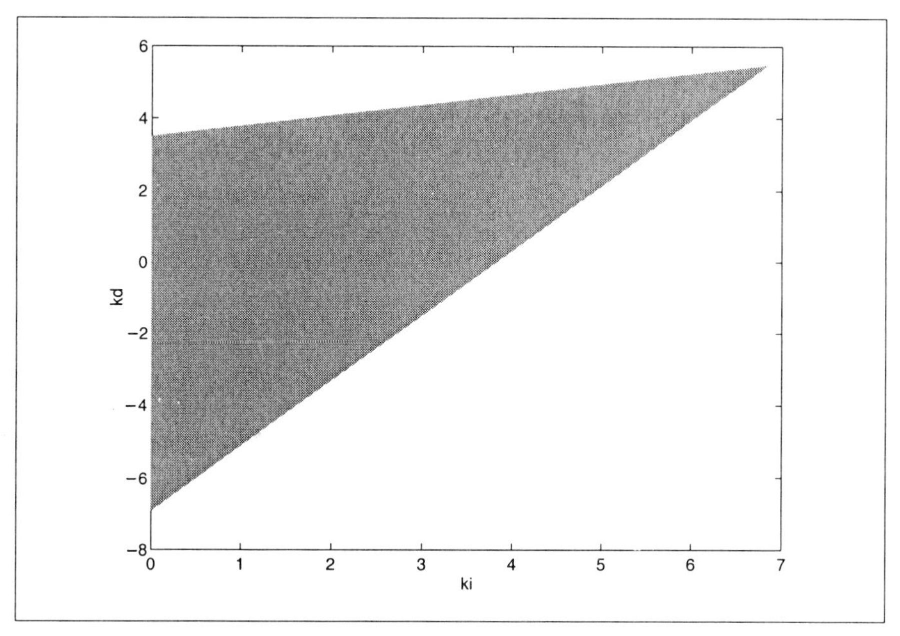

Fig. 4.5. The stabilizing set of (k_i, k_d) values when $k_p = 1$ (Example 4.4.2).

Since $n+m = 9$ which is odd, and $N^*(s)$ has no $j\omega$-axis roots, from Definition 4.4.1, the set A_5 becomes

$$A_5 = \left\{ \begin{array}{ll} \{-1,-1\} & \{1,1\} \\ \{-1,1\} & \{1,-1\} \end{array} \right\}$$

Since

$$l(N(s)) - r(N(s)) = -1$$

and

$$(-1)^{l-1} sgn[q(\infty, 5)] = 1,$$

it follows using Definition 4.4.3 that every string $\mathcal{I} = \{i_0, i_1\} \in F_5^*$ must satisfy

$$i_0 - 2i_1 = 7.$$

Hence

$$F_5^* = \phi$$

so that for $k_p = 5$, there are no stabilizing values for k_i and k_d.

Using the root locus ideas presented in Appendix A, the range of k_p values over which the sweeping needs to be carried out was narrowed down to

$$k_p \in (-8.5, 4.23337).$$

By sweeping over different k_p values in this interval and following the procedure illustrated above, we can generate the set of (k_p, k_i, k_d) values for which $\delta(s, k_p, k_i, k_d)$ is Hurwitz. This set is sketched in Fig. 4.6.

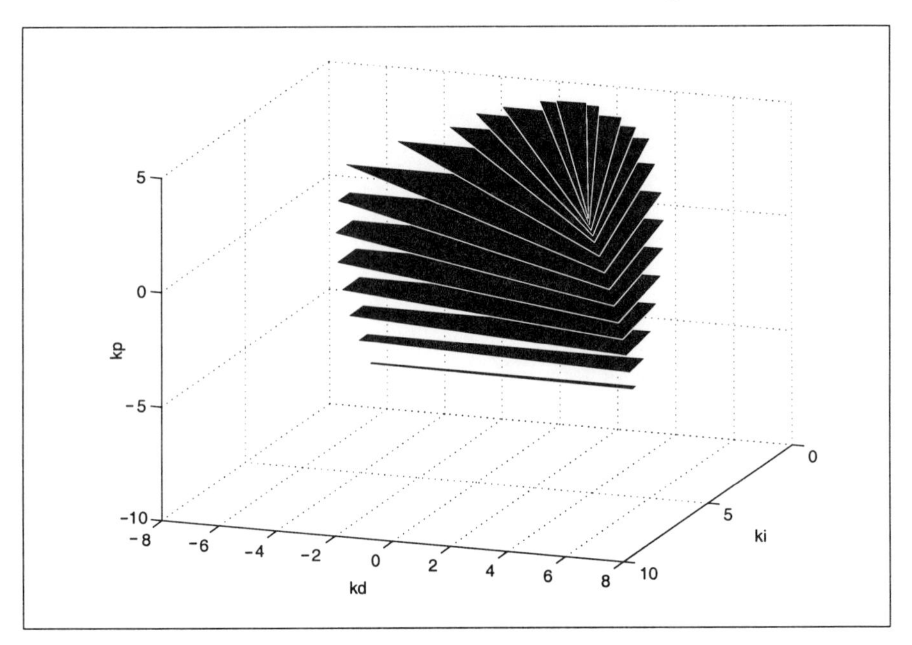

Fig. 4.6. The stabilizing set of (k_p, k_i, k_d) values (Example 4.4.2).

We now present another PID stabilization example. In this case, the set of all stabilizing PID gain values is the union of two disconnected sets.

Example 4.4.3. Consider the problem of choosing stabilizing PID gains for the plant $G(s) = \frac{N(s)}{D(s)}$ where

$$\begin{aligned} D(s) &= s^4 + 2s^3 + 3s^2 + 7s + 14 \\ N(s) &= s^3 + 3s^2 + s + 8 \end{aligned}$$

and

$$C(s) = k_p + \frac{k_i}{s} + k_d s = \frac{k_i + k_p s + k_d s^2}{s}.$$

The closed loop characteristic polynomial is

$$\delta(s, k_p, k_i, k_d) = sD(s) + (k_i + k_d s^2)N(s) + k_p sN(s).$$

We consider the even-odd decompositions for the polynomials $N(s)$ and $D(s)$, *i.e.*,

$$\begin{aligned} D(s) &= D_e(s^2) + sD_o(s^2) \\ N(s) &= N_e(s^2) + sN_o(s^2) \end{aligned}$$

where

$$\begin{aligned} D_e(s^2) &= s^4 + 3s^2 + 14 \\ D_o(s^2) &= 2s^2 + 7 \\ N_e(s^2) &= 3s^2 + 8 \\ N_o(s^2) &= s^2 + 1. \end{aligned}$$

Now

$$\begin{aligned} N^*(s) &= N(-s) \\ &= N_e(s^2) - sN_o(s^2) \\ &= (3s^2 + 8) - s(s^2 + 1). \end{aligned}$$

Therefore, from (4.20) we obtain

$$\begin{aligned} \delta(s, k_p, k_i, k_d)N^*(s) = & [s^2(-s^6 + 2s^4 + 20s^2 + 42) + (k_i + k_d s^2) \\ & (-s^6 + 7s^4 + 47s^2 + 64)] + s[(s^6 + 8s^4 \\ & +59s^2 + 112) + k_p(-s^6 + 7s^4 + 47s^2 + 64)] \end{aligned}$$

so that

$$\delta(j\omega, k_p, k_i, k_d)N^*(j\omega) = [p_1(\omega) + (k_i - k_d\omega^2)p_2(\omega)] + j[q_1(\omega) + k_p q_2(\omega)]$$

where

$$\begin{aligned} p_1(\omega) &= -\omega^8 - 2\omega^6 + 20\omega^4 - 42\omega^2 \\ p_2(\omega) &= \omega^6 + 7\omega^4 - 47\omega^2 + 64 \\ q_1(\omega) &= -\omega^7 + 8\omega^5 - 59\omega^3 + 112\omega \\ q_2(\omega) &= \omega^7 + 7\omega^5 - 47\omega^3 + 64\omega. \end{aligned}$$

Using the root locus ideas presented in Appendix A, the range of k_p values over which the sweeping needs to be carried out was narrowed down to

$$k_p \in (-3.27212, -1.75) \cup (0.52172, 1.55064).$$

By sweeping over different k_p values in these two intervals and using Theorem 4.4.1 repeatedly, we obtained two disconnected sets of $(k_p,\ k_i,\ k_d)$ values for which $\delta(s, k_p, k_i, k_d)$ is Hurwitz. These sets are sketched in Fig. 4.7.

4.5 PID Controllers Without Pure Derivative Action

The ideal PID controller

$$C(s) = k_p + \frac{k_i}{s} + k_d s \tag{4.24}$$

includes a pure derivative term $k_d s$. This pure derivative action is undesirable because of two reasons: (1) it can result in the amplification of high frequency

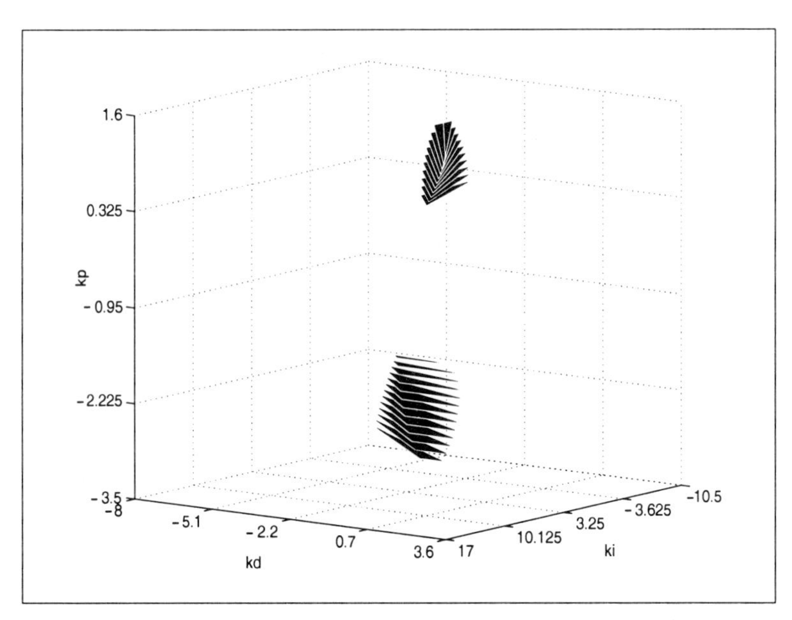

Fig. 4.7. Two disconnected stabilizing sets of $(k_p,\ k_i,\ k_d)$ values. (Example 4.4.3)

sensor noise; and (2) the closed loop system may not be internally stable, e.g., for a relative degree one plant, the transfer function from the command signal to the control input is improper.

This problem is usually avoided by replacing the pure derivative term $k_d s$ in (4.24) by

$$\frac{k_d s}{1+Ts}$$

where T is a small positive number typically chosen in the range from 0.01 to 0.1. With this modification, the new PID controller becomes

$$C(s) = k_p + \frac{k_i}{s} + \frac{k_d s}{1+Ts} = \frac{(k_d + k_p T)s^2 + (k_p + k_i T)s + k_i}{s(1+Ts)}. \tag{4.25}$$

An important question that now arises is whether the pure PID stabilization results of the last section can be suitably modified to handle this case. The aim of this section is to provide an affirmative answer to this question.

Accordingly we choose a fixed T and consider the problem of determining the values of $(k_p,\ k_i,\ k_d)$ in (4.25) for which the closed loop characteristic polynomial is Hurwitz. As we now show, this problem is equivalent to the problem of stabilizing a *modified* plant using an ideal PID controller.

Suppose that

$$G(s) = \frac{N(s)}{D(s)}$$

is the plant to be stabilized using the modified PID controller (4.25). The closed loop characteristic polynomial is given by

$$\delta(s, k_p, k_i, k_d) = s(1+Ts)D(s) + [(k_d + k_pT)s^2 + (k_p + k_iT)s + k_i]N(s).$$

Let us now define $k_p^{'}$, $k_i^{'}$, $k_d^{'}$ by

$$\begin{cases} k_p^{'} & := & k_p + k_iT \\ k_i^{'} & := & k_i \\ k_d^{'} & := & k_d + k_pT \end{cases} \tag{4.26}$$

Then the closed loop characteristic polynomial can be rewritten as

$$\delta(s, k_p, k_i, k_d) = s(1+Ts)D(s) + (k_d^{'}s^2 + k_p^{'}s + k_i^{'})N(s).$$

Comparing the above expression with (4.17), it is clear that the problem of determining (k_p, k_i, k_d) such that $\delta(s, k_p, k_i, k_d)$ is Hurwitz is equivalent to the problem of determining the gain values of the ideal PID controller

$$C^{'}(s) = k_p^{'} + \frac{k_i^{'}}{s} + k_d^{'}s$$

which stabilize the modified plant

$$\frac{N(s)}{(1+Ts)D(s)}. \tag{4.27}$$

The values of $(k_p^{'}, k_i^{'}, k_d^{'})$ which stabilize (4.27) can, of course, be determined using the results of the last section. The set of stabilizing gain values (k_p, k_i, k_d) for the original problem can then be obtained using the relationship

$$\begin{bmatrix} k_p \\ k_i \\ k_d \end{bmatrix} = \begin{bmatrix} 1 & -T & 0 \\ 0 & 1 & 0 \\ -T & T^2 & 1 \end{bmatrix} \begin{bmatrix} k_p^{'} \\ k_i^{'} \\ k_d^{'} \end{bmatrix} \tag{4.28}$$

which is an immediate consequence of the definitions in (4.26).

From the results of the last section, we know that for a fixed $k_p^{'}$, the stabilizing set of $(k_i^{'}, k_d^{'})$ values is a union of either polygons or intersections of half planes. Since the transformation (4.28) is linear, a polygon or a half plane in the $(k_p^{'}, k_i^{'}, k_d^{'})$ space will map into a polygon or a half plane in the (k_p, k_i, k_d) space. Moreover the vertex of a polygon in the $(k_p^{'}, k_i^{'}, k_d^{'})$ space maps into the vertex of the corresponding polygon in the (k_p, k_i, k_d) space. It, therefore, follows that some short cuts can be taken while using the linear transformation (4.28) to determine the stabilizing (k_p, k_i, k_d) values from the stabilizing $(k_p^{'}, k_i^{'}, k_d^{'})$ values determined using the results of the last section.

We now present an example where the controller is given by (4.25) and the approach discussed above is used to determine the stabilizing values of (k_p, k_i, k_d).

Example 4.5.1. Consider the problem of choosing stabilizing PID gains for the plant

$$G(s) = \frac{N(s)}{D(s)}$$

where

$$\begin{aligned} D(s) &= s^4 + 5s^3 + 10s^2 - s + 1 \\ N(s) &= s - 100 \end{aligned}$$

and the PID controller is given by (4.25) with $T = 0.1$. Defining

$$\begin{cases} k_p^{'} &:= k_p + k_i T \\ k_i^{'} &:= k_i \\ k_d^{'} &:= k_d + k_p T \end{cases}$$

we used the results of the last section to determine the set of all gain values of the PID controller

$$C^{'}(s) = k_p^{'} + \frac{k_i^{'}}{s} + k_d^{'} s$$

which stabilize the plant

$$\frac{N(s)}{(1+Ts)D(s)}.$$

This stabilizing set is sketched in Fig. 4.8.

Now using the linear transformation (4.28), we obtained the corresponding stabilizing set of $(k_p,\ k_i,\ k_d)$ values for the original plant $\frac{N(s)}{D(s)}$. This set is sketched in Fig. 4.9.

4.6 Some Applications of the Stabilizing PID Characterization

In this section, we describe some possible ways in which one could make use of the characterization of stabilizing PID controllers that we developed in the last two sections.

4.6.1 Assessing the Stability of a Ziegler-Nichols Design

We present below an example to show how the characterization of all stabilizing PID controllers can be used to assess the stability of PID controllers designed using other approaches. In particular, we focus on a PID controller designed using the classical Ziegler-Nichols frequency response method.

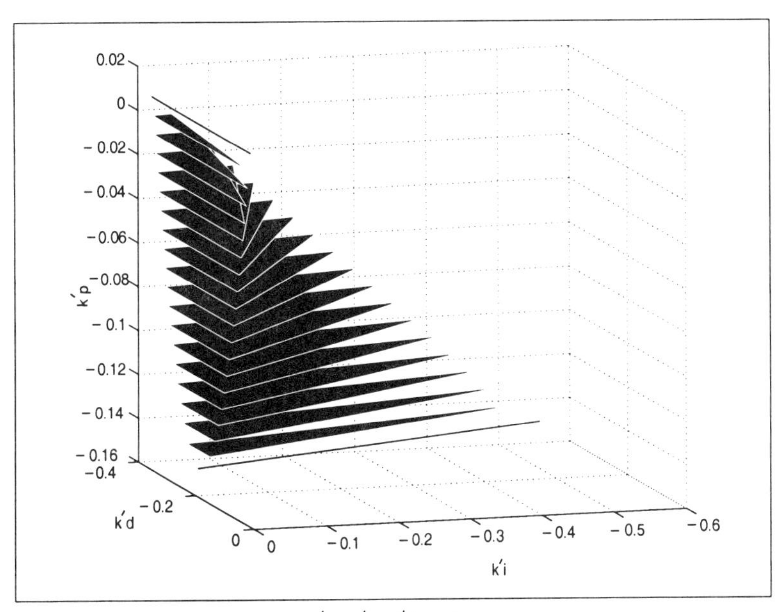

Fig. 4.8. The stabilizing set of $(k_p^{'}, k_i^{'}, k_d^{'})$ values.

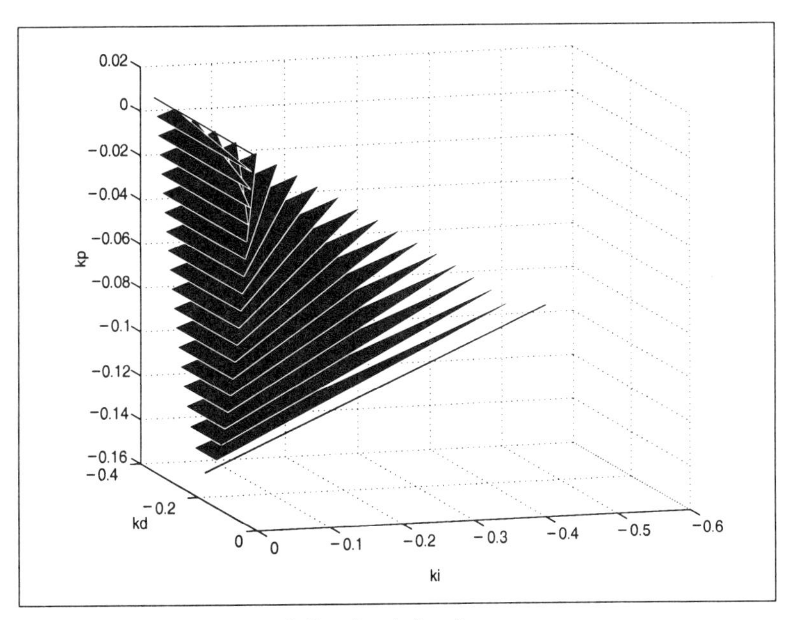

Fig. 4.9. The stabilizing set of (k_p, k_i, k_d) values.

Example 4.6.1. Consider the plant

$$G(s) = \frac{N(s)}{D(s)} = \frac{1}{(s+1)^8} \tag{4.29}$$

and the PID controller

$$C(s) = k_p + \frac{k_i}{s} + k_d s.$$

Using the results of Section 4.4, the the set of all stabilizing $(k_p,\ k_i,\ k_d)$ values for this plant can be determined as follows.

The closed loop characteristic polynomial is

$$\delta(s, k_p, k_i, k_d) = s(s+1)^8 + (k_d s^2 + k_p s + k_i)$$

and

$$N^*(s) = N(-s) = N(s) = 1.$$

Thus

$$\begin{aligned} &\delta(j\omega, k_p, k_i, k_d) N^*(j\omega) \\ &\quad = [8\omega^8 - 56\omega^6 + 56\omega^4 - 8\omega^2 \\ &\quad +(k_i - k_d\omega^2)] + j\omega(\omega^8 - 28\omega^6 \\ &\quad +70\omega^4 - 28\omega^2 + 1 + k_p). \end{aligned}$$

Since $N^*(s)$ is Hurwitz, it follows using the Hermite-Biehler Theorem that for any $(k_p,\ k_i,\ k_d)$, such that $\delta(s, k_p, k_i, k_d)$ is Hurwitz,

$$(\omega^8 - 28\omega^6 + 70\omega^4 - 28\omega^2 + 1 + k_p)$$

must have four and only four distinct positive real roots. Using the root locus ideas presented in Appendix A, we conclude that this is only possible for

$$k_p \in (-1,\ 2.0751).$$

By sweeping over

$$k_p \in (-1,\ 2.0751)$$

and using the results of the last section, we obtained the stabilizing set of $(k_p,\ k_i,\ k_d)$ values sketched in Fig. 4.10.

Now let us examine where in this plot, the parameters obtained from a Ziegler-Nichols frequency response design would be located. For the plant of this example, the ultimate gain $K_u = 1.883$ and the ultimate period $T_u = 15.16$. Hence, using the Ziegler-Nichols frequency response formulas, we obtain $k_p = 1.13$, $k_i = 0.149$, and $k_d = 2.147$. Now for k_p fixed at 1.13, the set of stabilizing (k_i, k_d) values can be obtained from Fig. 4.10. This set is sketched in Fig. 4.11. From Fig. 4.11 it is clear that for this example, the PID controller parameters obtained by the Ziegler-Nichols frequency response method are dangerously close to the stability boundary.

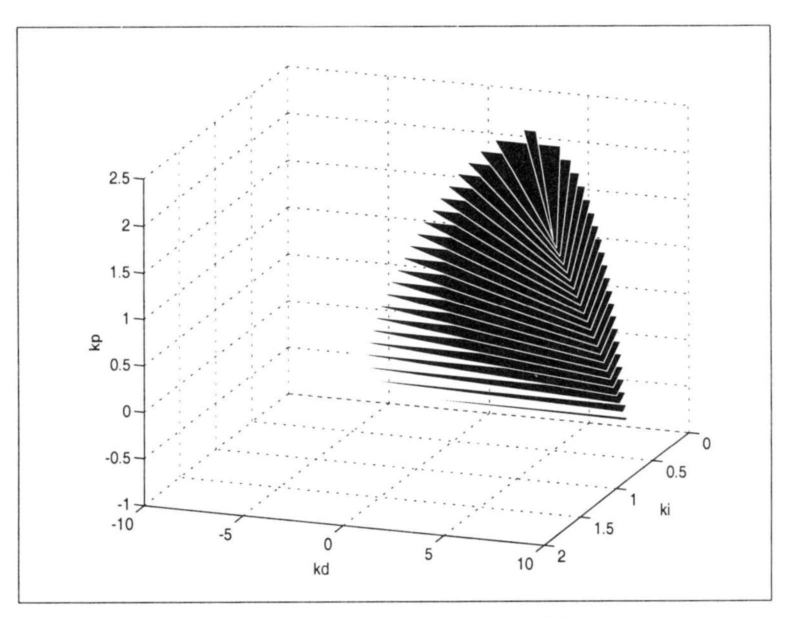

Fig. 4.10. The stabilizing region of (k_p, k_i, k_d) values (Example 4.6.1)

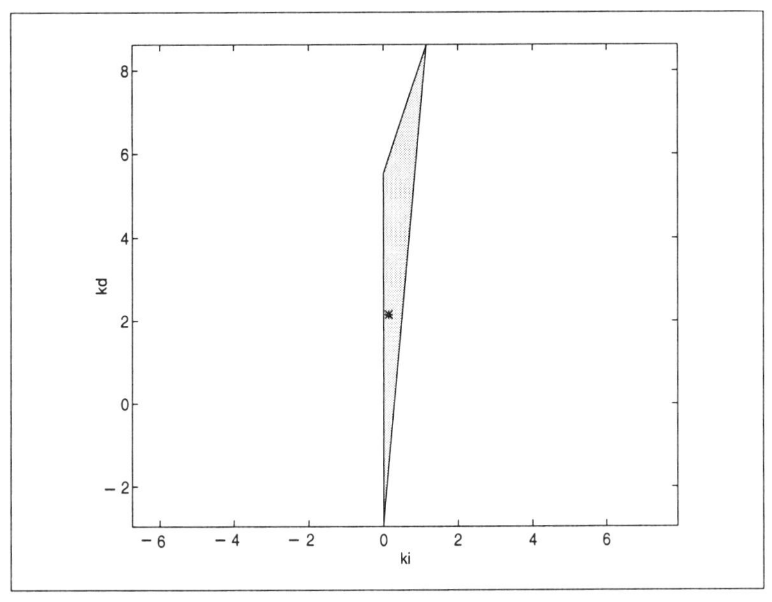

Fig. 4.11. The stabilizing set of (k_i, k_d) values when $k_p = 1.13$; $*$ denotes the parameters corresponding to the Ziegler-Nichols frequency response method. (Example 4.6.1)

4.6.2 A Possible Approach for Redesigning the PID Parameters

From Example 4.6.1, we know that the classical tuning methods may give PID controller parameters that are dangerously close to instability in the k_i-k_d plane. To avoid such an undesirable scenario, one could exploit the shape of the stability regions in the k_i-k_d plane to maximize the l_2 stability margin. This can be done as follows. Since for a fixed k_p, the stabilizing regions in the $(k_i,\ k_d)$ plane are either convex polygons or half-planes, one can choose the $(k_i,\ k_d)$ value which is at the center of the largest circle inscribed inside the stabilizing region. This requires that given a convex polygon, we be able to calculate the coordinates of the center of the largest inscribed circle. Accordingly, we next describe a systematic procedure for characterizing the largest circle inside a convex polygon.

Inscribing the Largest Circle Inside a Convex Polygon. Consider the general m-sided polygon illustrated in Fig. 4.12. This polygon can be represented analytically by a set of linear inequalities:

$$\mathcal{P} = \{x | {a_i}^T x \leq b_i, i = 1, ..., m\}$$

where each inequality generates the half plane containing one side of the polygon. Now the circle $\mathcal{C}$ inscribed in a polygon $\mathcal{P}$ can be expressed as:

$$\mathcal{C} = \{x_c + u |\ \|u\| \leq r\}$$

where x_c is the center of the circle (also called the Chebychev center), u is a vector with origin at the center of the circle, and r is the radius of the inscribed circle.

The circle $\mathcal{C}$ will lie in $\mathcal{P}$ if and only if:

$$sup\{{a_i}^T x_c + {a_i}^T u |\ \|u\| \leq r\} \leq b_i,\ (i = 1, ..., m).$$

This condition can be rewritten as follows:

$${a_i}^T x_c + r\|a_i\| \leq b_i,\ (i = 1, ..., m).$$

Hence, the problem of finding the largest circle $\mathcal{C}$ inside a polygon $\mathcal{P}$ is a Linear Programming (LP) problem:

$$\begin{aligned} &\text{maximize } r \\ &\text{subject to } {a_i}^T x_c + r\|a_i\| \leq b_i,\ (i = 1, ..., m). \end{aligned}$$

The following example illustrates the detailed calculations involved in using the above procedure to find the center x_c and the radius r of the largest inscribed circle.

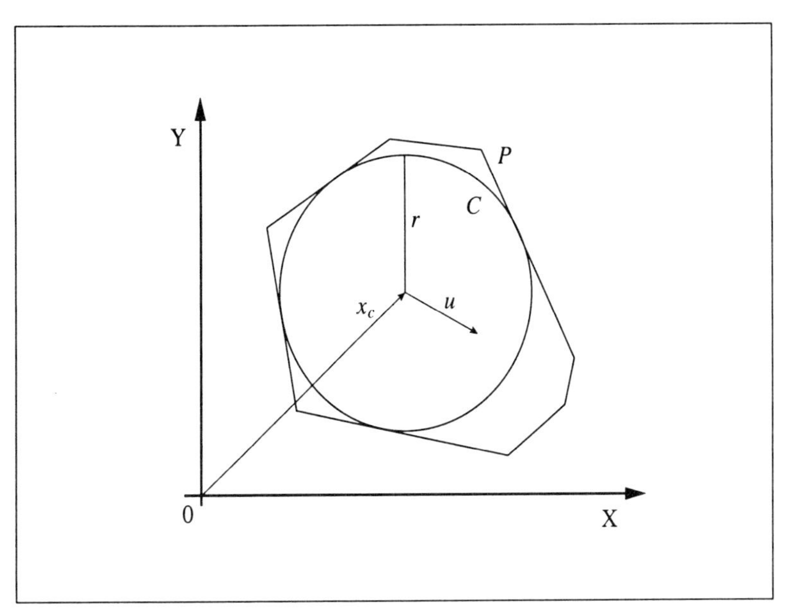

Fig. 4.12. The largest circle $\mathcal{C}$ inside a polygon $\mathcal{P}$.

Example 4.6.2. Consider a set of linear inequalities:

$$\left\{\begin{array}{rcl} [-1.3333,\ -1]x & \leq & -4 \\ [-1.3333,\ 1]x & \leq & 4 \\ [0,\ 1]x & \leq & 8 \\ [1.1429,\ 1]x & \leq & 15.4286 \\ [1,\ -1]x & \leq & 6 \\ [0.25,\ -1]x & \leq & 0.75 \end{array}\right. \tag{4.30}$$

The admissible solutions of the linear inequalities (4.30) represent the polygon $\mathcal{P}$ sketched in Fig. 4.13.

Now, in order to find the largest circle inside the polygon, we have to solve the LP problem:

Maximize r

subject to the following constraints:

$$\left\{\begin{array}{rcl} [-1.3333,\ -1]x_c + 1.6667r & \leq & -4 \\ [-1.3333,\ 1]x_c + 1.6667r & \leq & 4 \\ [0,\ 1]x_c + r & \leq & 8 \\ [1.1429,\ 1]x_c + 1.5186r & \leq & 15.4286 \\ [1,\ -1]x_c + 1.4142r & \leq & 6 \\ [0.25,\ -1]x_c + 1.0308r & \leq & 0.75 \end{array}\right.$$

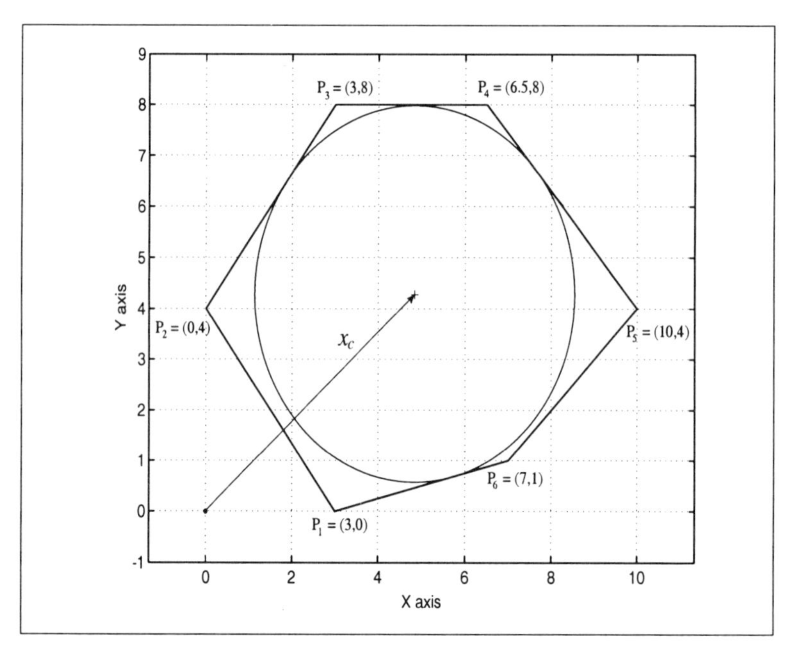

Fig. 4.13. Largest circle $\mathcal{C}$ inside a polygon $\mathcal{P}$ for Example 4.6.2.

The solution of this LP problem leads us to the conclusion that the largest circle inside the polygon $\mathcal{P}$ has a radius $r = 3.7035$ and its center is located at $x_c = (4.8368, 4.2767)$. This circle is indicated in Fig. 4.13.

We now return to the plant considered in Example 4.6.1 and apply the above procedure to it.

Example 4.6.3. Consider the same plant as the one in Example 4.6.1, *i.e.*,

$$G(s) = \frac{1}{(s+1)^8}.$$

Now, from Example 4.6.1, we know that for the existence of stabilizing (k_i, k_d) values k_p must necessarily lie in $(-1,\ 2.0751)$. So, by sweeping over $k_p \in (-1,\ 2.0751)$, we found the largest circle inscribed in the stabilizing (k_i, k_d) region at each stage. Thereafter we chose that value of k_p that gave us the circle with the largest radius in the k_i-k_d plane. This value of k_p was found to be $k_p = 1.32759$ and the largest circle was centered at

$$k_i = 0.42563,\ k_d = 5.15291$$

and had a radius of 0.42563. The stabilizing region in the k_i-k_d plane for $k_p = 1.32759$ along with the maximum l_2 circle is sketched in Fig. 4.14.

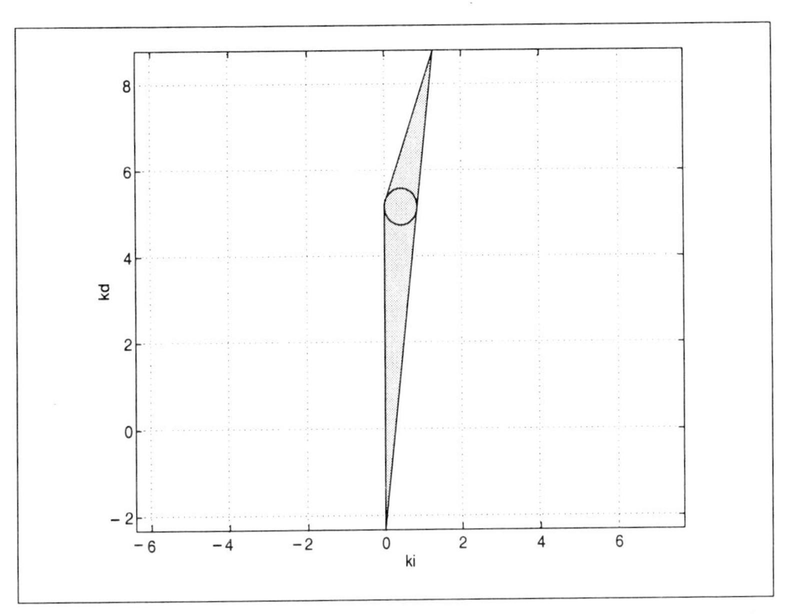

Fig. 4.14. The stabilizing region for $k_p = 1.32759$ and the maximum l_2 circle (Example 4.6.3).

4.7 Notes and References

The characterization of all stabilizing PID controllers for a given linear time invariant plant was developed by Ho, Datta and Bhattacharyya [21]. Given a linear time-invariant system, the YJBK parametrization [46] provides a complete characterization of all stabilizing controllers. This parametrization is central to many of the modern optimal control approaches such as H_2, H_∞ [17] and L_1 [14] since it isolates the problem of optimizing the performance index from the problem of ensuring stability. The YJBK parametrization, however, cannot accomodate constraints on the controller order and/or structure and this motivated the results presented here. The characterization of the center and radius of the largest circle inscribed inside a given convex polygon is due to Boyd and Vandenberghe [10].

CHAPTER 5
OPTIMAL DESIGN USING PID CONTROLLERS

In this chapter, we show how the characterization of all stabilizing PID controllers obtained in the last chapter can be used to design PID controllers which optimize various performance indices. The design procedure, which is essentially a search over the stabilizing set, is demonstrated using numerical examples. The performance of the optimal PIDs is also compared, wherever appropriate, to that of the corresponding unconstrained order optimal designs.

5.1 Introduction

In the last chapter, we obtained characterizations of all stabilizing P, PI and PID controllers for a given linear time-invariant plant. These characterizations make it possible to carry out optimal controller designs subject to constraints on the controller order and/or structure. In this chapter, our objective is to show, via examples, how the results of the last chapter can be used to carry out constant gain, PI and PID designs to optimize various performance indices.

The chapter is organized as follows. First, in Section 5.2, we present standard (unconstrained order) H_2 and H_∞ optimal design examples from the literature. These examples are included mainly to serve as benchmarks against which the constrained order designs to be presented later can be compared. In Section 5.3, we show how the characterization of all stabilizing gain values for a given plant can be used to carry out H_2 and H_∞ optimal designs when the compensator is constrained to be a constant gain. In Section 5.4, we show how the results of Section 4.3 can be used to carry out PI designs which globally optimize various performance indices. Section 5.5 explores the possibility of carrying out globally optimal PID designs based on the results of Section 4.4.

The results of this chapter and the preceding one take on added significance when viewed in the context of practical applications. Most of the currently available optimization techniques of modern optimal control including H_2, H_∞ [17] and L_1 Optimal [14] and μ cannot be directly used in applications because they cannot accommodate fixed structure controllers such as PID. At the same time PID controllers are the most important class of controllers since 98% of the controllers in use in applications worldwide are of this category. Thus the results of this chapter and the previous one are expected to significantly impact controller design for industrial applications.

5.2 Unconstrained Order Optimal Designs

In this section, we present examples of standard H_2 and H_∞ optimal designs from the literature. In these designs, the controller order and/or structure are not constrained *a priori*. These designs serve as benchmarks against which to compare the performance of controllers that are optimal subject to constraints on the controller order.

5.2.1 H_∞ Robust Controller Design Using the YJBK Parametrization

The following example is taken from ([16], p. 192). In this example, we design an optimal H_∞ robust controller that minimizes $\|W_2(s)T(s)\|_\infty$ where $T(s)$ is the complementary sensitivity function and the weight $W_2(s)$ is chosen as the high-pass function

$$W_2(s) = \frac{s+0.1}{s+1}.$$

The transfer function of the plant in question is:

$$G(s) = \frac{s-1}{s^2+0.8s-0.2}.$$

This plant is a bad plant to control since it has RHP poles and zeros. The YJBK parametrization is used to parametrize all proper feedback controllers that stabilize the plant and the YJBK parameter $Q(s)$ is selected optimally to minimize $\|W_2(s)T(s)\|_\infty$. The optimal value of the cost function is

$$\begin{aligned}\gamma_{opt} &= \inf_{Q(s) stable} \|W_2(s)T(s)\|_\infty \\ &= 0.375\end{aligned}$$

with the infimizing $Q(s)$ being given by

$$Q(s) = \frac{-5(s+1)(0.075s-0.195)}{s+0.1}.$$

Since this infimizing $Q(s)$ is not proper, we divide it by the factor $(\tau s+1)$ to make it proper, where τ is chosen to be a small positive number. For instance, choosing $\tau = 0.01$ we come up with the sub-optimally robust controller given by:

$$\begin{aligned} C(s) &= \frac{q_3^0 s^3 + q_2^0 s^2 + q_1^0 s + q_0^0}{s^3 + p_2^0 s^2 + p_1^0 s + p_0^0} \\ &= \frac{-39.3s^3 - 114.48s^2 - 112.68s - 37.5}{s^3 + 141.6s^2 + 275.4s + 137.5}. \end{aligned}$$

The level of performance achieved by this sub-optimally robust controller is given by

$$\|W_2(s)T(s)\|_\infty = 0.3911$$

which is a reasonable approximation to the unattainable infimum.

5.2.2 H_2 Optimal Controller Design using the YJBK Parametrization

Consider the H_2 optimal control problem shown in Fig. 5.1. Here $G(s) = \frac{N(s)}{D(s)}$

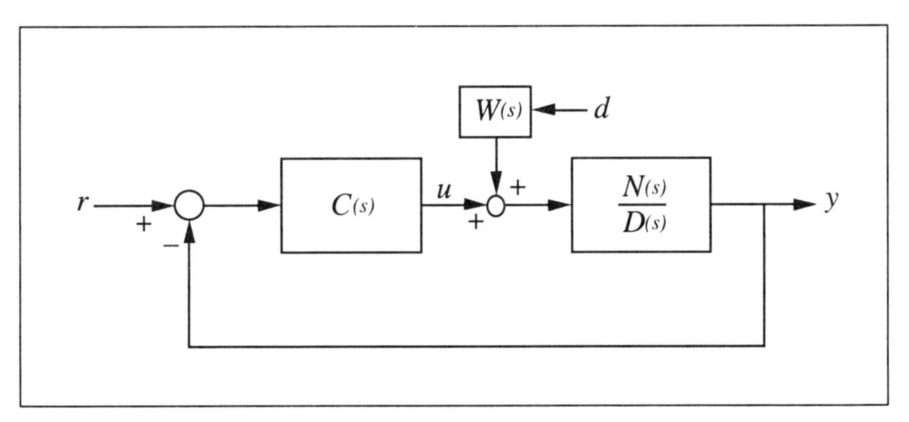

Fig. 5.1. The H_2 optimal control problem.

is the plant to be controlled, $C(s)$ is the compensator to be designed, r is the command signal, d is an input disturbance and $W(s)$ is a weighting function. The plant $\frac{N(s)}{D(s)}$ is the same as before, *i.e.*,

$$\frac{N(s)}{D(s)} = \frac{s-1}{s^2 + 0.8s - 0.2}.$$

Let us assume that the plant disturbance d is random white noise. The control objective here is to design a stabilizing compensator $C(s)$ which minimizes the H_2 norm of the weighted transfer function from d to y. This transfer function is given by $W(s)G(s)S(s)$, where $S(s)$ is the sensitivity function. Assume that the weight $W(s)$ is chosen as

$$W(s) = \frac{s+0.1}{s+1}.$$

As in the H_∞ optimal control problem, we use the YJBK parametrization to parametrize all proper feedback controllers that stabilize the given plant. Thereafter the YJBK parameter $Q(s)$ is selected optimally to minimize $\|W(s)G(s)S(s)\|_2$.

The optimal value of the cost function in this case is

$$\begin{aligned}\gamma_{opt} &= \inf_{Q(s) stable} \|W(s)G(s)S(s)\|_2 \\ &= 0.9722718241\end{aligned}$$

with the infimizing $Q(s)$ being given by

$$Q(s) = \frac{-5(s+1)(0.075s-0.195)}{s+0.1}.$$

Since this infimizing $Q(s)$ is not proper, we divide it by the factor $(\tau s+1)$ to make it proper, where τ is chosen to be a small positive number. For instance, choosing $\tau = 0.01$ we get the sub-optimal H_2 controller given by:

$$C(s) = \frac{-39.3s^3 - 114.48s^2 - 112.68s - 37.5}{s^3 + 141.6s^2 + 275.4s + 137.5}$$

which is the same controller as the H_∞ robust controller obtained earlier. The level of performance achieved by this sub-optimal controller is given by

$$\|W(s)G(s)S(s)\|_2 = 0.97264$$

which is a good approximation to the unattainable infimum.

5.3 Design Using a Constant Gain

In this section, we assume that the plant of the last two subsections is to be optimally controlled using a constant gain. This may reflect hardware constraints or the requirement of simplicity. The constant gain controller now must stabilize the given plant and minimize the same performance indices considered in the last two subsections. In other words, we try to minimize $\|W_2(s)T(s)\|_\infty$ and $\|W(s)G(s)S(s)\|_2$ when the controller $C(s)$ is restricted to be a constant gain, *i.e.*, $C(s) = k$. Using the results of the last chapter, we conclude that the set of all stabilizing k values for the given plant is given by $k \in (-0.8,\ -0.2)$.

The plot of $\|W_2(s)T(s)\|_\infty$ versus $k \in (-0.8,\ -0.2)$ is shown in Fig. 5.2. From Fig. 5.2, we see that

$$\min_{k\in(-0.8,\ -0.2)} \|W_2(s)T(s)\|_\infty = 0.5205$$

and this occurs at $k = -0.2510$. Thus the optimal H_∞ norm attained by this constant gain controller is fairly good, especially when we take into account

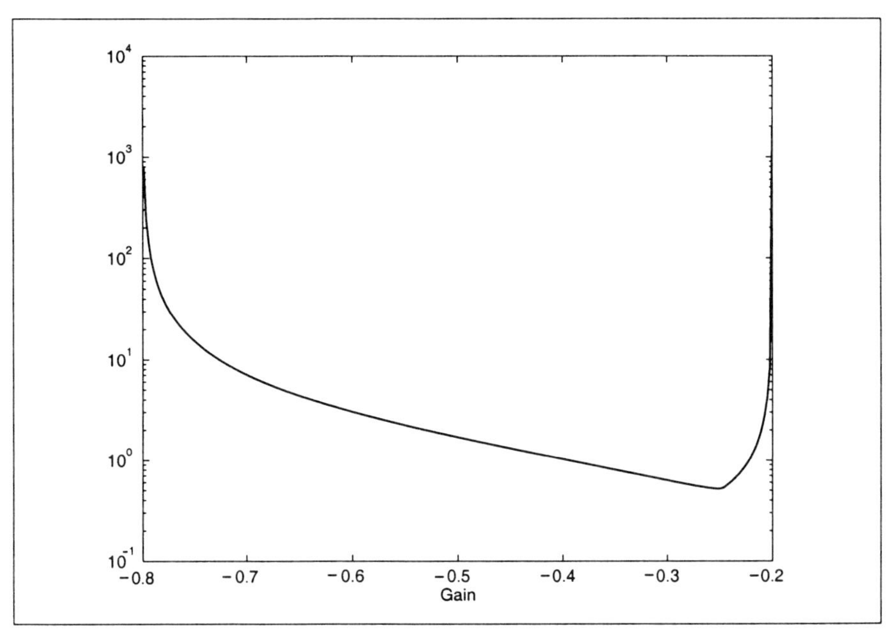

Fig. 5.2. H_∞ norm of weighted complementary sensitivity function (Constant Gain Stabilization).

the fact that the order of this controller is much lower than that of the unconstrained order H_∞ optimal one designed in Section 5.2.

The plot of $\|W(s)G(s)S(s)\|_2$ versus $k \in (-0.8,\ -0.2)$ is shown in Fig. 5.3.

From Fig. 5.3, we see that

$$\min_{k\in(-0.8,\ -0.2)} \|W(s)G(s)S(s)\|_2 \quad = \quad 1.03830$$

and this occurs at $k = -0.26811$. Thus the optimal H_2 norm attained by this constant gain controller is also quite good, especially when the order reduction is taken into account.

5.4 Design Using a PI Controller

In this section, we use the characterization of all stabilizing PI controllers developed in Section 4.3, to design stabilizing PI controllers that minimize certain performance indices. The plant is chosen to be the same as the one in Sections 5.2 and 5.3, namely

$$G(s) = \frac{s-1}{s^2 + 0.8s - 0.2}.$$

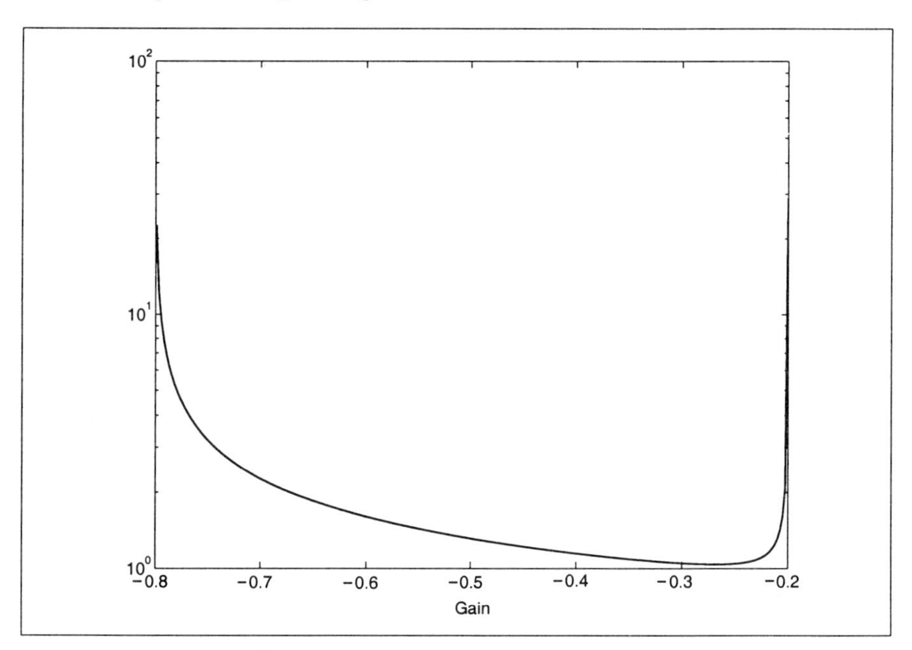

Fig. 5.3. H_2 norm of $\|W(s)G(s)S(s)\|_2$ (Constant Gain Stabilization).

The performance indices considered are also the same as before, namely $\|W_2(s)T(s)\|_\infty$ and $\|W(s)G(s)S(s)\|_2$ where

$$W_2(s) = W(s) = \frac{s+0.1}{s+1}$$

and $T(s)$, $S(s)$ denote the complementary sensitivity and sensitivity functions respectively. The only difference is that this time the controller $C(s)$ is restricted to be a PI controller, *i.e.*,

$$C(s) = \frac{k_p s + k_i}{s}.$$

The motivation for using PI controllers is their ready availability and the guarantee that they provide of robust tracking and disturbance rejection for constant inputs. The set of all stabilizing (k_p, k_i) values for the given plant is determined as follows by using the results of Sections 4.2 and 4.3. Now, the closed loop characteristic polynomial is given by

$$\delta(s, k_p, k_i) = s(s^2 + 0.8s - 0.2) + (k_p s + k_i)(s - 1)$$

and

$$N^*(s) = N'(-s) = s + 1.$$

Thus

$$\delta(s, k_p, k_i)N^*(s) = s^4 + 1.8s^3 + 0.6s^2 - 0.2s + k_p s(s^2 - 1) + k_i(s^2 - 1)$$

so that

$$\delta(j\omega, k_p, k_i)N^*(j\omega) = [\omega^4 - 0.6\omega^2 + k_i(-\omega^2 - 1)] + j\omega[-1.8\omega^2 - 0.2 + k_p(-\omega^2 - 1)]$$

Since $N^*(s)$ is Hurwitz, it follows using the Hermite-Biehler Theorem[1] that for any $(k_p,\ k_i)$ such that $\delta(s, k_p, k_i)$ is Hurwitz,

$$[-1.8\omega^2 - 0.2 + k_p(-\omega^2 - 1)]$$

must have one and only one positive real root. This implies that $k_p \in (-1.8,\ -0.2)$. Now for every fixed $k_p \in (-1.8,\ -0.2)$, the PI controller stabilization problem becomes a constant gain stabilization problem. Thus by sweeping over $k_p \in (-1.8,\ -0.2)$ and using the constant gain stabilization results of Section 4.2, we obtain the stabilizing region of $(k_p,\ k_i)$ values sketched in Fig. 5.4.

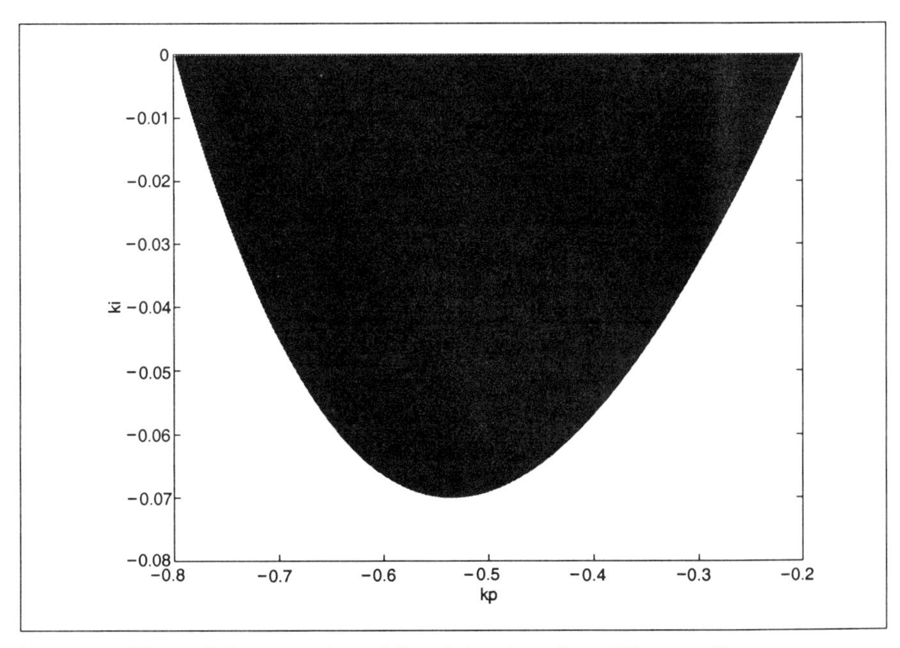

Fig. 5.4. The stabilizing region of $(k_p,\ k_i)$ values for a PI controller .

The plot of $\|W_2(s)T(s)\|_\infty$ versus stabilizing $(k_p,\ k_i)$ values is shown in Fig. 5.5 while the plot of $\|W(s)G(s)S(s)\|_2$ versus stabilizing $(k_p,\ k_i)$ values is shown in Fig. 5.6.

[1] Note that, for this example, since $N^*(s)$ is Hurwitz, the *Generalized* Hermite-Biehler Theorem does not have to be invoked.

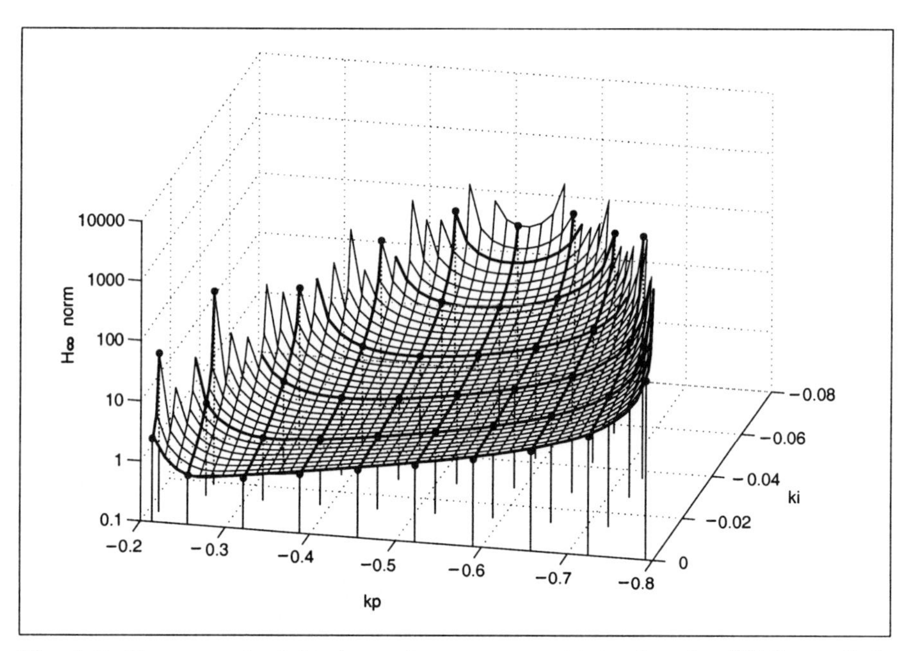

Fig. 5.5. H_∞ norm of weighted complementary sensitivity function (PI Controller).

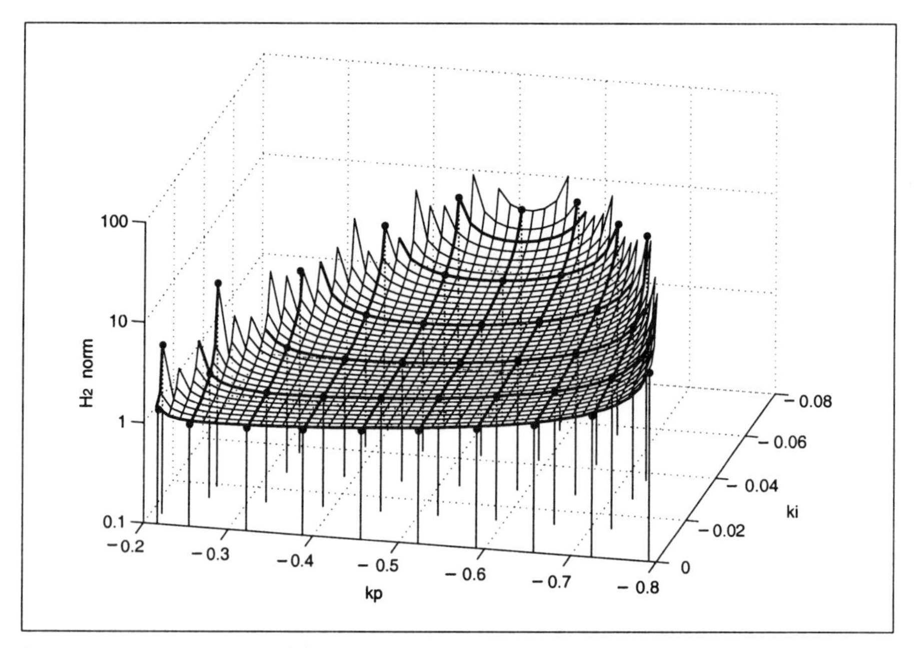

Fig. 5.6. H_2 norm of $W(s)G(s)S(s)$ (PI Controller).

From Fig. 5.5, we see that $\min \|W_2(s)T(s)\|_\infty = 0.5205$ and this occurs at $k_p = -0.2510$ and $k_i = 0$, which is identical to the H_∞ optimal constant gain solution obtained in Section 5.3. Also from Fig. 5.6, we see that $\min \|W(s)G(s)S(s)\|_2 = 1.03830$ and this occurs at $k_p = -0.26811$ and $k_i = 0$ which is once again identical to the H_2 optimal constant gain solution.

In both the above H_∞ and H_2 designs using PI controllers, we obtain optimal solutions with $k_i = 0$. Mathematically, at $k_i = 0$ the PI controller becomes

$$C(s) = \frac{k_p s}{s} = k_p$$

which is a constant gain controller. However, the main purpose for using PI controllers is to use the integral term to achieve zero steady state errors to step inputs. To preserve the zero steady state property at $k_i = 0$, the optimal PI controllers would need to be implemented as

$$C(s) = \frac{k_p s}{s}.$$

Unfortunately, this hardware implementation will destabilize the closed loop system and cause unbounded steady state errors. To avoid such an illogical implementation, it is reasonable to impose a constraint on k_i. From the stabilizing region in Fig. 5.4, a possible design constraint could be $k_i \leq -0.01$. With this parametric constraint, we obtained the level curves shown in Fig. 5.7. From Fig. 5.7 we see that

$$\min \|W(s)T(s)\|_\infty = 0.93322$$

and this occurs at

$$k_p = -0.31723 \text{ and } k_i = -0.01.$$

Similar level curves for $\|W(s)G(s)S(s)\|_2$, subject to $k_i \leq -0.01$ are shown in Fig. 5.8. From Fig. 5.8, we see that

$$\min \|W(s)G(s)S(s)\|_2 = 1.18342$$

and this occurs at

$$k_p = -0.32655 \text{ and } k_i = -0.01.$$

The purpose of the integral term in a PI controller is to achieve zero steady state errors when tracking step inputs. Thus, when using a PI controller, we can employ several time domain performance specifications such as settling time, maximum overshoot, minimum undershoot, etc. to quantify the performance of the closed loop system. The characterization of all stabilizing (k_p, k_i) values in Fig. 5.4 enables us to graphically display the variation of these performance indices over the entire stabilizing region in the parameter space. A few instances are given below.

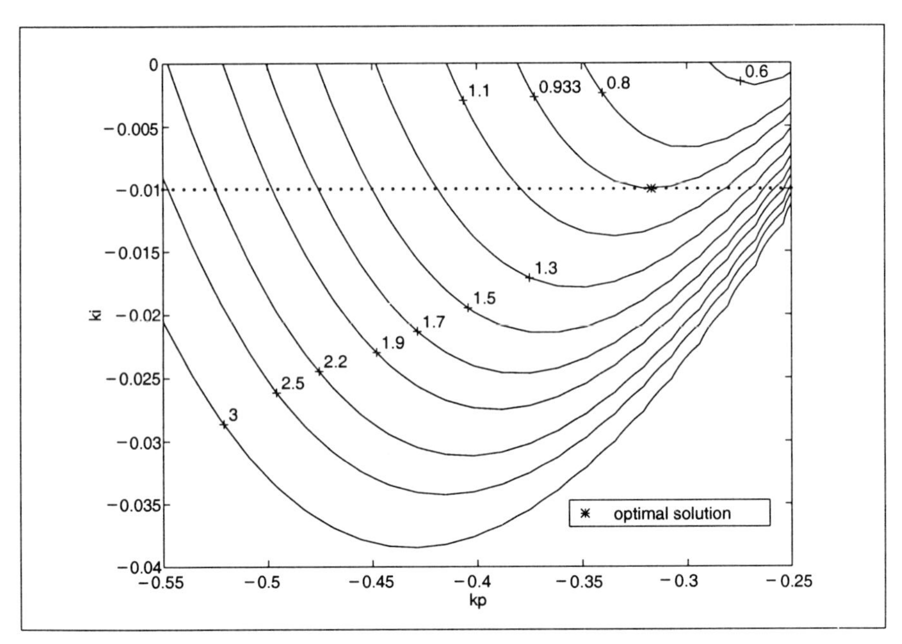

Fig. 5.7. Level curves of H_∞ norm weighted complementary sensitivity function with $k_i \leq -0.01$ (PI Controller).

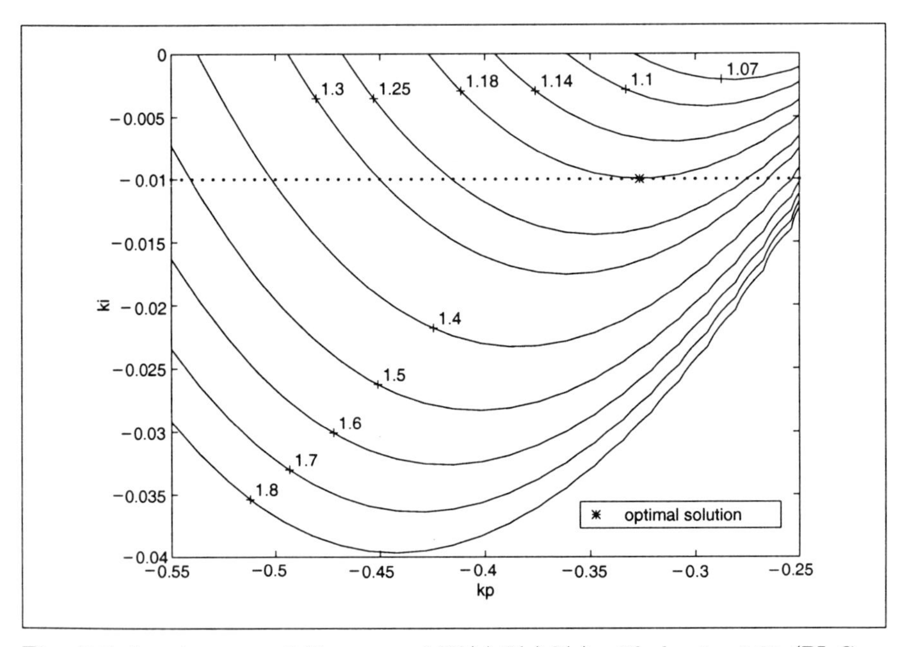

Fig. 5.8. Level curves of H_2 norm of $W(s)G(s)S(s)$ with $k_i \leq -0.01$ (PI Controller).

(1) The plot of settling time (not exceeding 70 secs) versus stabilizing (k_p, k_i) values is shown in Fig. 5.9 while the corresponding level curves are sketched in Fig. 5.10. From Fig. 5.10, we see that the minimum settling time is 21.44 secs and this occurs at $k_p = -0.34793$ and $k_i = -0.01369$.
(2) The plot of overshoot versus stabilizing (k_p, k_i) values is shown in Fig. 5.11 while the corresponding level curves are sketched in Fig. 5.12. From Fig. 5.12, we see that the minimum overshoot is 1.54026 and this occurs at $k_p = -0.46803$ and $k_i = -0.01$.
(3) The plot of undershoot versus stabilizing (k_p, k_i) values is shown in Fig. 5.13 while the corresponding level curves are sketched in Fig. 5.14. From Fig. 5.14, we see that the minimum undershoot is 0.09969 and this occurs at $k_p = -0.26483$ and $k_i = -0.01$.

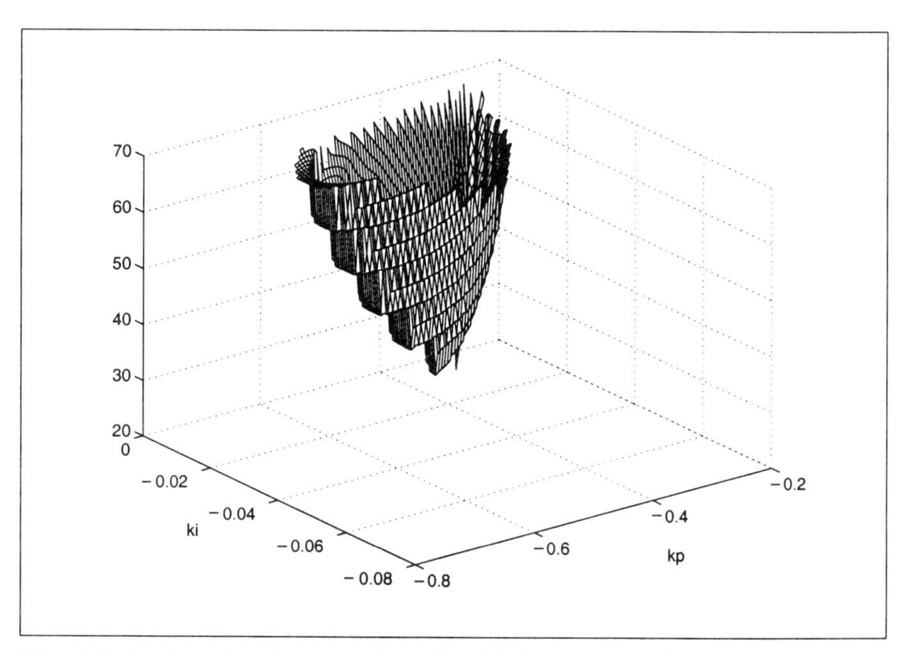

Fig. 5.9. Settling time ($\leq$ 70 secs) vs. (k_p, k_i) (PI Controller).

The plots in Figs. 5.7, 5.8, 5.10, 5.12, and 5.14 can be used to design PI controllers to meet given performance specifications. For example, if we are given the following unit step response specifications:

1. Settling time $\leq$ 30 secs;
2. Overshoot $\leq$ 2,

we can superimpose Figs. 5.10 and 5.12 to obtain Fig. 5.15. From Fig. 5.15, we can then choose stabilizing (k_p, k_i) values for guaranteeing the perfor-

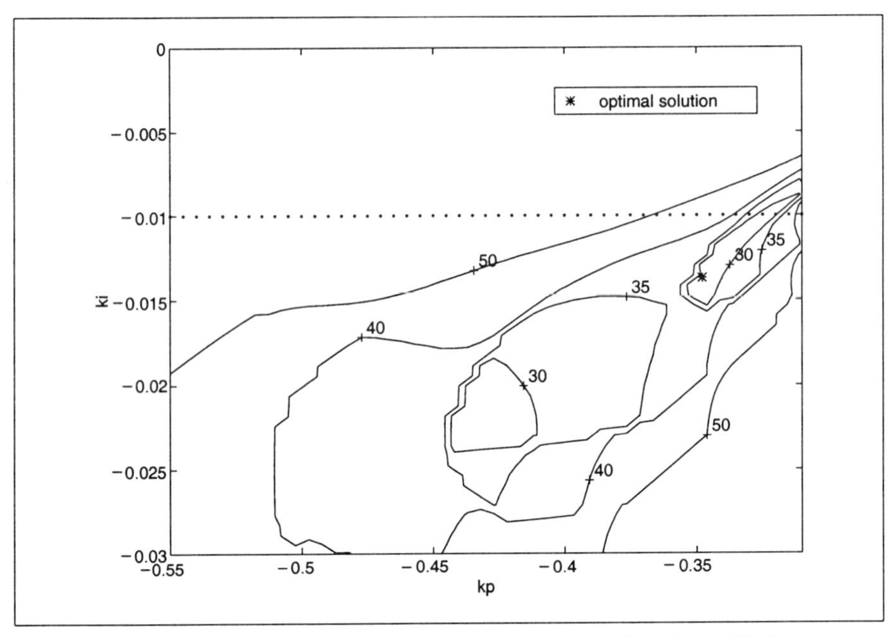

Fig. 5.10. Level curves of settling time with $k_i \leq -0.01$ (PI Controller).

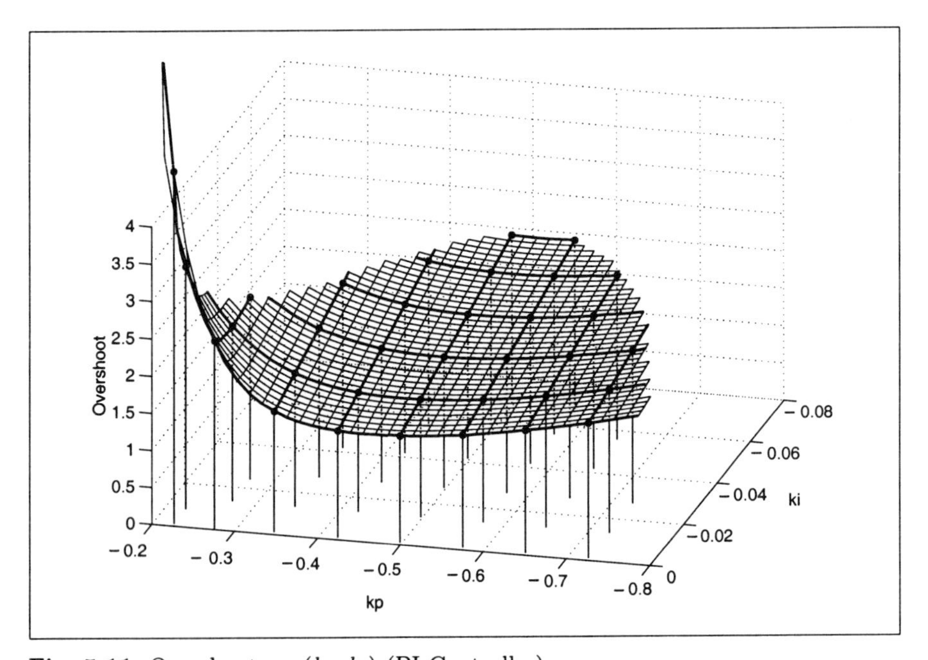

Fig. 5.11. Overshoot vs. (k_p, k_i) (PI Controller).

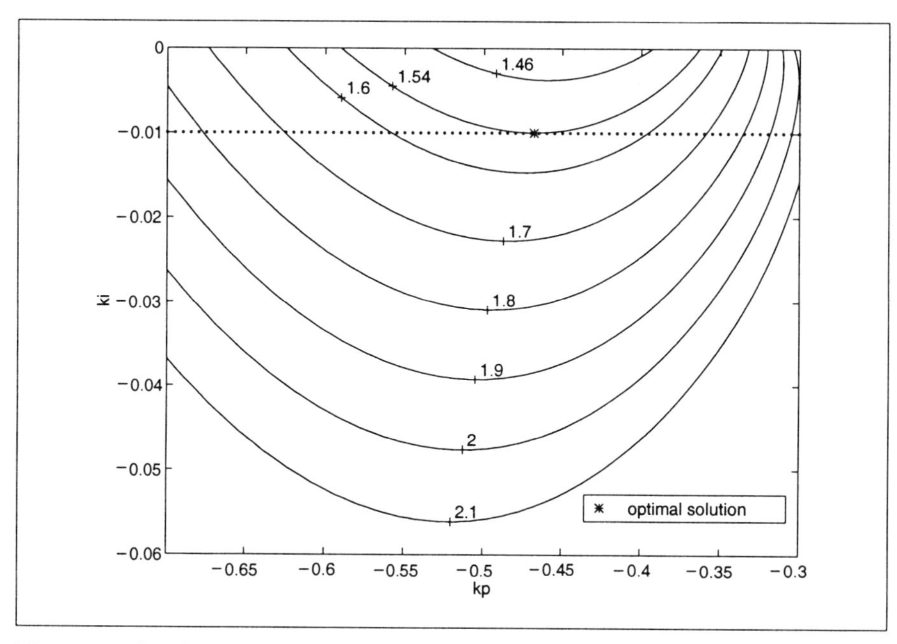

Fig. 5.12. Level curves of overshoot with $k_i \leq -0.01$ (PI Controller).

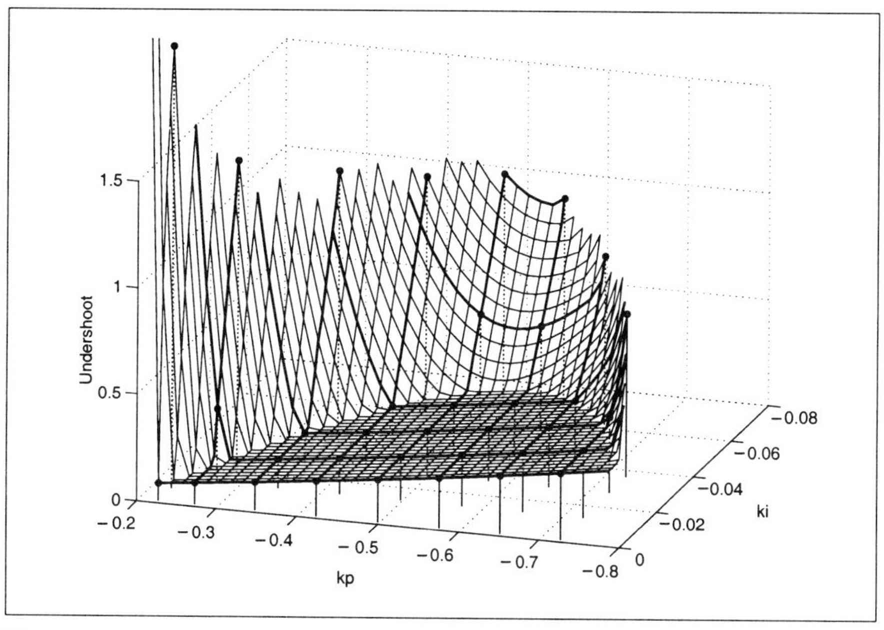

Fig. 5.13. Undershoot vs. (k_p, k_i) (PI Controller).

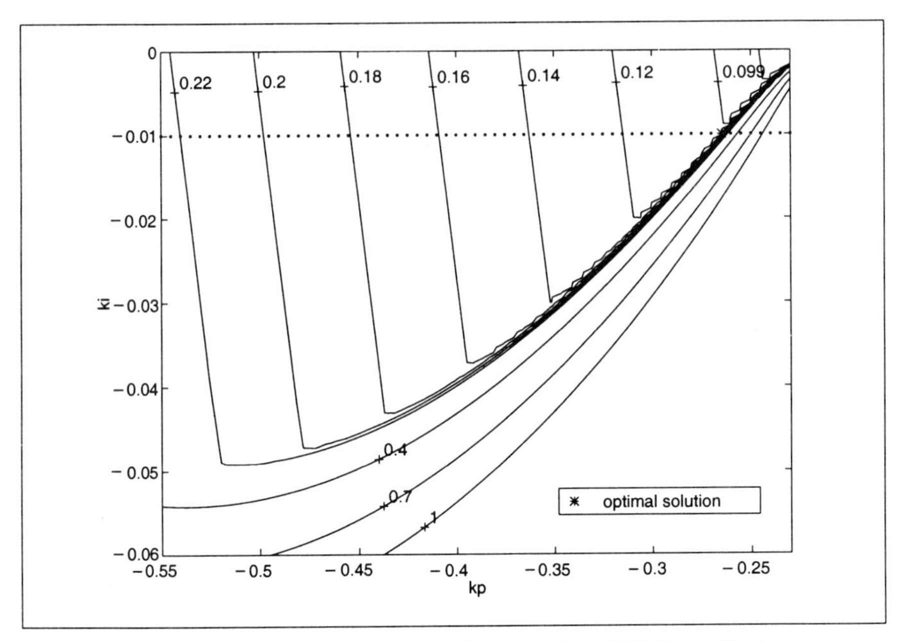

Fig. 5.14. Level curves of undershoot with $k_i \leq -0.01$ (PI Controller).

mance requirements. For instance, at $k_p = -0.43$ and $k_i = -0.02$, the given performance specifications are satisfied. The closed loop step response using the designed PI controller is given in Fig. 5.16 which shows that the desired performance specifications have indeed been satisfied. It should be pointed out that the design specification here of 200 % on the overshoot is abnormally high. However, from Fig. 5.12, it is clear that for the given plant, one cannot do much better since the minimum possible overshoot using a PI controller is about 146 %.

5.5 Design Using a PID Controller

In this section, we use the characterization of all stabilizing PID controllers developed in Section 4.4 to design PID controllers that minimize certain performance indices. We consider the same plant and performance indices considered in the last section. The only difference is that the controller $C(s)$ is now restricted to be a PID, *i.e.*,

$$C(s) = \frac{k_d s^2 + k_p s + k_i}{s}.$$

The set of all stabilizing (k_p, k_i, k_d) values for the given plant is determined as follows. The closed loop characteristic polynomial is

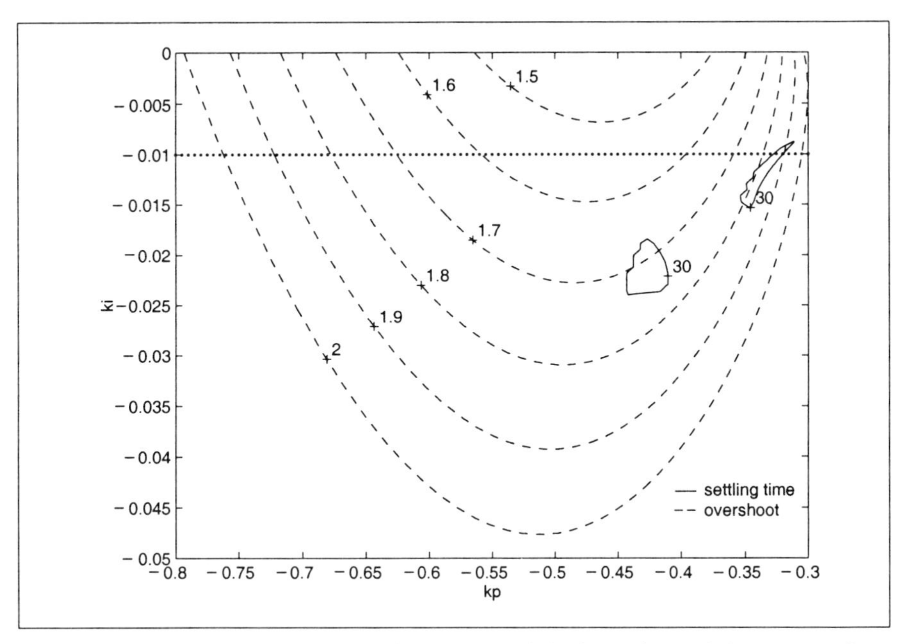

Fig. 5.15. The level curves of overshoot ≤ 2 and the boundary of the region where settling time ≤ 30 secs with $k_i \leq -0.01$ (PI Controller).

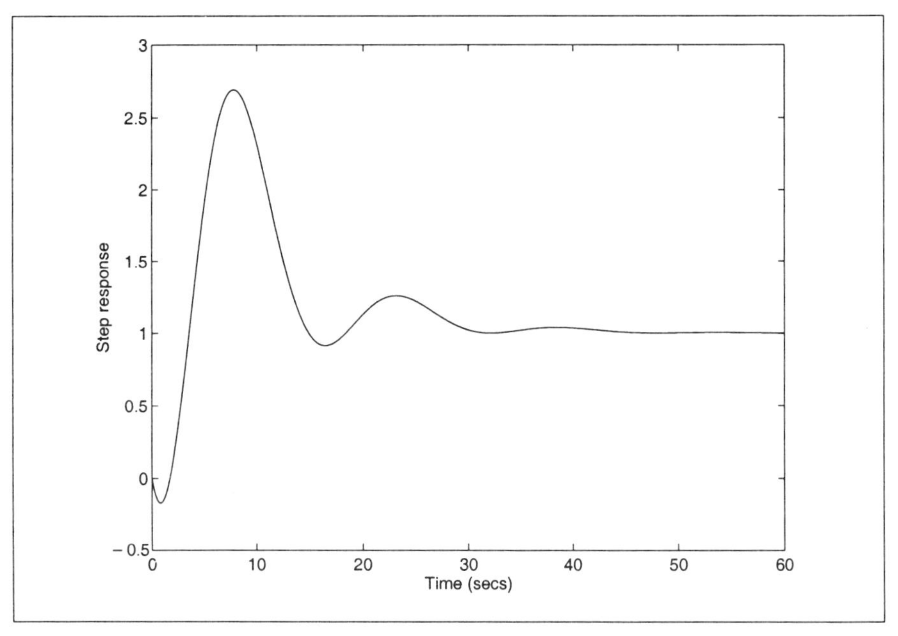

Fig. 5.16. Step response using the designed PI Controller.

$$\delta(s, k_p, k_i, k_d) = s(s^2 + 0.8s - 0.2) + (k_d s^2 + k_p s + k_i)(s - 1)$$

$$\text{and } N^*(s) = N'(-s) = s + 1.$$

$$\begin{aligned} \text{Thus } \delta(s, k_p, k_i, k_d)N^*(s) &= s^4 + 1.8s^3 + 0.6s^2 - 0.2s + k_d s^2(s^2 - 1) \\ &\quad + k_p s(s^2 - 1) + k_i(s^2 - 1) \end{aligned}$$

so that

$$\begin{aligned} \delta(j\omega, k_p, k_i, k_d)N^*(j\omega) = {} & [\omega^4 - 0.6\omega^2 + k_i(-\omega^2 - 1) + k_d(\omega^4 + \omega^2)] \\ & + j\omega[-1.8\omega^2 - 0.2 + kp(-\omega^2 - 1)] \end{aligned}$$

Since $N^*(s)$ is Hurwitz, it follows using the Hermite-Biehler Theorem that for any $(k_p,\ k_i,\ k_d)$ such that $\delta(s, k_p, k_i, k_d)$ is Hurwitz,

$$[-1.8\omega^2 - 0.2 + kp(-\omega^2 - 1)]$$

must have one and only one positive real root. This implies that

$$k_p \in (-1.8,\ -0.2).$$

Now for every fixed $k_p \in (-1.8,\ -0.2)$ and using the results of Section 4.4, the PID controller stabilization problem becomes a linear programming problem. Thus by sweeping over

$$k_p \in (-1.8,\ -0.2),$$

and using Theorem 4.4.1 repeatedly, we obtain the stabilizing region of $(k_p,\ k_i,\ k_d)$ values sketched in Fig. 5.17.

Now for a fixed k_p, the stabilizing region in the $(k_i,\ k_d)$ plane is a union of either convex polygons or intersections of half-planes. By sweeping over k_p, we can search for

$$\min_{(k_p,\ k_i,\ k_d)} \|W_2(s)T(s)\|_\infty$$

by first determining the minimum over each stabilizing $(k_i,\ k_d)$ region. Using such an approach, we found that the minimum value of $\|W_2(s)T(s)\|_\infty$ is 0.37884 and this occurs at $k_p = -0.2718$, $k_i = 0$, and $k_d = -0.27475$. Clearly, the H_∞ performance achieved with this PID controller is very close to the lower bound determined in Section 5.2. As in the case of the H_∞ norm, we can determine

$$\min_{(k_p,\ k_i,\ k_d)} \|W(s)G(s)S(s)\|_2$$

by first determining the minimum over each stabilizing (k_i, k_d) region corresponding to a fixed k_p, and then sweeping over k_p. Using such an approach, we found that the minimum value of

$$\|W(s)G(s)S(s)\|_2 \text{ is } 0.9722774567$$

and this occurs at

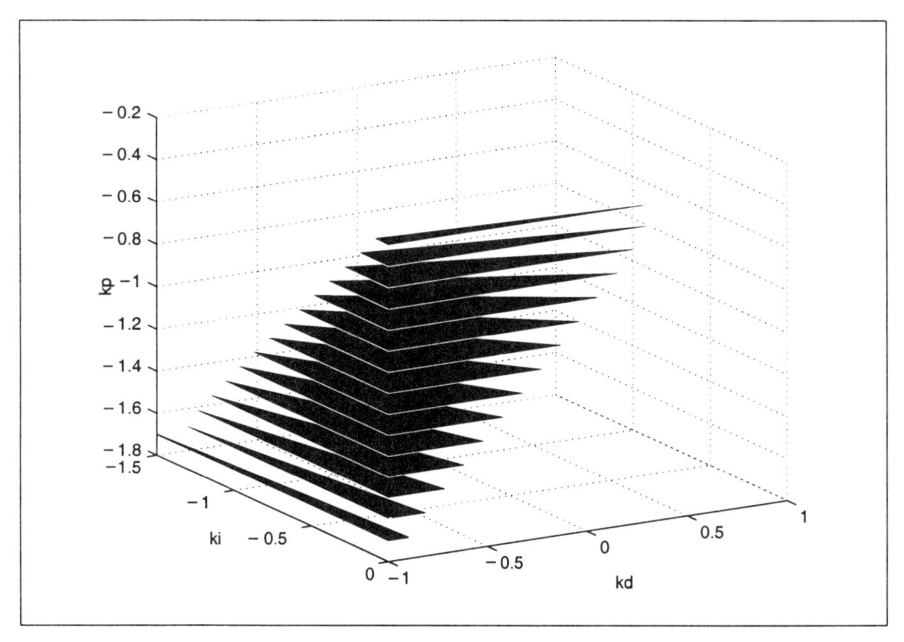

Fig. 5.17. The stabilizing region of $(k_p,\ k_i,\ k_d)$ values for the PID controller.

$$k_p = -0.272611, k_i = 0, \text{ and } k_d = -0.275137.$$

Clearly, the H_2 performance achieved with this PID controller is also very close to the lower bound determined in Section 5.2.

In both the above H_∞ and H_2 designs using PID controllers, we obtain optimal solutions with $k_i = 0$. At $k_i = 0$, the PID controller becomes

$$C(s) = \frac{k_d s^2 + k_p s}{s}.$$

Once again, this implementation will destabilize the closed loop system and cause unbounded steady state errors to ramp inputs. The situation can be remedied as before by choosing $k_i \leq -0.01$ as a design constraint.

By keeping k_p fixed and searching for

$$\min_{(k_i \leq -0.01,\ k_d)} \|W_2(s)T(s)\|_\infty,$$

and then sweeping over k_p, we found that

$$\min_{(k_p,\ k_i \leq -0.01,\ k_d)} \|W_2(s)T(s)\|_\infty = 0.51028$$

and this occurs at

$$k_p = -0.33889, k_i = -0.01, \text{ and } k_d = -0.33136.$$

The stabilizing region in the (k_i, k_d) plane corresponding to $kp = -0.33889$ is sketched in Fig. 5.18 while the plot of $\|W_2(s)T(s)\|_\infty$ on this plane is given in Fig. 5.19. The level curves in Fig. 5.20 give a clearer picture of the occurrence of this minimum value. Following a similar procedure as in the case of the H_∞ norm, we found that

$$\min_{(k_p,\ k_i \leq -0.01,\ k_d)} \|W(s)G(s)S(s)\|_2 = 1.02785$$

and this occurs at $k_p = -0.316$, $k_i = -0.01$, and $k_d = -0.34603$. The stabilizing region in the $(k_i,\ k_d)$ plane corresponding to

$$k_p = -0.316$$

is sketched in Fig. 5.21 while the plot of $\|W(s)G(s)S(s)\|_2$ on this plane is given in Fig. 5.22. The level curves in Fig. 5.23 give a clearer picture of the occurrence of this minimum value.

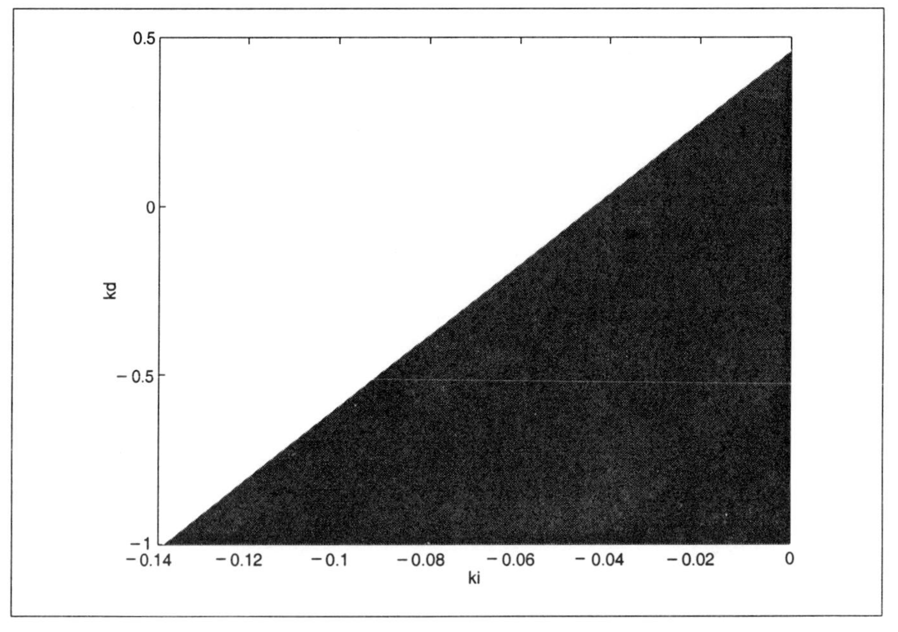

Fig. 5.18. The stabilizing region of $(k_i,\ k_d)$ values for $k_p = -0.33889$.

As in the last section, the characterization of all stabilizing (k_p, k_i, k_d) values in Fig. 5.17 can be also used to generate graphical plots showing the variation of various time domain performance indices over the entire stabilizing region in the parameter space. A few instances are given below.

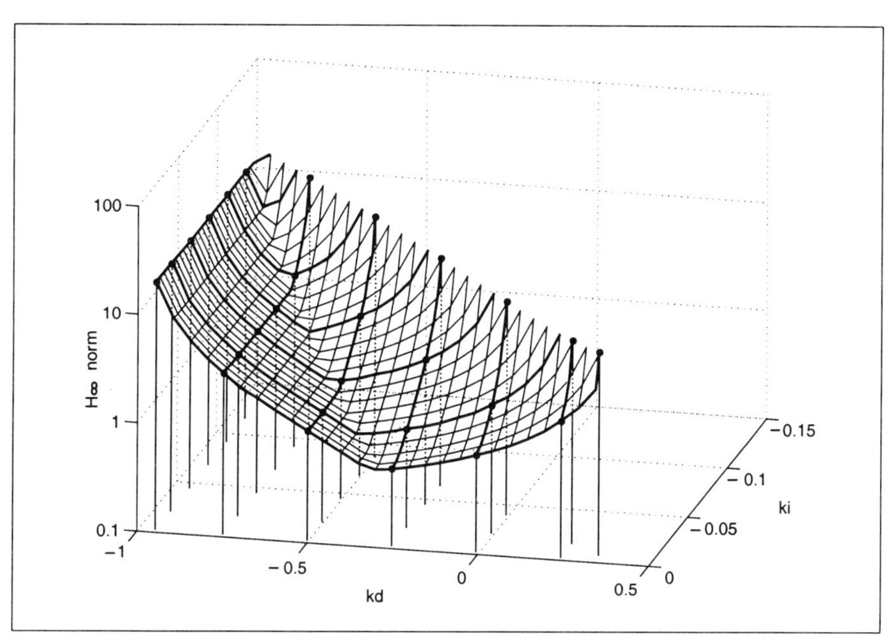

Fig. 5.19. H_∞ norm of weighted complementary sensitivity function vs. (k_i, k_d) values for $k_p = -0.33889$.

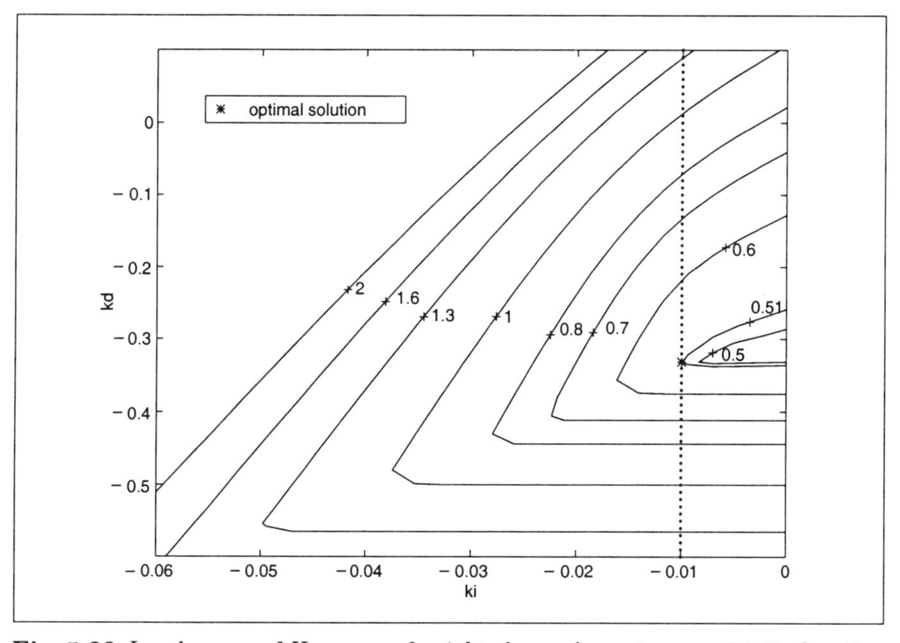

Fig. 5.20. Level curves of H_∞ norm of weighted complementary sensitivity function vs. (k_i, k_d) values for $k_p = -0.33889$ with $k_i \leq -0.01$.

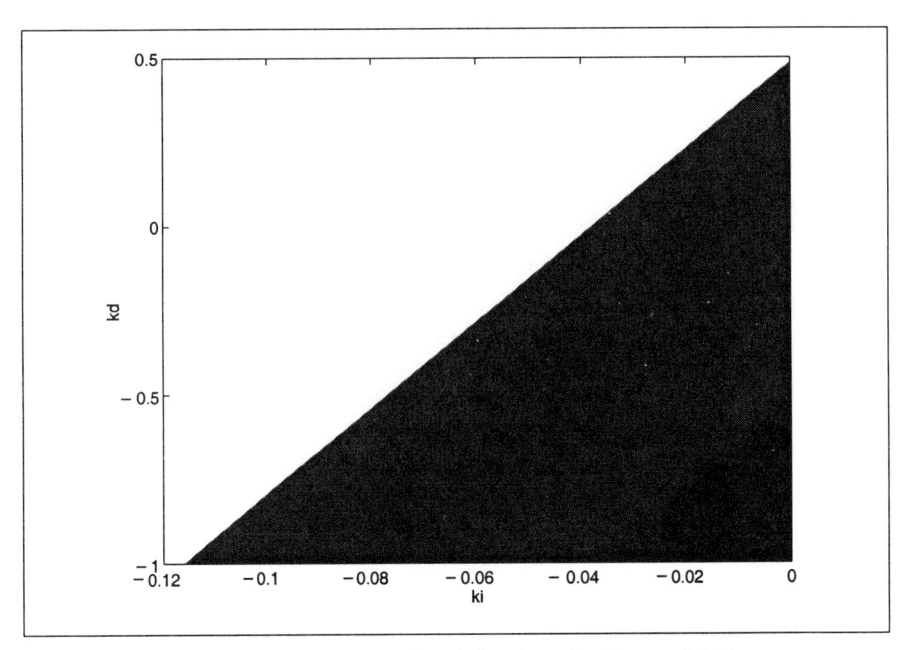

Fig. 5.21. The stabilizing region of (k_i, k_d) values for $k_p = -0.316$.

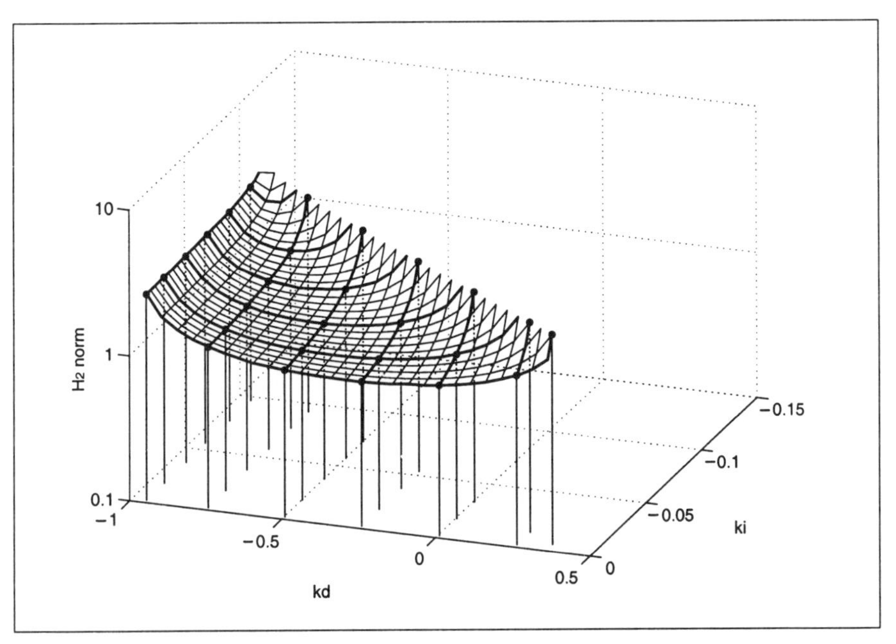

Fig. 5.22. H_2 norm of $W(s)G(s)S(s)$ vs. (k_i, k_d) values for $k_p = -0.316$.

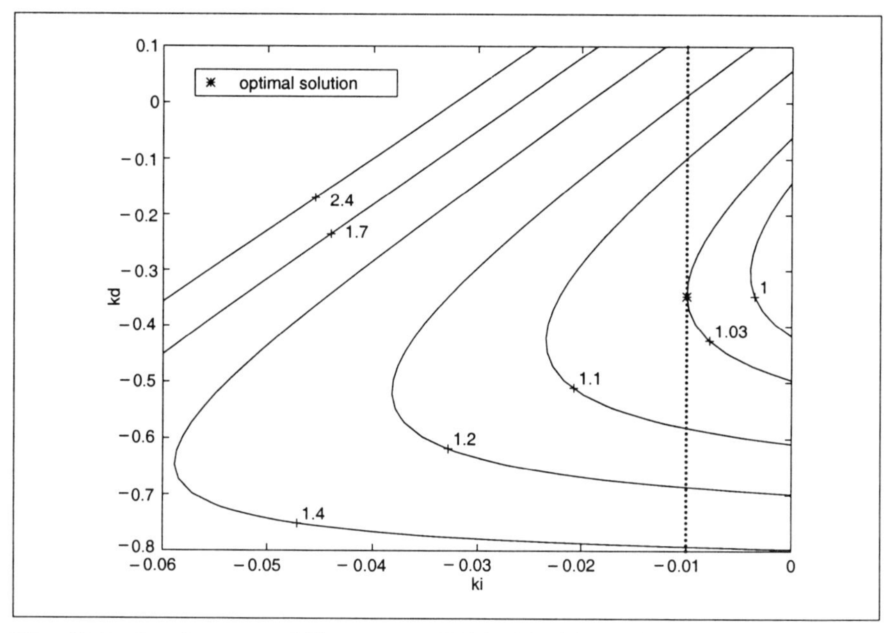

Fig. 5.23. Level curves of H_2 norm of $W(s)G(s)S(s)$ vs. (k_i, k_d) values for $k_p = -0.316$ with $ki \leq -0.01$.

(1) Settling time: Searching over the entire stabilizing set of $(k_p,\ k_i,\ k_d)$ values, with the constraint $k_i \leq -0.01$, we found the minimum settling time to be 0.679 secs and this occurs at

$$k_p = -1.71889, k_i = -0.72083, \text{ and } k_d = -0.99725.$$

The stabilizing region in the $(k_i,\ k_d)$ plane corresponding to $k_p = -1.71889$ is sketched in Fig. 5.24 while the corresponding plot of settling time values ($\leq$ 50 secs) on this region is given in Fig. 5.25. The level curves in Fig. 5.26 give a clearer picture of the occurrence of this minimum.

(2) Overshoot: By using the same search procedure, we determined the minimum overshoot to be 0.26582 and this occurs at

$$k_p = -0.98622, k_i = -0.01 \text{ and } k_d = -0.98621.$$

The stabilizing region in the $(k_i,\ k_d)$ plane corresponding to

$$k_p = -0.98622,$$

is sketched in Fig. 5.27 while the corresponding plot of overshoot values on this region is given in Fig. 5.28. The level curves in Fig. 5.29 give a clearer picture of the occurrence of this minimum.

(3) Undershoot: By using the same search procedure, we determined the minimum undershoot to be 0.09969 and this occurs at

$$k_p = -0.26483,\ k_i = -0.01 \text{ and } k_d = 0.$$

The stabilizing region in the $(k_i,\ k_d)$ plane corresponding to

$$k_p = -0.26483$$

is sketched in Fig. 5.30 while the corresponding plot of undershoot values on this region is given in Fig. 5.31. The level curves in Fig. 5.32 give a clearer picture of the occurrence of this minimum.

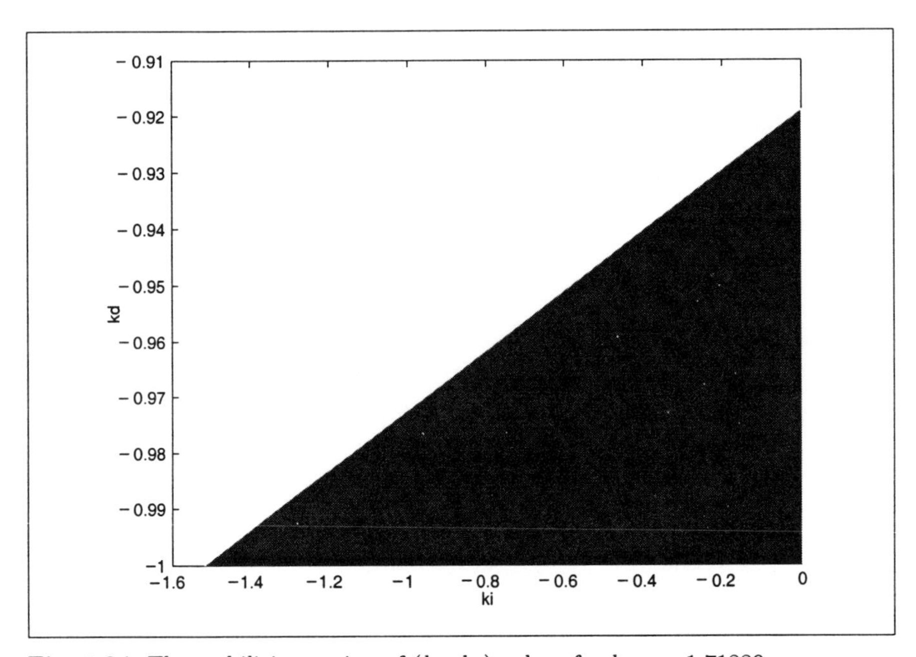

Fig. 5.24. The stabilizing region of $(k_i,\ k_d)$ values for $k_p = -1.71889$.

The plots generated above can be used to design PID controllers to meet given performance specifications. For instance, if we are given the following unit step response specifications:

(1) Settling time ≤ 10 secs.
(2) undershoot ≤ 1.
(3) and $|e_{ss_{ramp}}| \leq 1$, where $e_{ss_{ramp}}$ is the steady state error to a unit ramp input,

then we can proceed as follows. First, we note that

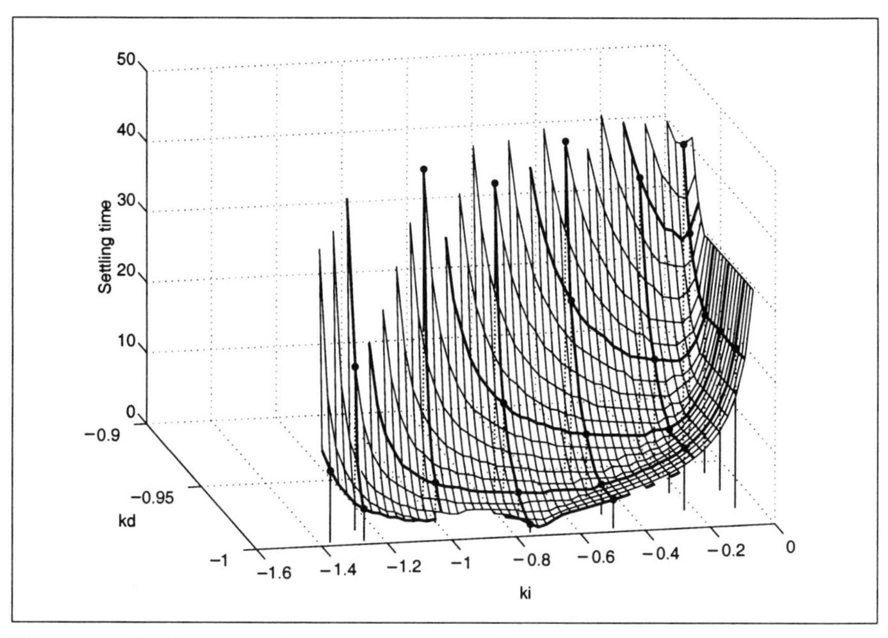

Fig. 5.25. Settling time ($\leq$ 50 secs) vs. (k_d, k_i) for $k_p = -1.71889$.

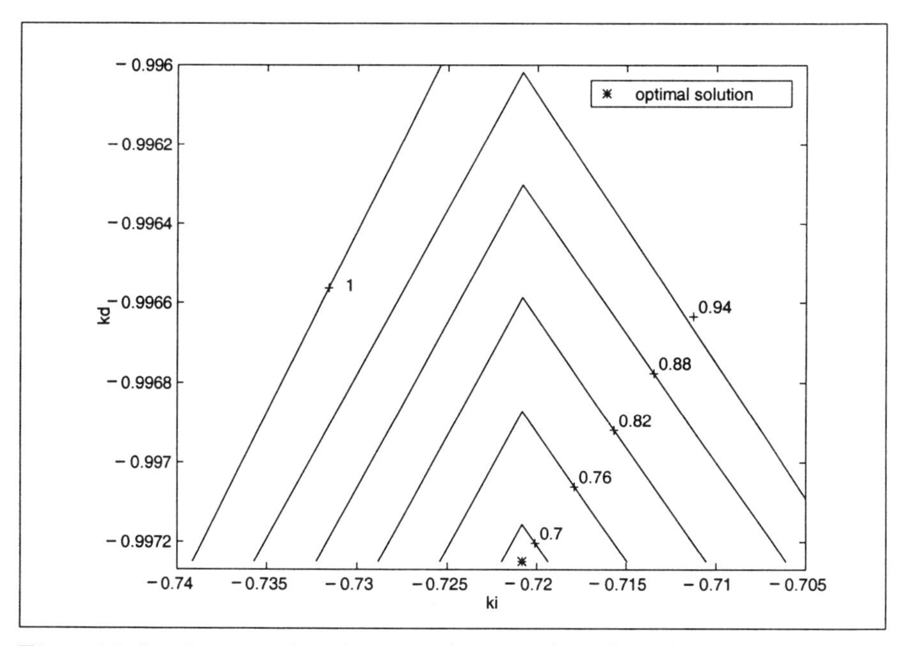

Fig. 5.26. Level curves of settling time ($\leq$ 50 secs) vs. (k_d, k_i) for $k_p = -1.71889$.

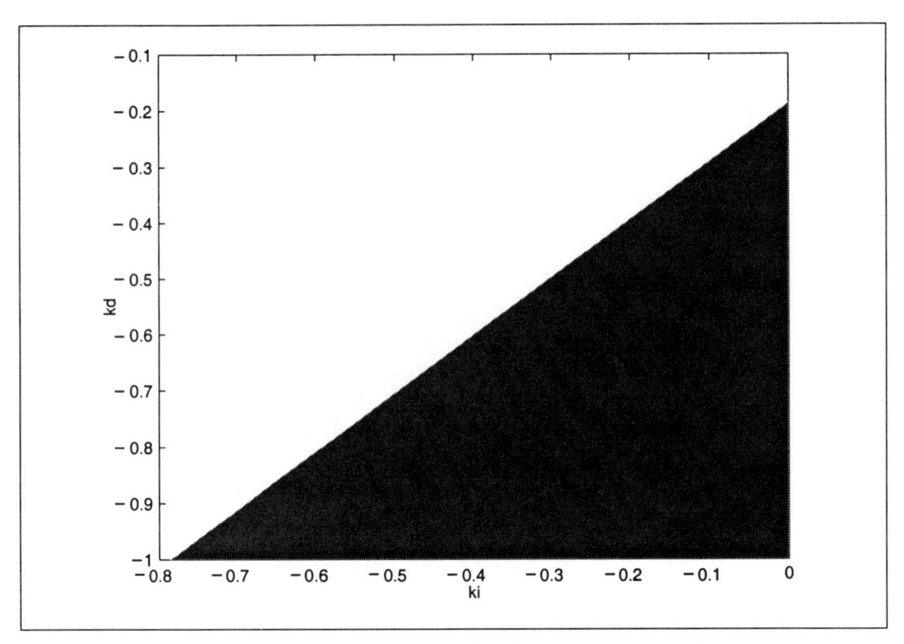

Fig. 5.27. The stabilizing region of $(k_i,\ k_d)$ values for $k_p = -0.98622$.

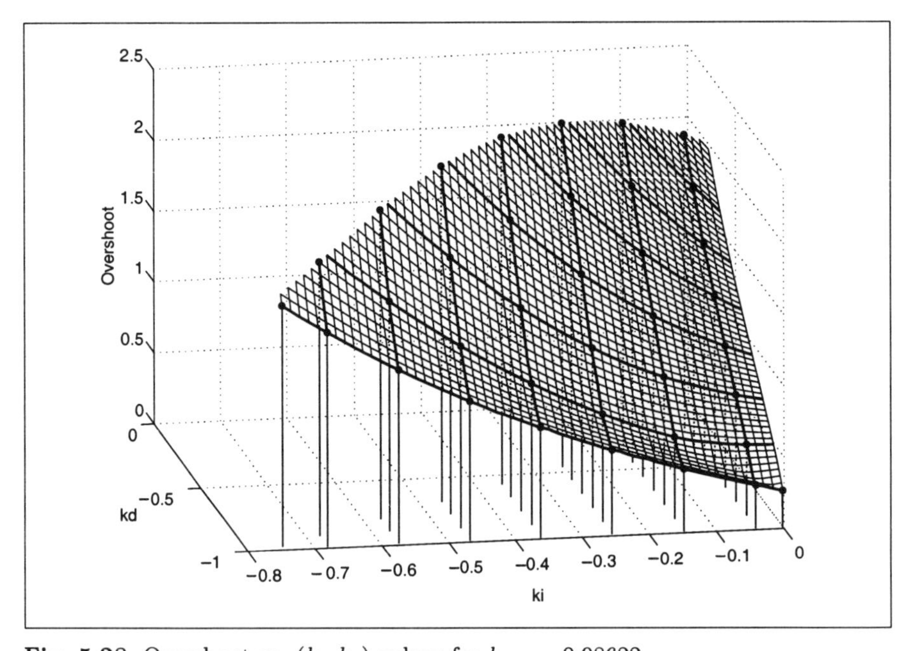

Fig. 5.28. Overshoot vs. (k_i, k_d) values for $k_p = -0.98622$.

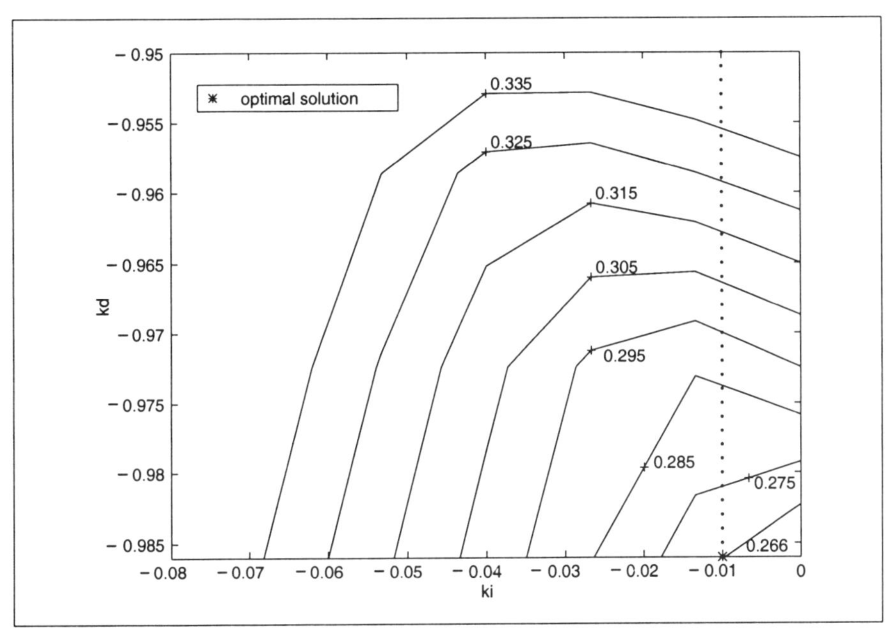

Fig. 5.29. Level curves of overshoot vs. (k_i, k_d) values for $k_p = -0.98622$ with $k_i \leq -0.01$.

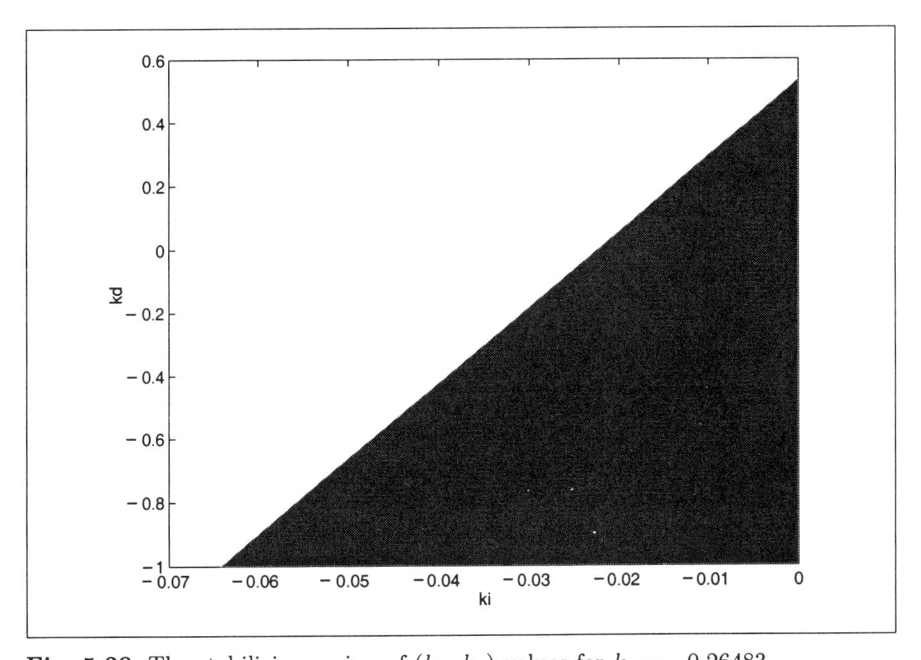

Fig. 5.30. The stabilizing region of $(k_i,\ k_d)$ values for $k_p = -0.26483$.

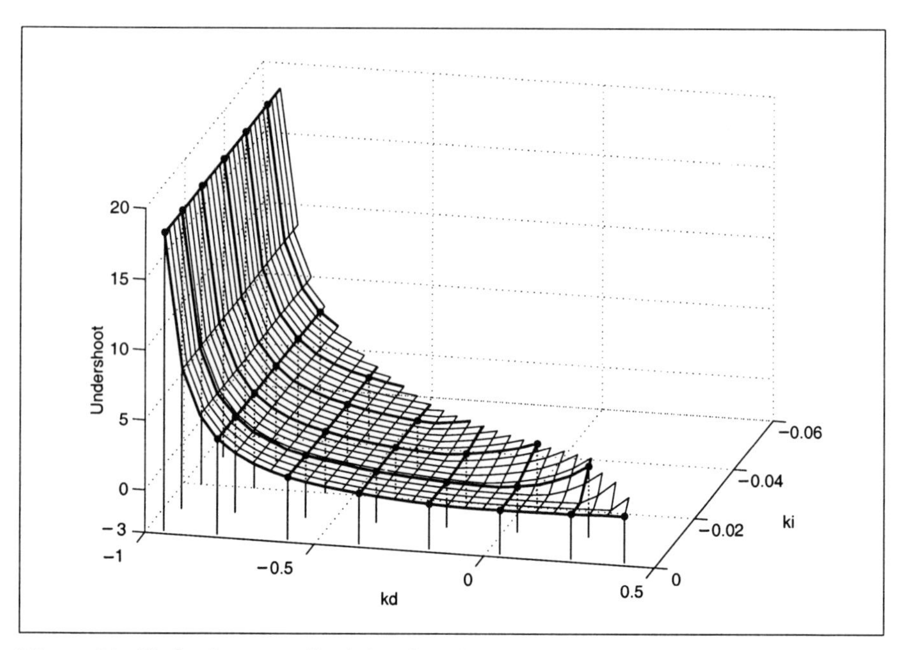

Fig. 5.31. Undershoot vs. (k_i, k_d) values for $k_p = -0.26483$.

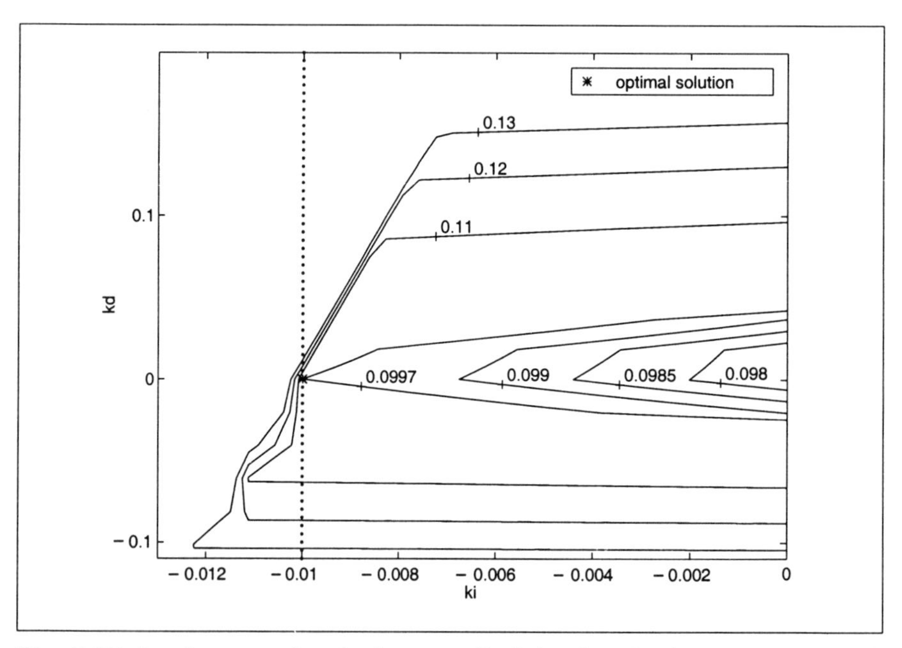

Fig. 5.32. Level curves of undershoot vs. (k_i, k_d) values for $k_p = -0.26483$ with $k_i \leq -0.01$.

$$|e_{ss_{ramp}}| = \left|\frac{0.2}{k_i}\right|$$

Also from Fig. 5.17, we require $k_i < 0$ to guarantee closed loop stability. Therefore, $|e_{ss_{ramp}}| \leq 1$ implies that $k_i \leq -0.2$ which narrows down the size of the admissible set of $(k_p,\ k_i,\ k_d)$ values. By searching over the entire set of stabilizing $(k_p,\ k_i,\ k_d)$ values subject to the constraint $k_i \leq -0.2$, we found that there is no PID controller which can meet the given performance specifications. Now we consider the same performance specifications except that we somewhat loosen up the undershoot constraint and require that the undershoot be less than 2. Then the performance specifications were met with

$$k_p = -0.857778,\ k_i = -0.200678,\ \text{and}\ k_d = -0.664633$$

and the corresponding overshoot was found to be 1.121412. The closed loop step response using the resulting PID controller is given in Fig. 5.33 which shows that the desired performance specifications have indeed been satisfied.

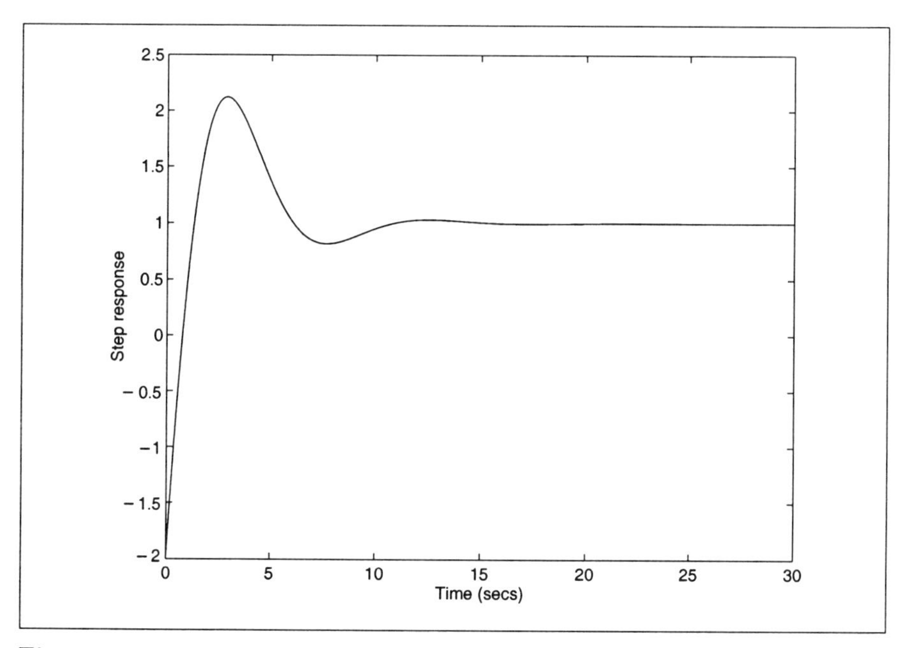

Fig. 5.33. Step response using the designed PID controller.

We conclude this section by presenting an example where the stabilizing PID gains form a *disconnected* set. In such a case, the global minimum of any performance index would have to be found by determining the *minimum over each of the connected components.*

Example 5.5.1. Consider the problem of designing an optimal H_∞ robust controller that minimizes $\|W(s)T(s)\|_\infty$ where $T(s)$ is the complementary sensitivity function and the weight $W(s)$ is chosen as the high-pass function

$$W(s) = \frac{s+0.1}{s+1}.$$

The transfer function of the plant in question is:

$$G(s) = \frac{s^3+3s^2+9}{s^4+2s^3+3s^2+7s+14}. \tag{5.1}$$

By using the root locus ideas in Appendix A, we determined that for the existence of stabilizing PID gains, k_p must lie in the union of the two disjoint intervals

$$(-1.8708, -1.5556)$$

and

$$(0.3157, 0.5333).$$

By sweeping over

$$k_p \in (-1.8708, -1.5556) \cup (0.3157, 0.5333),$$

we obtained the set of all stabilizing (k_p, k_i, k_d) values for the plant (5.1) as sketched in Fig. 5.34. Clearly, this set is disconnected and has two components. The two components are sketched separately in Figs. 5.35 and Fig. 5.36. For clarity of presentation, the stabilizing regions of Figs. 5.35 and 5.36 will be denoted in the rest of this example by S_1 and S_2 respectively.

As shown in Chapter 4, for a fixed k_p, the stabilizing region in the $(k_i,\ k_d)$ plane is a union of either convex polygons or intersections of half-planes. Thus, by sweeping over k_p, we can search for $\min_{(k_p,\ k_i,\ k_d)} \|W(s)T(s)\|_\infty$ by first determining the minimum over each stabilizing $(k_i,\ k_d)$ region. Using such an approach, we found that the minimum value of $\|W(s)T(s)\|_\infty$ in S_1 is 5.761 and this occurs at

$$k_p = -1.58367,\ k_i = 0,\ \text{ and } k_d = -1.28781.$$

Similarly, the minimum value of $\|W(s)T(s)\|_\infty$ in S_2 is 18.79598 and this occurs at

$$k_p = 0.42894,\ k_i = 2.4232,\ \text{ and } k_d = -0.46749.$$

Thus the globally optimal solution is

$$\min_{(k_p,\ k_i,\ k_d)\in S_1 \cup S_2} \|W(s)T(s)\|_\infty = 5.761$$

and this occurs at

$$k_p = -1.58367,\ k_i = 0,\ \text{ and } k_d = -1.28781.$$

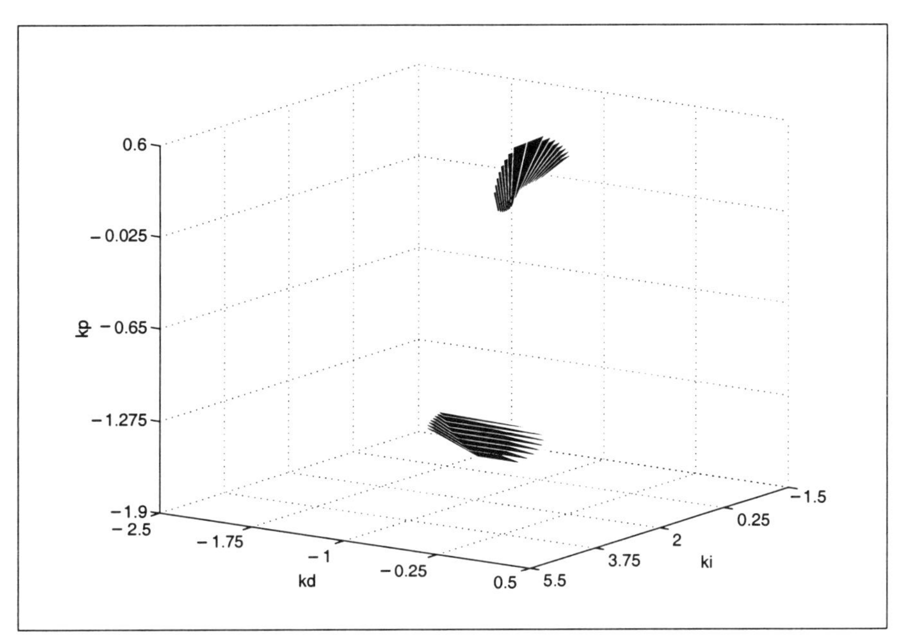

Fig. 5.34. Two disjoint stabilizing sets of (k_p, k_i, k_d) values (Example 5.5.1).

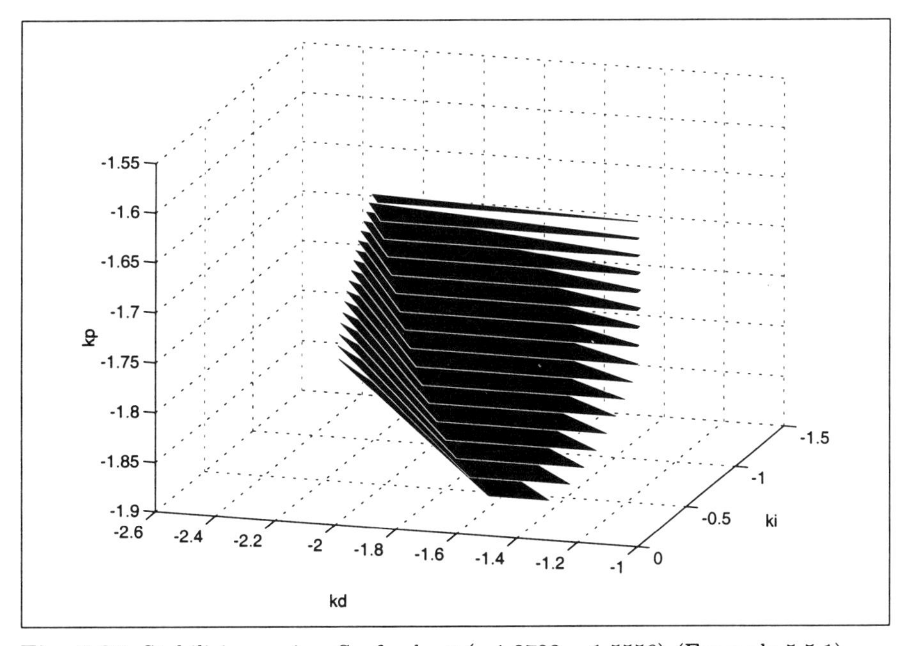

Fig. 5.35. Stabilizing region S_1, for $k_p \in (-1.8708, -1.5556)$ (Example 5.5.1).

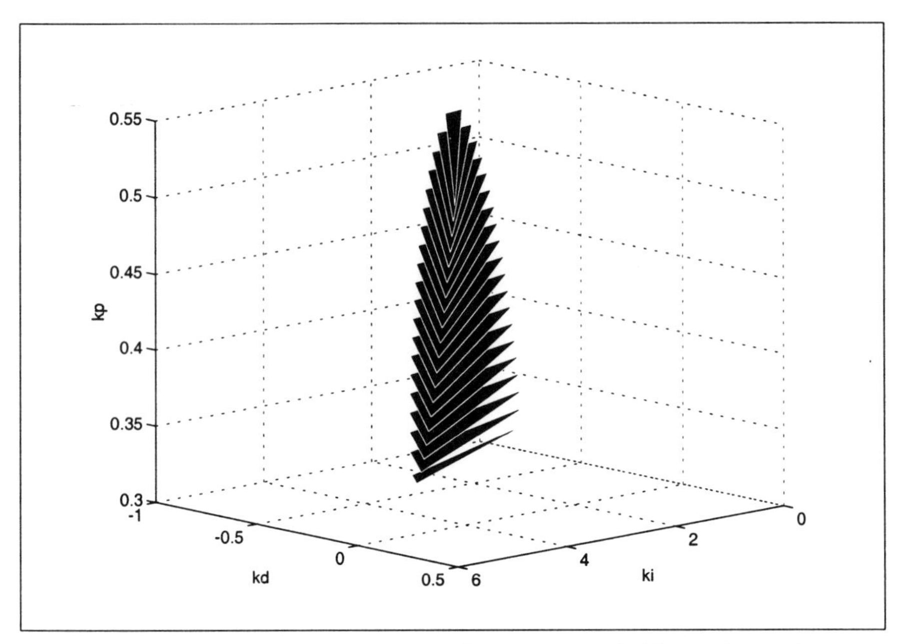

Fig. 5.36. Stabilizing region S_2, for $k_p \in (0.3157, 0.5333)$ (Example 5.5.1).

The optimal solution obtained above yields a zero value for the integral gain. Such a value is not desirable for reasons discussed in earlier examples. Consequently, we repeat the optimization procedure after imposing a reasonable constraint on the value of k_i. For the stabilizing region S_1, a reasonable constraint could be $k_i \leq -0.01$. By searching over S_1 subject to this design constraint, we found that

$$\min_{(k_p,\ k_i \leq -0.01,\ k_d)} \|W(s)T(s)\|_\infty = 6.26465$$

and this occurs at

$$k_p = -1.60364,\ k_i = -0.01, \quad \text{and } k_d = -1.31875.$$

Therefore, the globally optimal solution is

$$\min_{(k_p,\ k_i \leq -0.01,\ k_d) \in S_1 \cup S_2} \|W(s)T(s)\|_\infty = 6.26465$$

The stabilizing region in the (k_i, k_d) plane corresponding to $kp = -1.60364$ is sketched in Fig. 5.37 while the plot of $\|W(s)T(s)\|_\infty$ on this plane is given in Fig. 5.38. The level curves of $\|W(s)T(s)\|_\infty$ in Fig. 5.39 give a clearer picture of the occurrence of this minimum value.

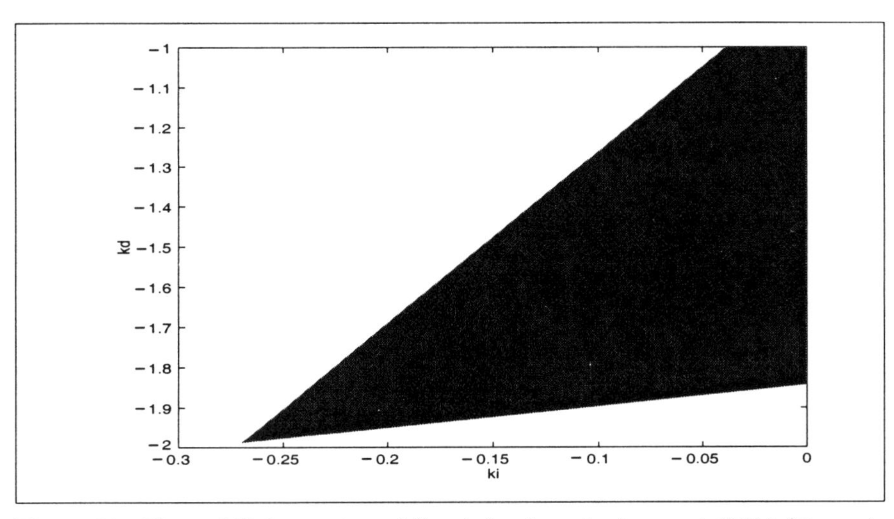

Fig. 5.37. The stabilizing region of $(k_i,\ k_d)$ values for $k_p = -1.60364$ (Example 5.5.1).

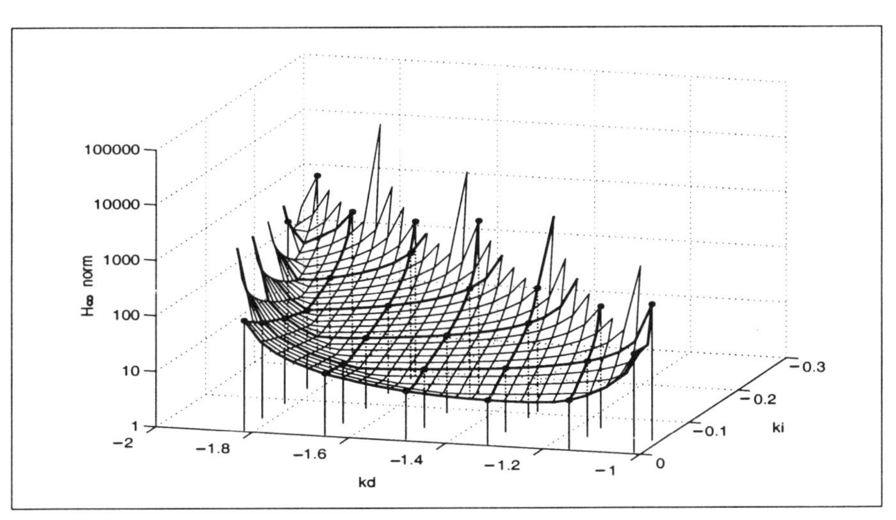

Fig. 5.38. H_∞ norm of weighted complementary sensitivity function vs. (k_i, k_d) values for $k_p = -1.60364$ (Example 5.5.1).

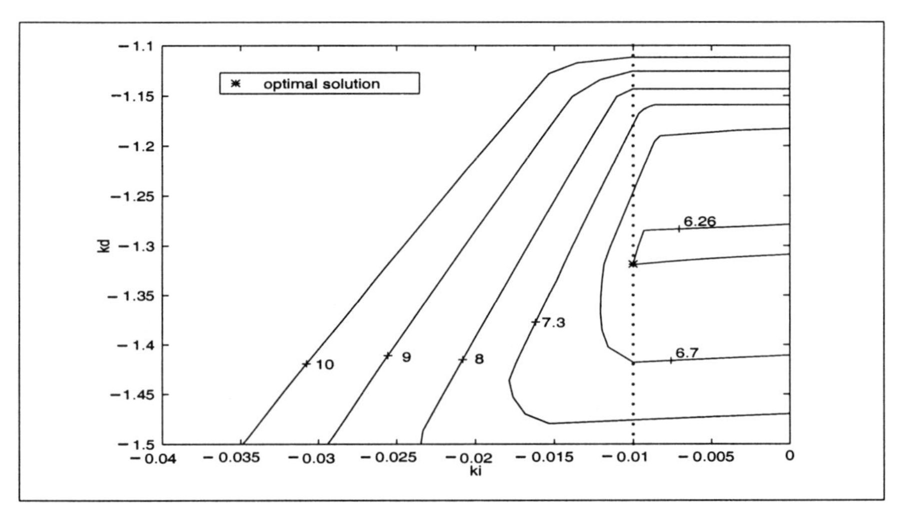

Fig. 5.39. Level curves of H_∞ norm of weighted complementary sensitivity function vs. (k_i, k_d) values for $k_p = -1.60364$ with $k_i \leq -0.01$ (Example 5.5.1).

5.6 Notes and References

The H_2 and H_∞ optimal design examples using P, PI and PID controllers are taken from Ho, Datta and Bhattacharyya [22]. Although the design procedure is essentially a brute force optimization search, nevertheless the fact that the results of the last chapter can be used to confine the search to the stabilizing region makes the problem orders of magnitude easier.

CHAPTER 6

ROBUST AND NON-FRAGILE PID CONTROLLER DESIGN

In this chapter, we provide a solution to the problem of *robustly* stabilizing a given interval plant family using a PID controller. In addition, we show how such a design can be made "non-fragile" in the space of the controller coefficients.

6.1 Introduction

The area of parametric robust control has made tremendous strides during the last decade. Ever since the spark generated by Kharitonov's celebrated Theorem, many important problems in this area have been formulated and elegantly solved. However, most of the existing results in the area of parametric robust control are of the analysis type. In other words, given a controller, they allow one to check, in a computationally efficient way, if the particular controller can simultaneously stabilize an entire plant family with real coefficient variations, or for that matter, if the particular controller can guarantee the satisfaction of a performance objective over the entire plant family. No systematic technique currently exists for *designing* a controller to stabilize an interval plant family.

The only synthesis type result for interval plants that we are aware of is the one in [8], where the parametric uncertainty in the plant is bounded using an H_∞ ball of perturbations, and results from H_∞ theory [33] are used to synthesize a stabilizing controller for the entire plant family. Such an approach is not satisfactory because of at least two reasons. First, the bounding of parametric uncertainty by an H_∞ ball of perturbations introduces conservatism into the design; and second, H_∞ controllers are typically of high order (same order as that of the plant) and such a design may suffer from extreme fragility in the controller coefficients [30]. Thus it seems that what is really needed in the parametric robust control area is a technique for synthesizing fixed and low order controllers that simultaneously stabilize a given interval plant family. The aim of this chapter is to do precisely that when the controller structure is restricted to being either a proportional gain (P), a proportional-integral (PI) or a proportional-integral-derivative (PID). This is done by combining analysis results from the area of parametric robust control with the results of Chapters 3 and 4 for designing fixed order and structure controllers for any given plant.

The chapter is organized as follows. In Section 6.2, we recall some relevant results from the area of parametric robust control. In Section 6.3, these results are used in conjunction with the results of Chapters 3 and 4 to provide a constructive procedure for obtaining all constant gain stabilizers for a given interval plant family. Similar procedures for PI and PID stabilization are presented in Sections 6.4 and 6.5 respectively. Illustrative examples are included in each case. Finally, in Section 6.6, we show how the robust PID controller of Section 6.5 can be made optimally "non-fragile" in the space of the PID parameters.

6.2 Kharitonov's Theorem and Its Generalization

In this section, we briefly review some results from the area of parametric robust control. We will only highlight those results which are most relevant to the subsequent development here. For an exhaustive treatment, the reader is referred to [7].

Definition 6.2.1. *Consider the set $\mathcal{F}$ of all real polynomials of degree n of the form*

$$p(s) = p_0 + p_1 s + p_2 s^2 + \cdots + p_n s^n$$

where the coefficients vary in independent intervals

$$p_0 \in [x_0,\, y_0],\; p_1 \in [x_1,\, y_1],\; \cdots,\; p_n \in [x_n,\, y_n],\; 0 \notin [x_n,\, y_n].$$

Such a set of polynomials is called an interval polynomial.

We next state Kharitonov's celebrated theorem which provides a necessary and sufficient condition for the Hurwitz stability of such an interval polynomial.

Theorem 6.2.1. *(Kharitonov's Theorem) Every polynomial in the interval family $\mathcal{F}$ is Hurwitz if and only if the following four polynomials called Kharitonov polynomials are Hurwitz:*

$$\begin{aligned}
K^1(s) &= x_0 + x_1 s + y_2 s^2 + y_3 s^3 + x_4 s^4 + x_5 s^5 + y_6 s^6 + \cdots \\
K^2(s) &= x_0 + y_1 s + y_2 s^2 + x_3 s^3 + x_4 s^4 + y_5 s^5 + y_6 s^6 + \cdots \\
K^3(s) &= y_0 + x_1 s + x_2 s^2 + y_3 s^3 + y_4 s^4 + x_5 s^5 + x_6 s^6 + \cdots \\
K^4(s) &= y_0 + y_1 s + x_2 s^2 + x_3 s^3 + y_4 s^4 + y_5 s^5 + x_6 s^6 + \cdots.
\end{aligned}$$

We now proceed to state a generalization of Kharitonov's Theorem which will play an important role in the sequel. However, before stating this theorem, we need to introduce some notation. Let m be an arbitrary integer and let $\bar{P}(s) = [P_1(s),\, P_2(s),\, \cdots,\, P_m(s)]$ be an m-tuple of real polynomials where each component $P_i(s)$ $i = 1, 2, \cdots, m$ is an interval polynomial. Consequently, with each $P_i(s)$, we can associate its four Kharitonov polynomials $K_i^1(s)$, $K_i^2(s)$, $K_i^3(s)$ and $K_i^4(s)$. We denote by $\mathcal{K}_m$ the set of m-tuples obtained as follows: For every fixed integer i between 1 and m set

$$P_i(s) = K_i^k(s), \text{ for some } k = 1,\ 2,\ 3,\ 4.$$

Clearly, there are at most 4^m distinct elements in $\mathcal{K}_m$. In addition, we define a family of m-tuples called generalized Kharitonov segments as follows: For any fixed integer l between 1 and m, set

$$P_i(s) = K_i^k(s), \text{ for } i \neq l \text{ and for some } k = 1, 2, 3, 4$$

and for $i = l$, suppose that $P_l(s)$ varies in one of the four segments

$$\begin{array}{l} \left[K_l^1(s),\ K_l^2(s)\right] \\ \left[K_l^1(s),\ K_l^3(s)\right] \\ \left[K_l^2(s),\ K_l^4(s)\right] \\ \left[K_l^3(s),\ K_l^4(s)\right]. \end{array}$$

By the segment $[K_l^1(s),\ K_l^2(s)]$, we mean the set of all convex combinations of the form

$$(1-\lambda)K_l^1(s) + \lambda K_l^2(s),\ \lambda \in [0,\ 1].$$

There are at most $m4^m$ distinct generalized Kharitonov segments and we will denote by S_m the family of all these m-tuples. With these preliminaries, we now state the theorem of Chapellat and Bhattacharyya [7].

Theorem 6.2.2. *(Generalized Kharitonov Theorem) [7]:*

1) *Given an m-tuple of fixed real or complex polynomials $[F_1(s),\ F_2(s),\ \cdots,\ F_m(s)]$, the polynomial family*

$$P_1(s)F_1(s) + P_2(s)F_2(s) + \cdots + P_m(s)F_m(s) \qquad (6.1)$$

is Hurwitz stable iff all the one-parameter polynomial families that result from replacing $[P_1(s),\ P_2(s),\ \cdots, P_m(s)]$ in the above expression by the elements of S_m are Hurwitz stable.

2) *If the polynomials $F_i(s)$ are real and of the form $F_i(s) = s^{t_i}(a_i s + b_i)U_i(s)Q_i(s)$ where $t_i \geq 0$ is an arbitrary integer, a_i and b_i are arbitrary real numbers, $U_i(s)$ is an anti-Hurwitz polynomial, and $Q_i(s)$ is an even or odd polynomial, then it is sufficient that (6.1) be Hurwitz stable with the P_i's replaced by the elements of $\mathcal{K}_m$.*

3) *If the F_i's are complex and*

$$\frac{d}{d\omega} arg[F_i(j\omega)] \ \leq\ 0 \ \ \forall i = 1,\ 2,\ \cdots,\ m$$

then it is sufficient that (6.1) be Hurwitz stable with the P_i's replaced by the elements of $\mathcal{K}_m$.

6.3 Robust Stabilization Using a Constant Gain

In this section, we consider the problem of characterizing all constant gain controllers that stabilize a given interval plant. The key idea is to use the results of Chapter 4 in conjunction with the Generalized Kharitonov Theorem.

Now let $\mathcal{G}(s)$ be an interval plant:

$$\mathcal{G}(s) = \frac{\mathcal{N}(s)}{\mathcal{D}(s)}$$

$$\begin{aligned} \mathcal{N}(s) &= a_0 + a_1 s + a_2 s^2 + \cdots + a_m s^m \\ \mathcal{D}(s) &= b_0 + b_1 s + b_2 s^2 + \cdots + b_n s^n \end{aligned}$$

where $n \geq m$, $a_m \neq 0$, $b_n \neq 0$ and the coefficients of $\mathcal{N}(s)$ and $\mathcal{D}(s)$ vary in independent intervals, *i.e.*,

$$\begin{aligned} &a_0 \in [\underline{a}_0, \, \overline{a}_0], \; a_1 \in [\underline{a}_1, \, \overline{a}_1], \cdots, \; a_m \in [\underline{a}_m, \, \overline{a}_m] \\ &b_0 \in [\underline{b}_0, \, \overline{b}_0], \; b_1 \in [\underline{b}_1, \, \overline{b}_1], \cdots, \; b_n \in [\underline{b}_n, \, \overline{b}_n]. \end{aligned}$$

The controller $C(s)$ in question is a constant gain *i.e.*,

$$C(s) = k.$$

The family of closed loop characteristic polynomials $\Delta(s, k)$ is given by:

$$\Delta(s, k) = \mathcal{D}(s) + k\mathcal{N}(s).$$

The problem of characterizing all constant gain stabilizers for an interval plant is to determine all the values of k for which the family of closed loop characteristic polynomials $\Delta(s, k)$ is Hurwitz.

Let $\mathcal{N}^i(s)$, $\mathcal{D}^i(s)$, $i = 1, 2, 3, 4$ be the Kharitonov polynomials corresponding to $N(s)$ and $D(s)$ respectively and let $\mathcal{G}_K(s)$ denote the family of 16 vertex plants defined as:

$$\mathcal{G}_K(s) = \{G_{ij}(s) \mid G_{ij}(s) = \frac{\mathcal{N}^i(s)}{\mathcal{D}^j(s)}, i = 1, 2, 3, 4, \; j = 1, 2, 3, 4\}.$$

The closed-loop characteristic polynomial for each of these vertex plants $G_{ij}(s)$ is denoted by $\delta_{ij}(s, k)$ and is defined as

$$\delta_{ij}(s, k) = \mathcal{D}^j(s) + k\mathcal{N}^i(s).$$

We can now state the following Theorem which characterizes all constant gain stabilizers for an interval plant.

Theorem 6.3.1. *Let $\mathcal{G}(s)$ be an interval plant as defined above. Then the entire family $\mathcal{G}(s)$ is stabilizable by a constant gain k if and only if the following conditions hold:*
(i) Each $G_{ij}(s) \in \mathcal{G}_K(s)$ is stabilizable by a constant gain.
and
(ii) $\mathcal{K} \neq \emptyset$, where $\mathcal{K} = \cap_{i=1,\cdots,4,\ j=1,\cdots,4} K_{ij}$ and K_{ij} is the set of all stabilizing gain values for $G_{ij}(s)$[1].
Furthermore the set of all stabilizing gain values for the entire family $\mathcal{G}(s)$ is precisely given by $\mathcal{K}$.

Proof. Now we have

$$\begin{aligned}\Delta(s,\,k) &= \mathcal{D}(s) + k\mathcal{N}(s)\\ &= F_1(s)\mathcal{D}(s) + F_2(s)\mathcal{N}(s)\end{aligned}$$

where $F_1(s) = 1$ and $F_2(s) = k$. Using Theorem 6.2.2 2), it follows that the entire family $\Delta(s,k)$ is Hurwitz stable iff $\delta_{ij}(s,k)$, $i = 1,2,3,4$, $j = 1,2,3,4$ are all Hurwitz. Therefore, the entire family $\mathcal{G}(s)$ is stabilized by a constant gain k if and only if every element of $\mathcal{G}_K(s)$ is simultaneously stabilized by k. Thus one can use the results of Section 4.2 to find out all the constant gain stabilizers for each member of $\mathcal{G}_K(s)$ and then take their intersection to obtain all constant gain stabilizers for the given interval plant. This completes the proof. ♣

We now present a simple example to illustrate the detailed calculations involved in coming up with all constant gain stabilizers for an interval plant.

Example 6.3.1. Consider the interval plant

$$\mathcal{G}(s) = \frac{\mathcal{N}(s)}{\mathcal{D}(s)},$$

$$\begin{aligned}\text{where } \mathcal{N}(s) &= a_0 + a_1 s + a_2 s^2\\ \mathcal{D}(s) &= b_0 + b_1 s + b_2 s^2 + b_3 s^3 + b_4 s^4\end{aligned}$$

with

$$\begin{aligned}&a_2 \in [1,\ 1],\ a_1 \in [1,\ 2],\ a_0 \in [1,\ 2]\\ \text{and}\quad &b_4 \in [1,\ 1],\ b_3 \in [3,\ 4],\ b_2 \in [4,\ 4],\ b_1 \in [5,\ 8],\ b_0 \in [6,\ 7].\end{aligned}$$

Then the Kharitonov polynomials corresponding to $\mathcal{N}(s)$ and $\mathcal{D}(s)$ are

[1] Note that for any particular $G_{ij}(s)$, the set of all stabilizing feedback gains K_{ij} can be determined using the results of Section 4.2, specifically Theorem 4.2.1.

$$\mathcal{N}^1(s) = s^2 + s + 1,\ \mathcal{N}^2(s) = s^2 + 2s + 1,$$
$$\mathcal{N}^3(s) = s^2 + s + 2,\ \mathcal{N}^4(s) = s^2 + 2s + 2,$$
$$\mathcal{D}^1(s) = s^4 + 4s^3 + 4s^2 + 5s + 6,\ \mathcal{D}^2(s) = s^4 + 3s^3 + 4s^2 + 8s + 6,$$
$$\mathcal{D}^3(s) = s^4 + 4s^3 + 4s^2 + 5s + 7,\ \mathcal{D}^4(s) = s^4 + 3s^3 + 4s^2 + 8s + 7.$$

Furthermore,

$$\mathcal{G}_K(s) = \{G_{ij}(s) \mid G_{ij}(s) = \frac{\mathcal{N}^i(s)}{\mathcal{D}^j(s)}, i = 1,2,3,4,\ j = 1,2,3,4\}.$$

Using the results of Section 4.2, specifically Theorem 4.2.1, the sets K_{ij} corresponding to $G_{ij}(s)$, $i = 1,2,3,4,\ j = 1,2,3,4$ are

$$\begin{aligned}
&K_{11} = (2.3885, \infty),\ K_{12} = (1.5584, \infty),\\
&K_{13} = (3.0, \infty),\ K_{14} = (2.0523, \infty),\\
&K_{21} = (1.7749, \infty),\ K_{22} = (2.0, \infty),\\
&K_{23} = (2.2720, \infty),\ K_{24} = (2.5584, \infty),\\
&K_{31} = (-3.0, -2.8297) \cup (4.8297, \infty),\ K_{32} = (2.8541, \infty),\\
&K_{33} = (-3.5, -3.4721) \cup (5.4721, \infty),\ K_{34} = (3.4686, \infty),\\
&K_{41} = (3.2016, \infty),\ K_{42} = (-3.0, -2.8541) \cup (3.8541, \infty),\\
&K_{43} = (3.7749, \infty),\ K_{44} = (-3.5, -3.4686) \cup (4.4686, \infty).
\end{aligned}$$

Therefore, the set of all k values which stabilize the entire family $\mathcal{G}(s)$ is given by:

$$\begin{aligned}
\mathcal{K} &= \cap_{i=1,\cdots,4,\ j=1,\cdots,4} K_{ij}\\
&= (5.4721, \infty).
\end{aligned}$$

6.4 Robust Stabilization Using a PI Controller

In this section, we consider the problem of characterizing all PI controllers that stabilize a given interval plant $\mathcal{G}(s) = \frac{\mathcal{N}(s)}{\mathcal{D}(s)}$ where $\mathcal{N}(s)$, $\mathcal{D}(s)$ are as defined in Section 6.3. Since the controller $C(s)$ is now given by

$$C(s) = k_p + \frac{k_i}{s}$$

the family of closed loop characteristic polynomials $\Delta(s, k_p, k_i)$ becomes

$$\Delta(s, k_p, k_i) = s\mathcal{D}(s) + (k_i + k_p s)\mathcal{N}(s).$$

The problem of characterizing all stabilizing PI controllers is to determine all the values of k_p and k_i for which the entire family of closed loop characteristic polynomials $\Delta(s, k_p, k_i)$ is Hurwitz.

Let $\mathcal{N}^i(s),\ i=1,2,3,4$ and $\mathcal{D}^j(s),\ j=1,2,3,4$ be the Kharitonov polynomials corresponding to $\mathcal{N}(s)$ and $\mathcal{D}(s)$, respectively and let $\mathcal{G}_K(s)$ denote the family of 16 vertex plants:

$$\mathcal{G}_K(s) = \{G_{ij}(s) \mid G_{ij}(s) = \frac{\mathcal{N}^i(s)}{\mathcal{D}^j(s)}, i = 1,2,3,4,\ j = 1,2,3,4\}.$$

The closed-loop characteristic polynomial for each of these vertex plants $G_{ij}(s)$ is denoted by $\delta_{ij}(s,k_p,k_i)$ and is defined as

$$\delta_{ij}(s,k_p,k_i) = s\mathcal{D}^j(s) + (k_i + k_p s)\mathcal{N}^i(s).$$

We can now state the following Theorem on stabilizing an interval plant using a PI controller. The proof is essentially the same as that of Theorem 6.3.1 and is therefore omitted.

Theorem 6.4.1. *Let $\mathcal{G}(s)$ be an interval plant. Then the entire family $\mathcal{G}(s)$ is stabilized by a particular PI controller if and only if each $G_{ij}(s) \in \mathcal{G}_K(s)$ is stabilized by that same PI controller.*

In view of the above Theorem, we can now use the results of Section 4.3 to obtain a characterization of all PI controllers that stabilize the interval family $\mathcal{G}(s)$. As in Section 4.3, for any fixed k_p, we can solve the constant gain stabilization problem for each $G_{ij}(s)$ to determine the stabilizing set of k_i for that particular $G_{ij}(s)$. We denote $KI_{ij}(k_p)$ to be the set of stabilizing k_i corresponding to $G_{ij}(s)$ and a fixed k_p. With such a fixed k_p, the set of all stabilizing $(k_p,\ k_i)$ values for the entire $\mathcal{G}_K(s)$, denoted by $\mathcal{S}_{k_p}$ is given by

$$\mathcal{S}_{k_p} = \{(k_p,\ k_i) | k_i \in \cap_{i=1,\dots,4,\ j=1,\dots,4} KI_{ij}(k_p)\}.$$

The set of all stabilizing $(k_p,\ k_i)$ values for the entire family $\mathcal{G}(s)$ can now be found by simply sweeping over k_p. Now from the results of Section 4.3, we know that for a given fixed plant, the range of k_p values for which a stabilizing k_i may exist, can usually be narrowed down using some necessary conditions derived from roots locus ideas. Let KP_{ij} be the set of k_p values satisfying such a necessary condition for $G_{ij}(s)$. Define

$$\mathcal{KP} = \cap_{i=1,\dots,4,\ j=1,\dots,4} KP_{ij}.$$

Then $k_p \in \mathcal{KP}$ is a necessary condition that must be satisfied for any $(k_p,\ k_i)$ that stabilizes the entire family $\mathcal{G}(s)$. Thus, by sweeping over all $k_p \in \mathcal{KP}$ and solving constant gain stabilization problems for the 16 vertex plants at each stage, we can determine the set of all stabilizing (k_p, k_i) values for the entire family $\mathcal{G}(s)$.

We now present a simple example to illustrate the detailed calculations involved in determining all the stabilizing (k_p, k_i) values for a given interval plant $\mathcal{G}(s)$.

Example 6.4.1. Consider the interval plant

$$\mathcal{G}(s) = \frac{\mathcal{N}(s)}{\mathcal{D}(s)},$$

$$\begin{aligned} \text{where } \mathcal{N}(s) &= a_0 + a_1 s + a_2 s^2 + a_3 s^3 + a_4 s^4 \\ \mathcal{D}(s) &= b_0 + b_1 s + b_2 s^2 + b_3 s^3 + b_4 s^4 + b_5 s^5 \end{aligned}$$

with

$$\begin{array}{ll} & a_4 \in [1,\ 1],\ a_3 \in [2,\ 3],\ a_2 \in [39,\ 41],\ a_1 \in [48,\ 50],\ a_0 \in [-6,\ -3] \\ \text{and} & b_5 \in [1,\ 1],\ b_4 \in [2,\ 3],\ b_3 \in [31,\ 32],\ b_2 \in [35,\ 38],\ b_1 \in [49,\ 51], \\ & b_0 \in [97,\ 101]. \end{array}$$

Then the Kharitonov polynomials corresponding to $\mathcal{N}(s)$ and $\mathcal{D}(s)$ are

$$\begin{array}{l} \mathcal{N}^1(s) = s^4 + 3s^3 + 41s^2 + 48s - 6,\ \mathcal{N}^2(s) = s^4 + 2s^3 + 41s^2 + 50s - 6 \\ \mathcal{N}^3(s) = s^4 + 3s^3 + 39s^2 + 48s - 3,\ \mathcal{N}^4(s) = s^4 + 2s^3 + 39s^2 + 50s - 3 \\ \mathcal{D}^1(s) = s^5 + 2s^4 + 32s^3 + 38s^2 + 49s + 97, \\ \mathcal{D}^2(s) = s^5 + 2s^4 + 31s^3 + 38s^2 + 51s + 97, \\ \mathcal{D}^3(s) = s^5 + 3s^4 + 32s^3 + 35s^2 + 49s + 101, \\ \mathcal{D}^4(s) = s^5 + 3s^4 + 31s^3 + 35s^2 + 51s + 101. \end{array}$$

Now for a fixed k_p, for instance $k_p = 2$, we can solve constant gain stabilization problems to determine the set of stabilizing $KI_{ij}(2)$ corresponding to each $G_{ij}(s)$.

$$\begin{array}{l} KI_{11}(2) = (-1.4993, 0),\ KI_{12}(2) = (-1.5049, 0), \\ KI_{13}(2) = (-1.5673, 0),\ KI_{14}(2) = (-1.5734, 0), \\ KI_{21}(2) = (-1.4610, 0),\ KI_{22}(2) = (-1.4656, 0), \\ KI_{23}(2) = (-1.5281, 0),\ KI_{24}(2) = (-1.5332, 0), \\ KI_{31}(2) = (-1.7356, 0),\ KI_{32}(2) = (-1.7394, 0), \\ KI_{33}(2) = (-1.8100, 0),\ KI_{34}(2) = (-1.8142, 0), \\ KI_{41}(2) = (-1.6807, 0),\ KI_{42}(2) = (-1.6837, 0), \\ KI_{43}(2) = (-1.7534, 0),\ KI_{44}(2) = (-1.7567, 0). \end{array}$$

Since

$$\cap_{i=1,\ldots,4,\ j=1,\ldots,4}\, [KI_{ij}(2)] = (-1.4610,\ 0),$$

it follows that

$$\mathcal{S}_2 = \{(k_p,\ k_i) | kp = 2,\ k_i \in (-1.4610,\ 0)\}.$$

Furthermore, $KP_{ij}, i = 1,2,3,4,\ j = 1,2,3,4$ are given by:

$$\begin{aligned}
&KP_{11} = (0.0010, 16.1667),\ KP_{12} = (-0.0055, 16.1667),\\
&KP_{13} = (0, 16.8333),\ KP_{14} = (0, 16.8333),\\
&KP_{21} = (0, 16.1667),\ KP_{22} = (0, 16.1667),\\
&KP_{23} = (-1.7382, 16.8333),\ KP_{24} = (-1.6128, 17.9000),\\
&KP_{31} = (0.0022, 32.3333),\ KP_{32} = (0.0011, 32.3333),\\
&KP_{33} = (0, 33.6667),\ KP_{34} = (0, 33.6667),\\
&KP_{41} = (0, 32.3333),\ KP_{42} = (0, 32.3333),\\
&KP_{43} = (-2.0923, 33.6667),\ KP_{44} = (-1.9073, 33.6667).
\end{aligned}$$

Thus

$$\begin{aligned}
\mathcal{KP} &= \cap_{i=1,\ldots,4,\ j=1,\ldots,4} KP_{ij}\\
&= (0.0022, 16.1667).
\end{aligned}$$

Now by sweeping over all $k_p \in (0.0022, 16.1667)$ and solving the constant gain stabilization problems for each of the 16 vertex plants at each stage, we obtained the stabilizing (k_p, k_i) values for the entire family $\mathcal{G}(s)$. This is sketched in Fig. 6.1.

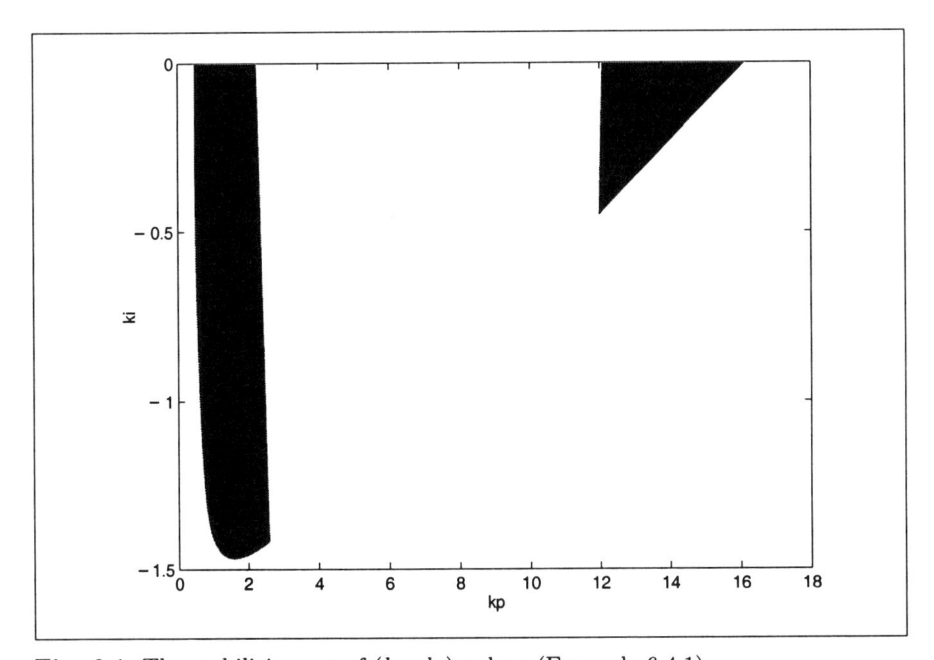

Fig. 6.1. The stabilizing set of (k_p, k_i) values (Example 6.4.1).

6.5 Robust Stabilization Using a PID Controller

In this section, we consider the problem of characterizing all PID controllers that stabilize a given interval plant $\mathcal{G}(s) = \frac{\mathcal{N}(s)}{\mathcal{D}(s)}$ where $\mathcal{N}(s)$, $\mathcal{D}(s)$ are as defined in Section 6.3. Since the controller $C(s)$ in question is now a PID controller, *i.e.*, $C(s) = \frac{k_d s^2 + k_p s + k_i}{s}$, the family of closed loop characteristic polynomials $\Delta(s, k_p, k_i, k_d)$ becomes

$$\Delta(s, k_p, k_i, k_d) = s\mathcal{D}(s) + (k_d s^2 + k_p s + k_i)\mathcal{N}(s).$$

The problem of characterizing all stabilizing PID controllers for the entire family $\mathcal{G}(s)$ is to determine all the values of k_p, k_i, and k_d for which the entire family of closed loop characteristic polynomials $\Delta(s, k_p, k_i, k_d)$ is Hurwitz.

Let $\mathcal{N}^i(s),\ i = 1, 2, 3, 4$ and $\mathcal{D}^j(s),\ j = 1, 2, 3, 4$ be the Kharitonov polynomials corresponding to $\mathcal{N}(s)$ and $\mathcal{D}(s)$, respectively. Furthermore, let $\mathcal{NS}^i(s),\ i = 1, 2, 3, 4$ be the four Kharitonov segments of $\mathcal{N}(s)$, where

$$\begin{aligned} \mathcal{NS}^1(s,\ \lambda) &= (1-\lambda)\mathcal{N}^1(s) + \lambda\mathcal{N}^2(s) \\ \mathcal{NS}^2(s,\ \lambda) &= (1-\lambda)\mathcal{N}^1(s) + \lambda\mathcal{N}^3(s) \\ \mathcal{NS}^3(s,\ \lambda) &= (1-\lambda)\mathcal{N}^2(s) + \lambda\mathcal{N}^4(s) \\ \mathcal{NS}^4(s,\ \lambda) &= (1-\lambda)\mathcal{N}^3(s) + \lambda\mathcal{N}^4(s) \end{aligned}$$

and $\lambda \in [0,\ 1]$. Let $\mathcal{G}_S(s)$ denote the family of 16 segment plants:

$$\begin{aligned} \mathcal{G}_S(s) &= \{G_{ij}(s,\ \lambda) \mid G_{ij}(s,\ \lambda) = \frac{\mathcal{NS}^i(s,\ \lambda)}{\mathcal{D}^j(s)}, i = 1, 2, 3, 4,\ j = 1, 2, 3, 4, \\ &\quad \lambda \in [0,\ 1]\}. \end{aligned}$$

The family of closed loop characteristic polynomials for each segment plant $G_{ij}(s,\ \lambda)$ is denoted by $\delta_{ij}(s, k_p, k_i, k_d, \lambda)$ and is given by

$$\delta_{ij}(s, k_p, k_i, k_d, \lambda) = s\mathcal{D}^j(s) + (k_i + k_p s + k_d s^2)\mathcal{NS}^i(s, \lambda).$$

We can now state the following result on stabilizing an interval plant using a PID controller.

Theorem 6.5.1. *Let $\mathcal{G}(s)$ be an interval plant. Then the entire family $\mathcal{G}(s)$ is stabilized by a particular PID controller, if and only if each segment plant $G_{ij}(s, \lambda) \in \mathcal{G}_S(s)$ is stabilized by that same PID controller.*

Proof. Now, we have

$$\begin{aligned} \Delta(s, k_p, k_i, k_d) &= s\mathcal{D}(s) + (k_i + k_p s + k_d s^2)\mathcal{N}(s) \\ &= F_1(s)\mathcal{D}(s) + F_2(s)\mathcal{N}(s) \end{aligned}$$

where

$$\begin{aligned} F_1(s) &= s \\ F_2(s) &= k_i + k_p s + k_d s^2. \end{aligned}$$

Using Theorem 6.2.2, it follows that the entire family $\Delta(s, k_p, k_i, k_d)$ is Hurwitz stable iff the entire family $\delta_{ij}(s, k_p, k_i, k_d, \lambda)$, $i = 1, 2, 3, 4$, $j = 1, 2, 3, 4$, $\lambda \in [0,\ 1]$ is Hurwitz. Hence, we can conclude that the entire family $\mathcal{G}(s)$ is stabilized by a PID controller iff every element of $\mathcal{G}_S(s)$ is simultaneously stabilized by such a PID controller. ♣

In view of the above Theorem, we can now use the results of Section 4.4 to obtain a computational characterization of all PID controllers that stabilize the interval family $\mathcal{G}(s)$. As in Section 4.4, for any fixed k_p and any fixed $\lambda^* \in [0, 1]$, we can solve a linear programming problem to determine the stabilizing set of (k_i, k_d) values for $G_{ij}(s,\ \lambda^*)$. Let this set be denoted by $\mathcal{S}_{(i,\ j,\ k_p,\ \lambda^*)}$. By keeping k_p fixed and letting λ^* vary in $[0, 1]$ we can determine the set of stabilizing (k_i, k_d) values for the entire segment plant $G_{ij}(s,\ \lambda)$. This set is denoted by $\mathcal{S}_{(i,\ j,\ k_p)}$ and is defined as

$$\mathcal{S}_{(i,\ j,\ k_p)} = \cap_{\lambda \in [0,\ 1]} \mathcal{S}_{(i,\ j,\ k_p,\ \lambda)}.$$

Now for a fixed k_p, the set of all stabilizing $(k_i,\ k_d)$ values for the entire set $\mathcal{G}_S(s)$, denoted by $\mathcal{S}_{k_p}$ is given by

$$\mathcal{S}_{k_p} = \cap_{i=1,2,3,4,\ j=1,2,3,4} \mathcal{S}_{(i,\ j,\ k_p)}$$

The set of all stabilizing (k_p, k_i, k_d) values for the interval plant $\mathcal{G}(s)$ can now be found by simply sweeping over k_p and determining $\mathcal{S}_{k_p}$ at each stage. Once again the range of k_p values over which the sweeping has to be carried out for the entire $\mathcal{G}_S(s)$ can be *a priori* narrowed down by using the root locus ideas presented in Appendix A. We now present a simple example to illustrate the detailed calculations involved in determining all the stabilizing (k_p, k_i, k_d) values for a given interval plant $\mathcal{G}(s)$.

Example 6.5.1. Consider the interval plant

$$\mathcal{G}(s) = \frac{\mathcal{N}(s)}{\mathcal{D}(s)},$$

$$\begin{aligned} \text{where } \mathcal{N}(s) &= a_0 + a_1 s + a_2 s^2 \\ \mathcal{D}(s) &= b_0 + b_1 s + b_2 s^2 + b_3 s^3 + b_4 s^4 + b_5 s^5 \end{aligned}$$

with

$$a_2 \in [1,\ 1],\ a_1 \in [-5,\ -4],\ a_0 \in [2,\ 4]$$

and $b_5 \in [1,\ 1]$, $b_4 \in [3,\ 4]$, $b_3 \in [5,\ 5]$, $b_2 \in [7,\ 9]$, $b_1 \in [8,\ 9]$, $b_0 \in [-2,\ -1]$.

The Kharitonov polynomials corresponding to $\mathcal{N}(s)$ and $\mathcal{D}(s)$ are

$$\begin{aligned}
\mathcal{N}^1(s) &= s^2 - 5s + 2, \mathcal{N}^2(s) = s^2 - 4s + 2,\\
\mathcal{N}^3(s) &= s^2 - 5s + 4, \mathcal{N}^4(s) = s^2 - 4s + 4,\\
\mathcal{D}^1(s) &= s^5 + 3s^4 + 5s^3 + 9s^2 + 8s - 2,\\
\mathcal{D}^2(s) &= s^5 + 3s^4 + 5s^3 + 9s^2 + 9s - 2,\\
\mathcal{D}^3(s) &= s^5 + 4s^4 + 5s^3 + 7s^2 + 8s - 1,\\
\text{and } \mathcal{D}^4(s) &= s^5 + 4s^4 + 5s^3 + 7s^2 + 9s - 1.
\end{aligned}$$

The Kharitonov segments and the family of sixteen segment plants are defined as at the beginning of this section.

Now for a fixed k_p, for instance $k_p = 1.05$, sweeping over $\lambda \in [0,1]$ and using the results of Section 4.4, we obtained the stabilizing $(k_i,\ k_d)$ values for $G_{11}(s,\ \lambda)$, *i.e.*, $\mathcal{S}_{(1,\ 1,\ 1.05)}$ sketched in Fig. 6.2. In Fig. 6.2, for different values of $\lambda \in [0,1]$, the boundaries of the stabilizing regions $\mathcal{S}_{(1,\ 1,\ 1.05,\ \lambda)}$ are indicated using solid lines. The shaded portion is $\mathcal{S}_{(1,\ 1,\ 1.05)}$ which is the intersection of $\mathcal{S}_{(1,\ 1,\ 1.05,\ \lambda)}$, as λ varies over the interval $[0,\ 1]$. Repeating

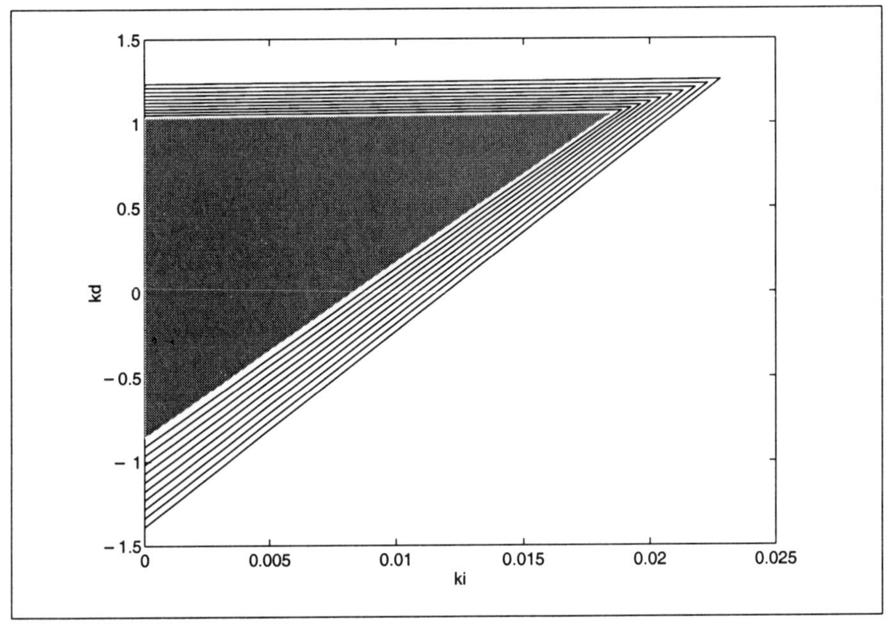

Fig. 6.2. The stabilizing set of $(k_i,\ k_d)$ values for $G_{11}(s,\ \lambda)$, $\lambda \in [0,\ 1]$ with $k_p = 1.05$ (Example 6.5.1).

the same procedure for the remaining fifteen segment plants, we obtained the regions $\mathcal{S}_{(i,\ j,\ 1.05)}$, $i = 1,2,3,4$, $j = 1,2,3,4$ as indicated by the solid lines in Fig. 6.3. The shaded portion, which is their intersection, is the region $\mathcal{S}_{1.05}$.

Now, using root locus ideas, it was determined that a necessary condition

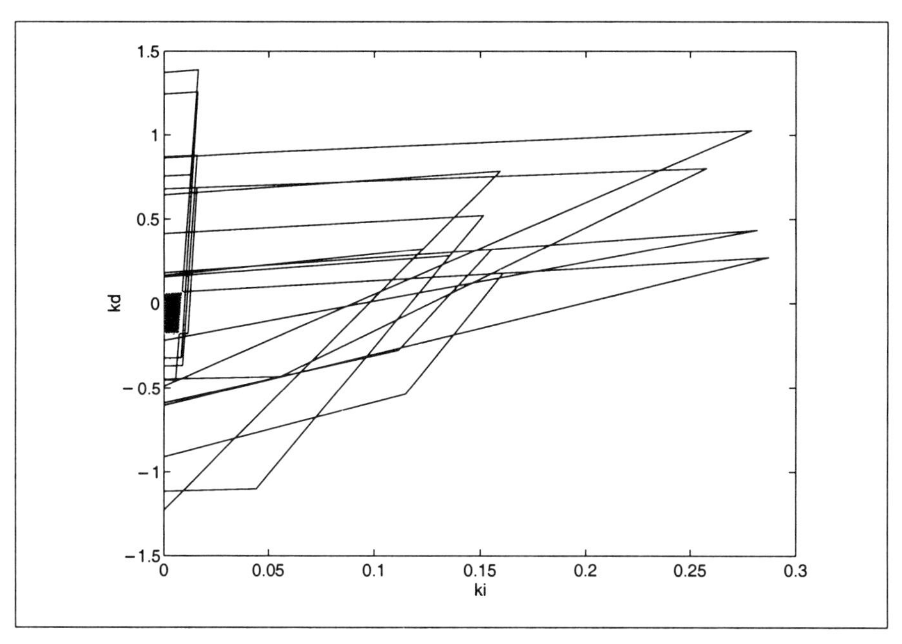

Fig. 6.3. The stabilizing set of $(k_i,\ k_d)$ values for $G_{ij}(s,\ \lambda)$, $\lambda \in [0,\ 1]$, $i = 1, 2, 3, 4$ and $j = 1, 2, 3, 4$ with $k_p = 1.05$ (Example 6.5.1).

for the existence of stabilizing (k_i, k_d) values for the entire family was that $k_p \in (1,\ 1.0869)$. Thus, by sweeping over $k_p \in (1,\ 1.0869)$, and repeating the above procedure, we obtained the stabilizing set of $(k_p,\ k_i,\ k_d)$ values sketched in Fig. 6.4.

6.6 Design of Robust and Non-Fragile PID Settings

In this section, we consider the problem of designing robust PID controllers for which the closed loop systems are not destabilized by small perturbations in the PID settings. This can be done by combining the results of the last section with the technique developed in Section 4.6.2 for fixed plants. A controller for which the closed loop system is destabilized by small perturbations in the controller coefficients is said to be "fragile." Any controller that is to be practically implemented must necessarily be non-fragile so that (1) round off errors during implementation do not destabilize the closed loop; and (2) the tuning of the parameters about the nominal design values is allowed.

Before proceeding further, we should note that for a given interval plant family and a fixed k_p, the set of stabilizing $(k_i,\ k_d)$ values is the union of

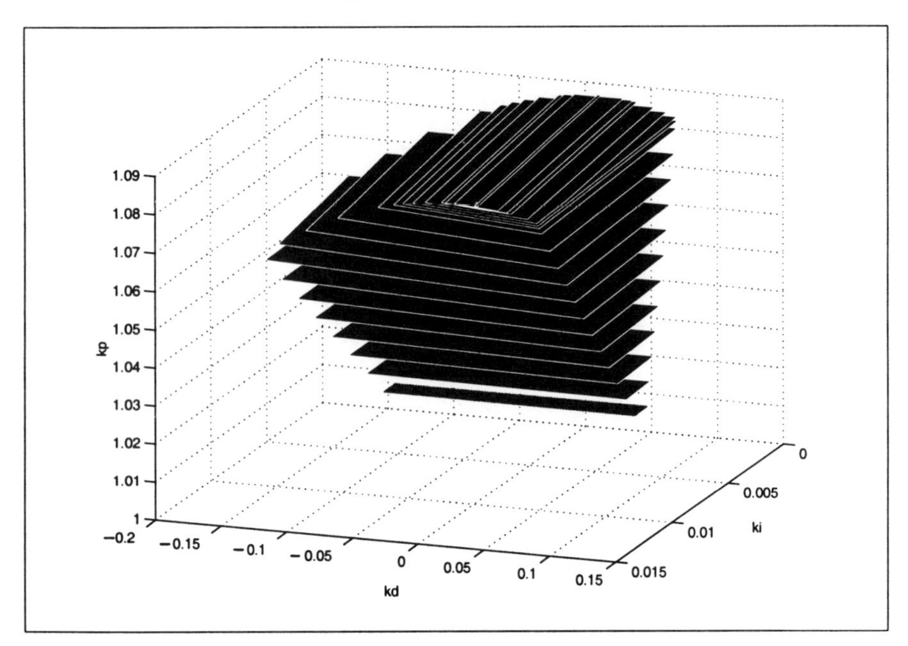

Fig. 6.4. The stabilizing set of $(k_p,\ k_i,\ k_d)$ values (Example 6.5.1).

the intersection of admissible solutions of sets of linear inequalities in terms of k_i and k_d. Thus, for a fixed k_p, the set of stabilizing $(k_i,\ k_d)$ values is a union of either convex polygons or intersections of half planes. Because of this parametric structure, for a fixed k_p, we can choose the pair $(k_i,\ k_d)$ to be at the center of the circle of largest radius inscribed inside the stabilizing region. The radius of this circle is the largest l_2 parametric stability margin in the space of $(k_i,\ k_d)$. The procedure for characterizing such a circle has already been presented in Section 4.6.2. Using this procedure, we can sweep over k_p and choose that value of k_p that gives us the inscribed circle with the largest radius. Clearly, setting the $(k_i,\ k_d)$ values at the center of this circle will yield the maximum l_2 parametric stability margin in the space of $(k_i,\ k_d)$. In the following example, we apply this procedure to the robust PID designed in Example 6.5.1

Example 6.6.1. Consider the same interval plant family as the one used in Example 6.5.1. From Example 6.5.1, we know that for the existence of stabilizing $(k_i,\ k_d)$ values for the entire family, k_p must lie in $(1,\ 1.0869)$. Accordingly, we sweep over $k_p \in (1,\ 1.0869)$ and find the largest circle inscribed in the stabilizing $(k_i,\ k_d)$ region at each stage. Thereafter we can choose the circle with the largest radius. Following this procedure we determined that the circle with the largest radius occurs at $k_p = 1.086$, its center is located at $k_i = 0.00684$, $k_d = 0.03738$ and its radius is $r = 0.00684$. Fig. 6.5 shows the stabilizing $(k_i,\ k_d)$ region along with the inscribed circle for $k_p = 1.086$.

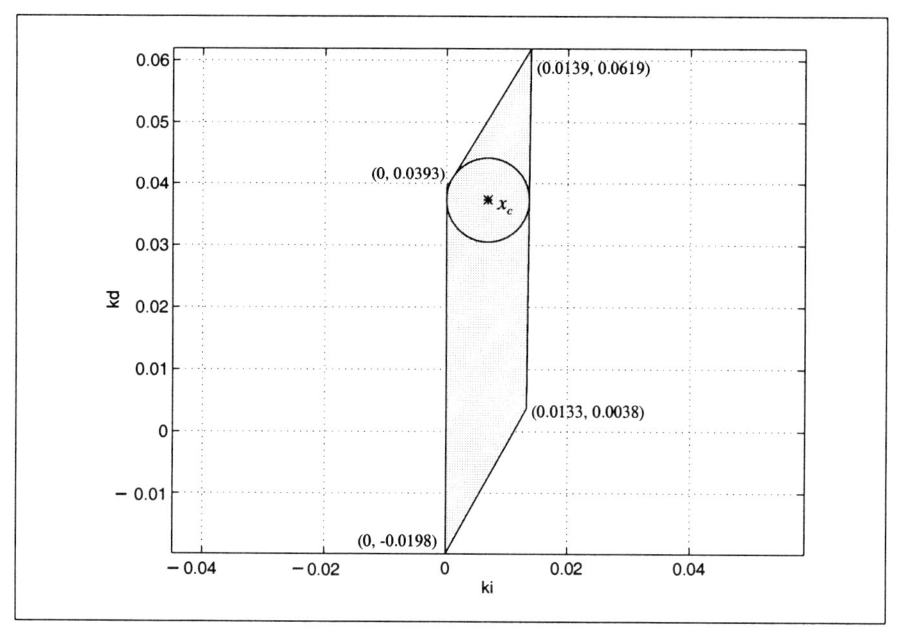

Fig. 6.5. (k_i, k_d) stabilizing region for $k_p = 1.086$ and the inscribed circle (Example 6.6.1).

6.7 Notes and References

The robust P, PI and PID stabilization results in this chapter are based on Ho, Datta and Bhattacharyya [24]. The approach for designing "non-fragile" PID controllers is due to Silva and Ho, and is unpublished. The issue of controller "fragility" was recently raised by Keel and Bhattacharyya [30]. A comprehensive treatment covering the area of Parametric Robust Control can be found in Bhattacharyya, Chapellat and Keel [7].

CHAPTER 7
STABILIZATION OF FIRST-ORDER SYSTEMS WITH TIME DELAY

In this chapter, we first consider the problem of stabilizing a first-order plant with dead time using a constant gain controller. Using a version of the Hermite-Biehler Theorem applicable to quasipolynomials, a complete analytical characterization of all stabilizing gain values is provided as a closed form solution. A similar approach is then used to tackle the problem of stabilizing a first-order plant with time delay using a PI controller.

7.1 Introduction

In earlier chapters we used the Hermite-Biehler Theorem to solve particular cases of the fixed order and structure stabilization problems. Specifically, in Chapter 3, a Generalization of the Hermite-Biehler Theorem was derived and in Chapter 4, we used this result to compute the set of all stabilizing P, PI and PID controllers.

Commonly in process control, the dynamic response of a plant is mathematically described by a first-order model with dead time [2]. These models are often used in practice to design and tune PID controllers. It is interesting to note that even though most of these tuning techniques provide satisfactory results, the set of all stabilizing PID controllers for these first-order models with dead time remains unknown. The synthesis results presented in the earlier chapters also cannot be applied directly since they were obtained for plants described by *rational transfer functions.* This provides us with the motivation to study the problem of characterizing the set of all PID controllers that stabilize a given first-order plant with time delay.

The chapter is organized as follows. In Section 7.2 we present an extension of the Hermite-Biehler Theorem applicable to quasipolynomials and some related results. This extension is used in Section 7.3 to obtain a closed-form solution to the constant gain stabilization problem for first-order plants with dead time. Section 7.4 deals with the problem of stabilizing a first-order plant with dead time using a pure integrator. Section 7.5 uses a similar approach to tackle the problem of stabilizing a first-order plant with dead time using a PI controller.

7.2 Extension of the Hermite-Biehler Theorem

Many control problems in process control engineering involve time delays. These time delays lead to dynamic models with characteristic equations of the form

$$\delta(s) = d(s) + e^{-sT_1}n_1(s) + e^{-sT_2}n_2(s) + ... + e^{-sT_m}n_m(s) \tag{7.2.1}$$

which are also known as quasipolynomials. It can be shown that the Hermite-Biehler Theorem for Hurwitz polynomials does not carry over to arbitrary functions $f(s)$ of the complex variable s. However, in [42] Pontryagin studied entire functions of the form $P(s, e^s)$, where $P(s,t)$ is a polynomial in two variables and is called a quasipolynomial. Based on Pontryagin's results, a suitable extension of the Hermite-Biehler Theorem can be developed (see [4], [7] and the references therein) to study the stability of certain classes of quasipolynomials characterized as follows. If in (7.2.1) we make the assumptions:

A1. $\deg[d(s)] = n$ and $\deg[n_i(s)] < n$ for $i = 1, 2, ..., m$;

A2. $0 < T_1 < T_2 < ... < T_m$,

then instead of (7.2.1) we can consider the quasipolynomial

$$\delta^*(s) = e^{sT_m}\delta(s) = e^{sT_m}d(s) + e^{s(T_m-T_1)}n_1(s) + e^{s(T_m-T_2)}n_2(s) + ... + n_m(s) \; . \tag{7.2.2}$$

Since e^{sT_m} does not have any finite zeros, the Hurwitz stability of $\delta(s)$ is equivalent to that of $\delta^*(s)$. The following theorem gives necessary and sufficient conditions for the Hurwitz stability of $\delta^*(s)$ [4].

Theorem 7.2.1. *Let $\delta^*(s)$ be given by (7.2.2), and write*

$$\delta^*(jw) = \delta_r(w) + j\delta_i(w) \; ,$$

where $\delta_r(w)$ and $\delta_i(w)$ represent respectively the real and imaginary parts of $\delta^*(jw)$. Under assumptions (A1) and (A2), $\delta^*(s)$ is Hurwitz stable if and only if

1. *$\delta_r(w)$ and $\delta_i(w)$ have only simple real roots and these interlace.*
2. *$\delta_i'(w_o)\delta_r(w_o) - \delta_i(w_o)\delta_r'(w_o) > 0$, for some w_o in $(-\infty, \infty)$*

where $\delta_r'(w)$ and $\delta_i'(w)$ denote the first derivative with respect to w of $\delta_r(w)$ and $\delta_i(w)$, respectively.

In the rest of this chapter, we will be making use of this theorem to provide solutions to the P and PI stabilization problems for first-order plants with dead time. A crucial step in applying the above theorem to check Hurwitz stability is to ensure that $\delta_r(w)$ and $\delta_i(w)$ have only *real* roots. Such a property can be ensured by using the following result, also due to Pontryagin [4].

Theorem 7.2.2. *Let M and N denote the highest powers of s and e^s respectively in $\delta^*(s)$. Let η be an appropriate constant such that the coefficients of terms of highest degree in $\delta_r(w)$ and $\delta_i(w)$ do not vanish at $w = \eta$. Then for the equations $\delta_r(w) = 0$ or $\delta_i(w) = 0$ to have only real roots, it is necessary and sufficient that in the interval*

$$-2l\pi + \eta \leq w \leq 2l\pi + \eta$$

$\delta_r(w)$ or $\delta_i(w)$ has exactly $4lN + M$ real roots starting with a sufficiently large l.

7.3 Stabilization using a Constant Gain

Systems with step responses like the one shown in Fig. 7.1 are commonly modeled as first order processes with a time delay [2] which can be mathematically described by

$$G(s) = \frac{k}{1 + Ts} e^{-Ls} \tag{7.3.1}$$

where k represents the steady-state gain of the plant, L represents the time delay, and T represents the time constant of the plant.

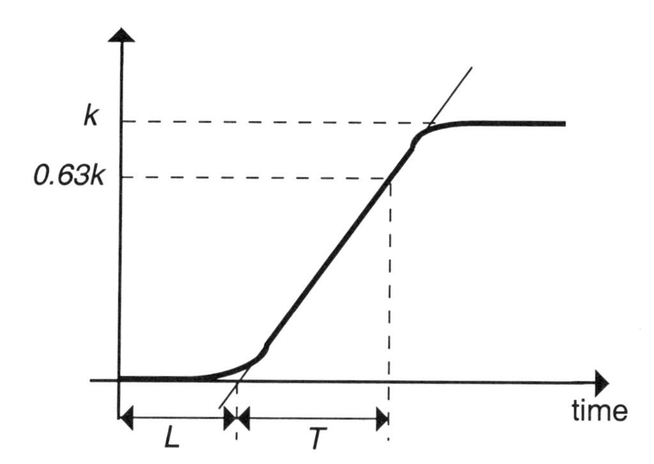

Fig. 7.1. Open-loop step response.

Consider now the feedback control system shown in Fig. 7.2 where r is the command signal, y is the output of the plant, $G(s)$ given by (7.3.1) is the plant to be controlled, and $C(s)$ is the controller. We first consider the case of a constant gain controller, *i.e.*,

$$C(s) = k_c \, .$$

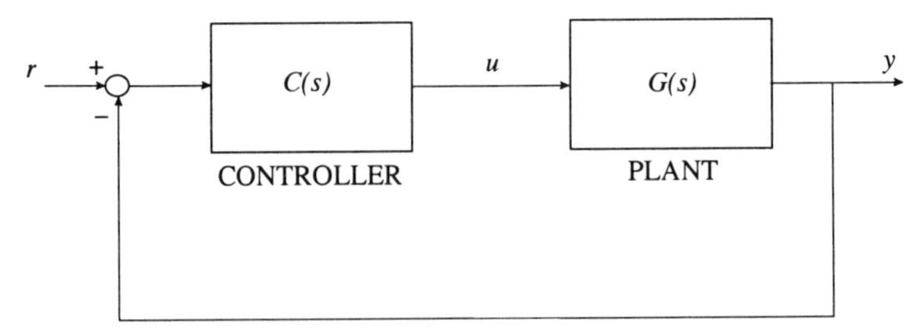

Fig. 7.2. Feedback control system.

Our objective is to analytically determine the values of the parameter k_c for which the closed-loop system is stable.

When the time delay of the plant model is zero, *i.e.*, $L = 0$, the closed-loop characteristic equation of the system is given by

$$\delta(s) = kk_c + 1 + Ts.$$

Clearly, this polynomial has a single root at $s = -\frac{1+k\cdot k_c}{T}$. Thus, for instance, if we assume that the steady state gain k is positive and $T > 0$ so that the plant is open-loop stable, then to ensure the stability of the closed-loop system we must have

$$k_c > -\frac{1}{k}\ .$$

If, on the other hand, the plant is open-loop unstable, *i.e.*, $T < 0$, then to ensure the stability of the closed-loop system we must have

$$k_c < -\frac{1}{k}\ . \tag{7.3.2}$$

Now, let us consider the case where the time delay of the model is different from zero and try to determine the set of all stabilizing gains. The closed-loop characteristic equation of the system is given by

$$\delta(s) = kk_c e^{-Ls} + 1 + Ts\ .$$

In order to study the stability of the closed-loop system, we need to determine if all the roots of the above expression lie in the open left half plane. Due to the presence of the exponential term e^{-Ls}, the number of roots of the expression $\delta(s)$ is infinite and this makes such a stability check extremely difficult. Instead, we can invoke Theorem 7.2.1 to determine the set of stabilizing gains k_c as follows.

First, we consider the quasipolynomial $\delta^*(s)$ defined by

$$\delta^*(s) = e^{Ls}\delta(s) = kk_c + (1 + Ts)e^{Ls}\ .$$

Substituting $s = jw$, we have

$$\delta^*(jw) = \delta_r(w) + j\delta_i(w)$$

where

$$\begin{aligned}\delta_r(w) &= cos(Lw) - Twsin(Lw) + kk_c \\ \delta_i(w) &= sin(Lw) + Twcos(Lw)\ .\end{aligned}$$

We now consider two different cases.

7.3.1 Open-loop Stable Plant

In this subsection we give a closed form solution to the constant gain stabilization problem for the case of an open-loop stable plant. This means that the time constant T of the plant satisfies $T > 0$. Moreover, we assume that $k > 0$ and $L > 0$.

Theorem 7.3.1. *Under the above assumptions on k and L, the set of all stabilizing gains k_c for a given open-loop stable plant with transfer function $G(s)$ as in (7.3.1) is given by*

$$-\frac{1}{k} < k_c < \frac{T}{kL}\sqrt{z_1^2 + \frac{L^2}{T^2}} \tag{7.3.3}$$

where z_1 is the solution of the equation

$$tan(z) = -\frac{T}{L}z$$

in the interval $(\frac{\pi}{2}, \pi)$.

Proof. With the change of variables $z = Lw$ the real and imaginary parts of $\delta^*(jw)$ can be rewritten as:

$$\delta_r(z) = cos(z) - \frac{T}{L}zsin(z) + kk_c \tag{7.3.4}$$

$$\delta_i(z) = sin(z) + \frac{T}{L}zcos(z)\ . \tag{7.3.5}$$

From Theorem 7.2.1, we need to check two conditions to ensure the stability of the quasipolynomial $\delta^*(s)$.

Step 1. We first check Condition 2 of Theorem 7.2.1:

$$E(w_o) = \delta_i'(w_o)\delta_r(w_o) - \delta_i(w_o)\delta_r'(w_o) > 0$$

for some w_o in $(-\infty, \infty)$. Let us take $w_o = z_o = 0$. Thus $\delta_i(z_o) = 0$ and $\delta_r(z_o) = 1 + kk_c$. We also have

$$\begin{aligned}\delta_i'(z) &= \left[\left(1 + \frac{T}{L}\right)cos(z) - \frac{T}{L}zsin(z)\right] \\ \Rightarrow E(z_o) &= \left(1 + \frac{T}{L}\right)(1 + kk_c)\ .\end{aligned}$$

By our initial assumption $T > 0$ and $L > 0$. Thus, if we pick $k_c > -\frac{1}{k}$ we have that $E(z_o) > 0$.

Step 2. We now check Condition 1 of Theorem 7.2.1: the interlacing of the roots of $\delta_r(z)$ and $\delta_i(z)$. From (7.3.5) we can compute the roots of the imaginary part, *i.e.*, $\delta_i(z) = 0$. This gives us the following equation

$$sin(z) + \frac{T}{L} zcos(z) = 0$$

$$\Leftrightarrow tan(z) = -\frac{T}{L} z \,. \tag{7.3.6}$$

An analytical solution of (7.3.6) is difficult to find. However, we can plot the two terms involved in this equation, *i.e.*, $tan(z)$ and $-\frac{T}{L}z$ to study the behavior of the roots of $\delta_i(z)$. Fig. 7.3 shows this plot. Clearly the non-negative real roots of the imaginary part are

$$z_o = 0, \; z_1 \epsilon (\frac{\pi}{2}, \pi), \; z_2 \epsilon (\frac{3\pi}{2}, 2\pi), \; z_3 \epsilon (\frac{5\pi}{2}, 3\pi),$$

and so on, where the roots z_i, for $i = 1, 2, 3, ...$ satisfy Equation 7.3.6.

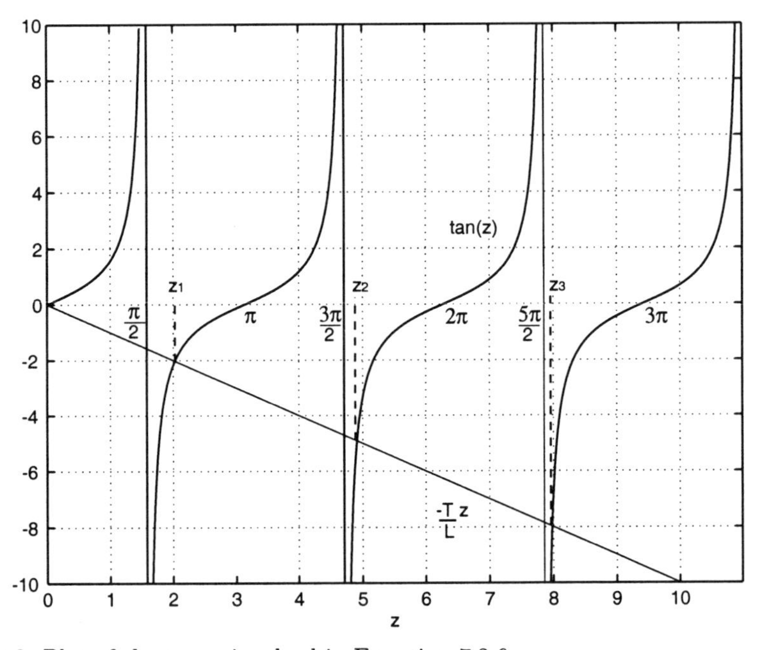

Fig. 7.3. Plot of the terms involved in Equation 7.3.6.

Let us now use Theorem 7.2.2 to check if $\delta_i(w)$ has only real roots. Substituting $s_1 = Ls$ in the expression for $\delta^*(s)$, we see that for the new quasipolynomial in s_1, $M = 1$ and $N = 1$. Next we choose $\eta = \frac{\pi}{4}$ to satisfy the requirement that $cos(\eta) \neq 0$. Now from Fig. 7.3, it is clear that in the interval $[0, 2\pi - \frac{\pi}{4}] = [0, \frac{7\pi}{4}]$, $\delta_i(z) = 0$ has three real roots including a root at the origin. Since $\delta_i(z)$ is an odd function it follows that in the interval $[-\frac{7\pi}{4}, \frac{7\pi}{4}]$, $\delta_i(z)$ will have 5 real roots. Also observe that $\delta_i(z)$ does not have any real roots in $(\frac{7\pi}{4}, \frac{9\pi}{4}]$. Thus $\delta_i(z)$ has $4N+M = 5$ real roots in the interval $[-2\pi + \frac{\pi}{4}, 2\pi + \frac{\pi}{4}]$. Moreover, $\delta_i(z)$ has two real roots in each of the intervals $[2l\pi + \frac{\pi}{4}, 2(l+1)\pi + \frac{\pi}{4}]$ and $[-2(l+1)\pi + \frac{\pi}{4}, -2l\pi + \frac{\pi}{4}]$ for $l = 1, 2, \ldots$. Hence it follows that $\delta_i(z)$ has exactly $4lN + M$ real roots in $[-2l\pi + \frac{\pi}{4}, 2l\pi + \frac{\pi}{4}]$ for $l = 1, 2, \ldots$, which by Theorem 7.2.2 implies that $\delta_i(z)$ has only real roots.

We now evaluate $\delta_r(z)$ at the roots of the imaginary part $\delta_i(z)$. For $z_o = 0$, using (7.3.4) we obtain

$$\begin{aligned} \delta_r(z_o) &= cos(0) - \frac{T}{L}0 \cdot sin(0) + kk_c \\ \Rightarrow \delta_r(z_o) &= 1 + kk_c \end{aligned} \tag{7.3.7}$$

For z_1, using (7.3.4) we obtain

$$\begin{aligned} \delta_r(z_1) &= cos(z_1) - \frac{T}{L}z_1 sin(z_1) + kk_c \\ &= cos(z_1) + tan(z_1)sin(z_1) + kk_c \quad \text{[using (7.3.6)]} \\ &= \frac{1}{cos(z_1)} + kk_c \end{aligned}$$

From Fig. 7.3 since $z_1 \epsilon (\frac{\pi}{2}, \pi)$ we obtain

$$\begin{aligned} cos(z_1) &= -\frac{L}{\sqrt{T^2 z_1^2 + L^2}} \\ \Rightarrow \delta_r(z_1) &= -\frac{T}{L}\sqrt{z_1^2 + \frac{L^2}{T^2}} + kk_c \end{aligned} \tag{7.3.8}$$

A similar analysis for $z_2 \epsilon (\frac{3\pi}{2}, 2\pi)$, $z_3 \epsilon (\frac{5\pi}{2}, 3\pi)$, and so on, gives us

$$\delta_r(z_2) = \frac{T}{L}\sqrt{z_2^2 + \frac{L^2}{T^2}} + kk_c \tag{7.3.9}$$

$$\delta_r(z_3) = -\frac{T}{L}\sqrt{z_3^2 + \frac{L^2}{T^2}} + kk_c \tag{7.3.10}$$

$$\vdots$$

Now, from Step 1 we have that $k_c > -\frac{1}{k}$. Thus, from (7.3.7) we see that $\delta_r(z_o) > 0$. Then, interlacing of the roots of $\delta_r(z)$ and $\delta_i(z)$ is equivalent to $\delta_r(z_1) < 0$, $\delta_r(z_2) > 0$, $\delta_r(z_3) < 0$, and so on . Using this fact and Equations 7.3.7-7.3.10 we obtain

$$\begin{aligned}
\delta_r(z_o) > 0 \quad &\Rightarrow \quad k_c > -\frac{1}{k} \\
\delta_r(z_1) < 0 \quad &\Rightarrow \quad k_c < \frac{T}{kL}\sqrt{z_1^2 + \frac{L^2}{T^2}} =: M_1 \\
\delta_r(z_2) > 0 \quad &\Rightarrow \quad k_c > -\frac{T}{kL}\sqrt{z_2^2 + \frac{L^2}{T^2}} =: M_2 \\
\delta_r(z_3) < 0 \quad &\Rightarrow \quad k_c < \frac{T}{kL}\sqrt{z_3^2 + \frac{L^2}{T^2}} =: M_3 \\
&\vdots
\end{aligned}$$

Since $z_1 < z_2 < z_3 < \ldots$, we conclude that $|M_1| < |M_2| < |M_3| < \ldots$. Thus intersecting the bounds previously found for k_c we conclude that for the interlacing property to hold, we must have

$$-\frac{1}{k} < k_c < \frac{T}{kL}\sqrt{z_1^2 + \frac{L^2}{T^2}} \; .$$

Note that for values of k_c in this range, the interlacing property and the fact that the roots of $\delta_i(z)$ are all real can be used in Theorem 7.2.2 to guarantee that $\delta_r(z)$ also has only real roots. Thus all the conditions of Theorem 7.2.1 are satisfied and this completes the proof. ♣

Remark 7.3.1. By explicitly evaluating the derivative with respect to L, it can be shown that the upper bound for k_c given in Theorem 7.3.1 is a monotonically decreasing function of the time delay L of the system. Thus, if the delay of the system is reduced from L_1 to L_2, with k_c fixed, the system remains stable since now a larger range of k_c can be tolerated.

We now present an example to illustrate the application of Theorem 7.3.1.

Example 7.3.1. Consider the constant gain stabilization problem where the plant parameters are $k = 1$, $L = 2$ sec, and $T = 1$ sec. Since the plant is open-loop stable we will use Theorem 7.3.1 to obtain the set of stabilizing gains. First, we compute $z_1 \epsilon (\frac{\pi}{2}, \pi)$ satisfying (7.3.6), *i.e.*,

$$tan(z) = -0.5z \; .$$

Solving this equation we obtain $z_1 = 2.2889$. Thus, from (7.3.3) the set of stabilizing gains is given by

$$-1 < k_c < 1.5198 \; .$$

We now set the controller parameter k_c to $\frac{1}{2}$ which is in the above stabilizing set. With this value of k_c, the characteristic quasipolynomial $\delta^*(s)$ of the system is given by

$$\delta^*(s) = \frac{1}{2} + (1+s)e^{2s} \; .$$

Substituting $s = jw$ we obtain

$$\delta^*(jw) = \left[\frac{1}{2} + cos(2w) - wsin(2w)\right] + j\left[sin(2w) + wcos(2w)\right] .$$

Fig. 7.4 shows the plot of the real and imaginary parts of $\delta^*(jw)$. As we can see the roots of the real and imaginary parts interlace. Fig. 7.5 shows the time response of the closed-loop system to a unit step input r which verifies that the closed-loop system is indeed stable.

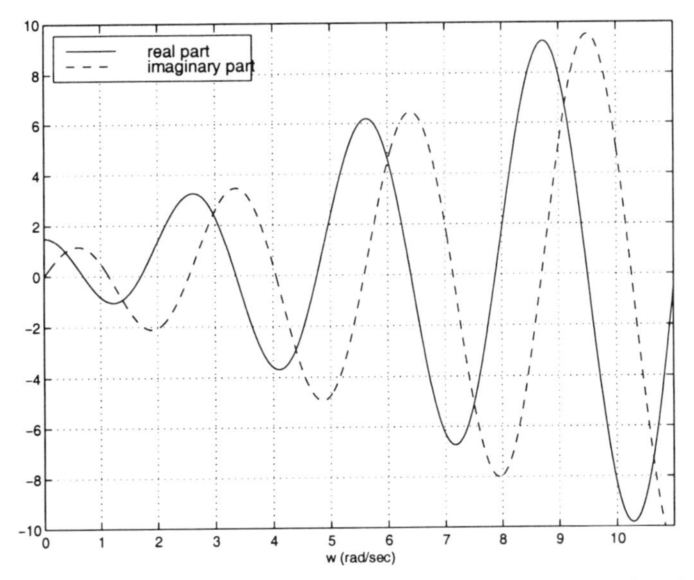

Fig. 7.4. Plot of the real and imaginary parts of $\delta^*(jw)$ for Example 7.3.1.

7.3.2 Open-loop Unstable Plant

In this subsection we present a Theorem that gives a closed form solution to the constant gain stabilization problem for the case of an open-loop unstable plant. This means that now $T < 0$. Of course, as before, we assume that $k > 0$ and $L > 0$.

Theorem 7.3.2. *Under the above assumptions on k and L, a necessary condition for a gain k_c to simultaneously stabilize the delay-free plant and the plant with delay is $|\frac{T}{L}| > 1$. If this necessary condition is satisfied, then the set of all stabilizing gains k_c for a given open-loop unstable plant with transfer function $G(s)$ as in (7.3.1) is given by*

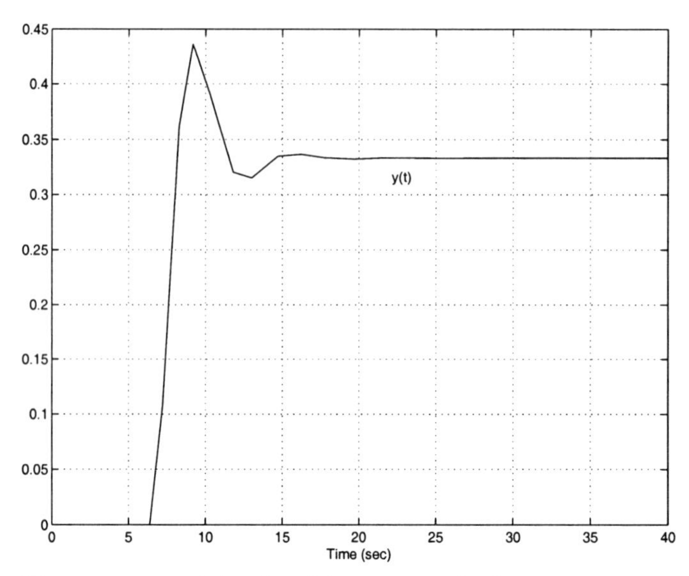

Fig. 7.5. Time response of the closed-loop system for Example 7.3.1.

$$\frac{T}{kL}\sqrt{z_1^2+\frac{L^2}{T^2}} < k_c < -\frac{1}{k} \tag{7.3.11}$$

where z_1 is the solution of the equation

$$tan(z) = -\frac{T}{L}z$$

in the interval $(0, \frac{\pi}{2})$.

Proof. The proof follows along the same lines as that of Theorem 7.3.1 and will be briefly sketched here. Again, the idea of the proof is to verify Conditions 1 and 2 of Theorem 7.2.1.

Step 1. First, we check Condition 2 of Theorem 7.2.1:

$$E(w_o) = \delta_i'(w_o)\delta_r(w_o) - \delta_i(w_o)\delta_r'(w_o) > 0$$

for some w_o in $(-\infty,\infty)$. Let us take $w_o = z_o = 0$. Thus $\delta_i(z_o) = 0$ and $\delta_r(z_o) = 1 + kk_c$. We also have

$$\begin{aligned} \delta_i'(z) &= \left[\left(1+\frac{T}{L}\right)cos(z) - \frac{T}{L}zsin(z)\right] \\ \Rightarrow E(z_o) &= \left(1+\frac{T}{L}\right)(1+kk_c)\,. \end{aligned}$$

From (7.3.2), it is clear that from the closed-loop stability of the delay-free system, we have $(1 + kk_c) < 0$. Hence to have $E(z_o) > 0$, we must have $1 + \frac{T}{L} < 0$ or $\frac{T}{L} < -1$,

$$\Rightarrow \left|\frac{T}{L}\right| > 1 \, .$$

Step 2. We now check Condition 1 of Theorem 7.2.1: the interlacing of the roots of $\delta_r(z)$ and $\delta_i(z)$. As in the previous case, the roots of $\delta_i(z)$ satisfy the following equation

$$tan(z) = -\frac{T}{L} z \, . \tag{7.3.12}$$

Again, since an analytical solution of (7.3.12) is difficult to find, we can plot the two terms involved in this equation, *i.e.*, $tan(z)$ and $-\frac{T}{L}z$ to study the behavior of the roots of $\delta_i(z)$. Fig. 7.6 shows this plot. Clearly the non-negative real roots of the imaginary part are

$$z_o = 0, \; z_1 \epsilon (0, \frac{\pi}{2}), \; z_2 \epsilon (\pi, \frac{3\pi}{2}), \; z_3 \epsilon (2\pi, \frac{5\pi}{2}),$$

and so on, where the roots z_i, for $i = 1, 2, 3, ...$ satisfy Equation 7.3.12. Arguing as in the proof of Theorem 7.3.1, we can show that $\delta_i(z)$ has only real roots.

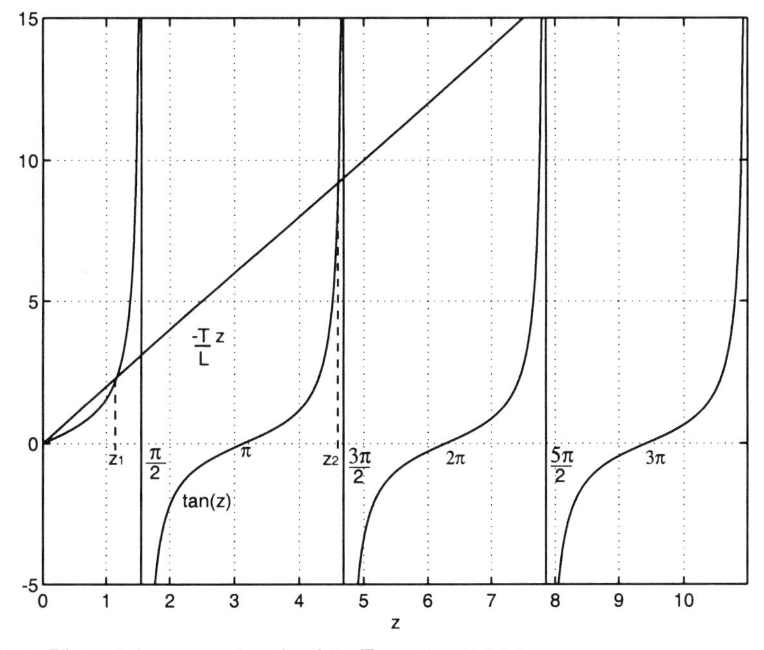

Fig. 7.6. Plot of the terms involved in Equation 7.3.12.

We now evaluate $\delta_r(z)$ at the roots of the imaginary part $\delta_i(z)$. For $z_o = 0$, using (7.3.4) we obtain

$$\delta_r(z_o) = 1 + kk_c \tag{7.3.13}$$

For z_1, using (7.3.4) and the fact that this root satisfies (7.3.12) we obtain

$$\delta_r(z_1) = \frac{1}{cos(z_1)} + kk_c$$

Recalling from Fig. 7.6 that $z_1 \epsilon (0, \frac{\pi}{2})$ we obtain

$$\begin{aligned} cos(z_1) &= \frac{L}{\sqrt{T^2 z_1^2 + L^2}} \\ \Rightarrow \delta_r(z_1) &= -\frac{T}{L}\sqrt{z_1^2 + \frac{L^2}{T^2}} + kk_c \end{aligned} \quad (7.3.14)$$

A similar analysis for $z_2 \epsilon (\pi, \frac{3\pi}{2})$, $z_3 \epsilon (2\pi, \frac{5\pi}{2})$, etc., gives us the following

$$\delta_r(z_2) = \frac{T}{L}\sqrt{z_2^2 + \frac{L^2}{T^2}} + kk_c \quad (7.3.15)$$

$$\delta_r(z_3) = -\frac{T}{L}\sqrt{z_3^2 + \frac{L^2}{T^2}} + kk_c \quad (7.3.16)$$

$$\vdots$$

To ensure the interlacing of the roots of $\delta_r(z)$ and $\delta_i(z)$ we need $\delta_r(z_o) < 0$ (which comes from Condition 7.3.2 for the closed-loop stability of the delay-free system), $\delta_r(z_1) > 0$, $\delta_r(z_2) < 0$, $\delta_r(z_3) > 0$, and so on. Using this fact and Equations 7.3.13-7.3.16 we obtain

$$\begin{aligned} \delta_r(z_o) < 0 &\Rightarrow k_c < -\frac{1}{k} \\ \delta_r(z_1) > 0 &\Rightarrow k_c > \frac{T}{kL}\sqrt{z_1^2 + \frac{L^2}{T^2}} =: M_1 \\ \delta_r(z_2) < 0 &\Rightarrow k_c < -\frac{T}{kL}\sqrt{z_2^2 + \frac{L^2}{T^2}} =: M_2 \\ \delta_r(z_3) > 0 &\Rightarrow k_c > \frac{T}{kL}\sqrt{z_3^2 + \frac{L^2}{T^2}} =: M_3 \\ &\vdots \end{aligned}$$

Since $z_1 < z_2 < z_3 < ...$, we conclude that $|M_1| < |M_2| < |M_3| < ...$. Thus intersecting the bounds previously found for k_c we see that the interlacing property holds provided that

$$\frac{T}{kL}\sqrt{z_1^2 + \frac{L^2}{T^2}} < k_c < -\frac{1}{k} \, .$$

As in the proof of Theorem 7.3.1, we can show that for values of k_c in the above set $\delta_r(z)$ has only real roots so that all conditions in Theorem 7.2.1 are satisfied. This completes the proof. ♣

Example 7.3.2. We now consider the constant gain stabilization problem for an open-loop unstable plant. Let $k = 1$, $L = 0.5$ sec, and $T = -2$ be the plant parameters. Since $\left|\frac{T}{L}\right| = 4 > 1$ we will use Theorem 7.3.2 to obtain the set of stabilizing gains. First, we compute $z_1 \epsilon (0, \frac{\pi}{2})$ satisfying (7.3.12), *i.e.*,

$$tan(z) = 4z \ .$$

Solving this equation we obtain $z_1 = 1.3932$. Thus, from (7.3.11) the set of stabilizing gains is given by

$$-5.6620 < k_c < -1 \ .$$

For the controller parameter k_c set to -3 the characteristic quasipolynomial $\delta^*(s)$ of the system is given by

$$\delta^*(s) = -3 + (1 - 2s)e^{0.5s} \ .$$

Substituting $s = jw$ we obtain

$$\delta^*(jw) = [-3 + cos(0.5w) + 2wsin(0.5w)] + j\,[sin(0.5w) - 2wcos(0.5w)] \ .$$

Fig. 7.7 shows the plot of the real and imaginary parts of $\delta^*(jw)$. It is clear from this plot that the roots of the real and imaginary parts interlace. Fig. 7.8 shows the time response of the closed-loop system to a unit step input r. Once again, we can see that the closed-loop system is stable.

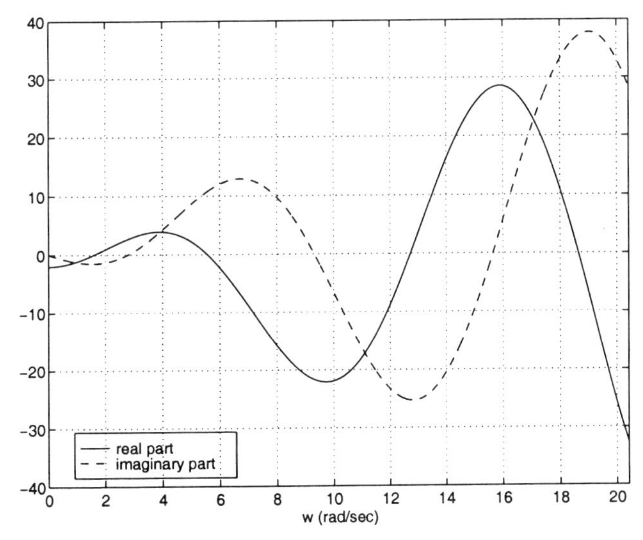

Fig. 7.7. Plot of the real and imaginary parts of $\delta^*(jw)$ for Example 7.3.2.

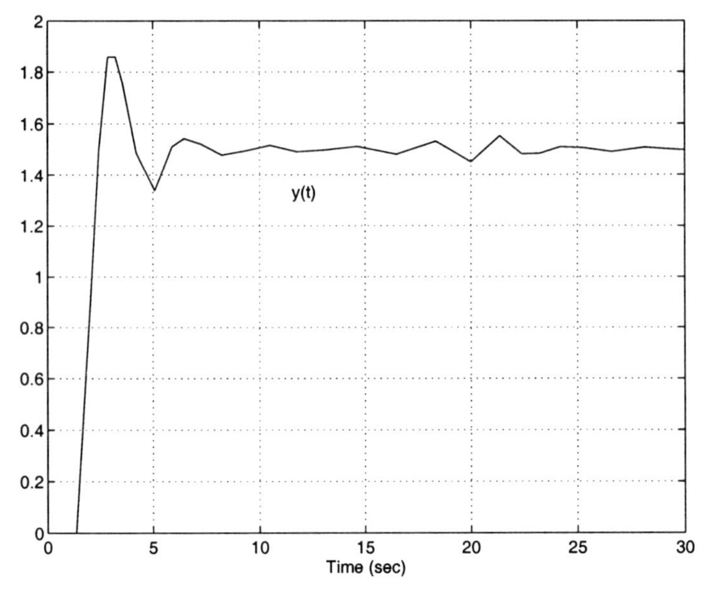

Fig. 7.8. Time response of the closed-loop system for Example 7.3.2.

7.4 Stabilization using a Pure Integrator

In this section we consider the stabilization problem shown in Fig. 7.2 where the controller is now a pure integrator, *i.e.*,

$$C(s) = \frac{k_i}{s} .$$

As in the previous section we want to determine conditions on the controller parameter k_i which ensure that the closed-loop system is stable.

When the time delay of the plant model is zero, *i.e.*, $L = 0$, the characteristic equation of the closed-loop system is given by

$$\delta(s) = Ts^2 + s + kk_i . \tag{7.4.1}$$

From this equation it is clear that if $T < 0$ then the delay-free closed-loop system cannot be stabilized using a pure integrator. Thus this case, corresponding to an unstable open loop plant, does not merit any further consideration. Accordingly, for the rest of this section, we restrict ourselves to only open loop stable plants, *i.e.*, plants for which $T > 0$. From (7.4.1), it is clear that for this case, the delay-free closed-loop system will be stable if and only if

$$k_i > 0 .$$

Thus the closed-loop stability requirement on the delay-free system imposes a lower bound of 0 on the controller parameter k_i.

When the time delay of the model is different from zero the closed-loop characteristic equation of the system is given by

$$\delta(s) = kk_i e^{-Ls} + (1+Ts)s\ .$$

In order to use Theorem 7.2.1 to find the set of stabilizing gains k_i we construct the quasipolynomial $\delta^*(s)$ as before, *i.e.*,

$$\delta^*(s) = e^{Ls}\delta(s) = kk_i + (1+Ts)se^{Ls}\ .$$

Substituting $s = jw$, we have

$$\delta^*(jw) = \delta_r(w) + j\delta_i(w)$$

where

$$\begin{aligned}\delta_r(w) &= kk_i - wsin(Lw) - Tw^2cos(Lw)\\ \delta_i(w) &= wcos(Lw) - Tw^2sin(Lw)\ .\end{aligned}$$

We can now state and prove the following theorem which gives a closed form solution to the stabilization problem using a pure integrator for the case of an open-loop stable plant with delay. Since the plant is open-loop stable, we have $T > 0$. In addition, we assume that $k > 0$ and $L > 0$.

Theorem 7.4.1. *Under the above assumptions on k and L, the set of all stabilizing gains k_i for a given open-loop stable plant with transfer function $G(s)$ as in (7.3.1) is given by*

$$0 < k_i < \frac{T}{kL^2} z_1 \sqrt{z_1^2 + \frac{L^2}{T^2}} \tag{7.4.2}$$

where z_1 is the solution of the equation

$$\frac{1}{tan(z)} = \frac{T}{L} z$$

in the interval $(0, \frac{\pi}{2})$.

Proof. With the change of variables $z = Lw$ the real and imaginary parts of $\delta^*(jw)$ can be rewritten as:

$$\delta_r(z) = kk_i - \frac{1}{L} zsin(z) - \frac{T}{L^2} z^2 cos(z) \tag{7.4.3}$$

$$\delta_i(z) = \frac{1}{L} zcos(z) - \frac{T}{L^2} z^2 sin(z)\ . \tag{7.4.4}$$

As in the proof of Theorem 7.3.1 we need to check two conditions to ensure the stability of the quasipolynomial $\delta^*(s)$.

Step 1. We check Condition 2 of Theorem 7.2.1:

$$E(w_o) = \delta_i^{'}(w_o)\delta_r(w_o) - \delta_i(w_o)\delta_r^{'}(w_o) > 0$$

for some w_o in $(-\infty, \infty)$. Let us take $w_o = z_o = 0$. Thus $\delta_i(z_o) = 0$ and $\delta_r(z_o) = kk_i$. We also have

$$\begin{aligned}
\delta_i'(z) &= \left(\frac{1}{L} - \frac{T}{L^2}z^2\right) cos(z) - \left(\frac{1}{L}z + \frac{2T}{L^2}z\right) sin(z) \\
\Rightarrow \delta_i'(z_o) &= \frac{1}{L} \\
\Rightarrow E(z_o) &= \frac{1}{L}kk_i \, .
\end{aligned}$$

Since $k > 0$ and $L > 0$ then to have $E(z_o) > 0$, we must pick $k_i > 0$. Note that this lower bound on k_i was also mandated by the stability requirement on the delay-free system.

Step 2. We now check Condition 1 of Theorem 7.2.1: the interlacing of the roots of $\delta_r(z)$ and $\delta_i(z)$. From (7.4.4) we can compute the roots of the imaginary part, *i.e.*, $\delta_i(z) = 0$. This gives us the following equation

$$\frac{z}{L}\left[cos(z) - \frac{T}{L}zsin(z)\right] = 0$$

Thus,

$$\begin{aligned}
z &= 0 \text{ or} \\
\frac{1}{tan(z)} &= \frac{T}{L}z \, .
\end{aligned} \tag{7.4.5}$$

Since the solutions of (7.4.5) are difficult to determine analytically, we plot the terms involved in this equation: $\frac{1}{tan(z)}$ and $\frac{T}{L}z$. In this way we can study the behavior of the roots of $\delta_i(z)$. Fig. 7.9 shows this plot. From (7.4.5) we see that one root of the imaginary part is $z_o = 0$. From Fig. 7.9 we find that the other positive real roots of the imaginary part are

$$z_1 \epsilon (0, \frac{\pi}{2}), \; z_2 \epsilon (\pi, \frac{3\pi}{2}), \; z_3 \epsilon (2\pi, \frac{5\pi}{2}),$$

and so on, where the roots z_i, for $i = 1, 2, 3, ...$ satisfy Equation 7.4.5.

Let us now use Theorem 7.2.2 to check if $\delta_i(z)$ has only real roots. Substituting $s_1 = Ls$ in the expression for $\delta^*(s)$, we see that for the new quasipolynomial in s_1, $M = 2$ and $N = 1$. Next we choose $\eta = \frac{\pi}{4}$ to satisfy the requirement that $sin(\eta) \neq 0$. Now from Fig. 7.9, it is clear that in the interval $[0, 2\pi - \frac{\pi}{4}] = [0, \frac{7\pi}{4}]$, $\delta_i(z)$ has three real roots, including a root at the origin. Since $\delta_i(z)$ is an odd function, it follows that in the interval $[-\frac{7\pi}{4}, \frac{7\pi}{4}]$, $\delta_i(z)$ will have five real roots. Also observe from Fig. 7.9 that $\delta_i(z)$ has a real root in the interval $(\frac{7\pi}{4}, \frac{9\pi}{4}]$. Thus $\delta_i(z)$ has $4N + M = 6$ real roots in the interval $[-2\pi + \frac{\pi}{4}, 2\pi + \frac{\pi}{4}]$. Moreover, $\delta_i(z)$ has two real roots in each of the intervals $[2l\pi + \frac{\pi}{4}, 2(l+1)\pi + \frac{\pi}{4}]$ and $[-2(l+1)\pi + \frac{\pi}{4}, -2l\pi + \frac{\pi}{4}]$ for $l = 1, 2, ...$. Hence, it follows that $\delta_i(z)$ has exactly $4lN + M$ real roots in

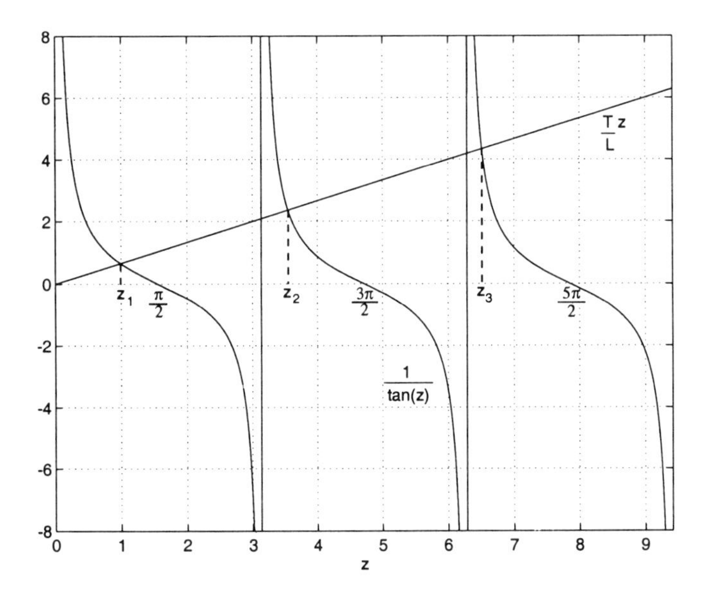

Fig. 7.9. Plot of the terms involved in Equation 7.4.5.

$[-2l\pi + \frac{\pi}{4}, 2l\pi + \frac{\pi}{4}]$, $l = 1, 2, ...$, which by Theorem 7.2.2 implies that $\delta_i(z)$ has only real roots.

We now evaluate $\delta_r(z)$ at the roots of the imaginary part $\delta_i(z)$. For $z_o = 0$, using (7.4.3) we obtain

$$\delta_r(z_o) = kk_i \tag{7.4.6}$$

For z_1, using (7.4.3) we obtain

$$\begin{aligned} \delta_r(z_1) &= kk_i - \frac{1}{L} z_1 sin(z_1) - \frac{T}{L^2} z_1^2 cos(z_1) \\ &= kk_i - \frac{sin(z_1)}{T tan(z_1)} - \frac{T}{L^2} z_1^2 cos(z_1) \text{ [using (7.4.5)]} \\ &= kk_i - \frac{T}{L^2}\left[z_1^2 + \frac{L^2}{T^2}\right] cos(z_1) \end{aligned}$$

From Fig. 7.9 since $z_1 \epsilon (0, \frac{\pi}{2})$ we obtain

$$\begin{aligned} cos(z_1) &= \frac{z_1}{\sqrt{z_1^2 + \frac{L^2}{T^2}}} \\ \Rightarrow \delta_r(z_1) &= kk_i - \frac{T}{L^2} z_1 \sqrt{z_1^2 + \frac{L^2}{T^2}} \end{aligned} \tag{7.4.7}$$

A similar analysis for $z_2 \epsilon (\pi, \frac{3\pi}{2})$, $z_3 \epsilon (2\pi, \frac{5\pi}{2})$, etc., gives us

$$\delta_r(z_2) = kk_i + \frac{T}{L^2} z_2 \sqrt{z_2^2 + \frac{L^2}{T^2}} \tag{7.4.8}$$

$$\delta_r(z_3) = kk_i - \frac{T}{L^2} z_3 \sqrt{z_3^2 + \frac{L^2}{T^2}} \tag{7.4.9}$$

$$\vdots$$

Interlacing of the roots of $\delta_r(z)$ and $\delta_i(z)$ is equivalent to $\delta_r(z_o) > 0$ (this is due to the fact that $k_i > 0$ as derived in the previous step), $\delta_r(z_1) < 0$, $\delta_r(z_2) > 0$, $\delta_r(z_3) < 0$, and so on . Using this fact and Equations 7.4.6-7.4.9 we obtain

$$\begin{aligned}
\delta_r(z_o) > 0 &\Rightarrow k_i > 0 \\
\delta_r(z_1) < 0 &\Rightarrow k_i < \frac{T}{kL^2} z_1 \sqrt{z_1^2 + \frac{L^2}{T^2}} =: M_1 \\
\delta_r(z_2) > 0 &\Rightarrow k_i > -\frac{T}{kL^2} z_2 \sqrt{z_2^2 + \frac{L^2}{T^2}} =: M_2 \\
\delta_r(z_3) < 0 &\Rightarrow k_i < \frac{T}{kL^2} z_3 \sqrt{z_3^2 + \frac{L^2}{T^2}} =: M_3 \\
&\vdots
\end{aligned}$$

Since $z_1 < z_2 < z_3 < ...$, we conclude that $|M_1| < |M_2| < |M_3| < ...$. Thus intersecting the bounds previously found for k_i we conclude that for the interlacing property to hold we must have

$$0 < k_i < \frac{T}{kL^2} z_1 \sqrt{z_1^2 + \frac{L^2}{T^2}} .$$

Note that for values of k_i in this range, the interlacing property and the fact that the roots of $\delta_i(z)$ are all real can be used in Theorem 7.2.2 to guarantee that $\delta_r(z)$ also has only real roots. Thus all the conditions of Theorem 7.2.1 are satisfied and this completes the proof. ♣

Example 7.4.1. Consider the pure integrator stabilization problem where the open-loop stable plant is defined by the parameters $k = 1$, $L = 1$ sec, and $T = 2$ sec. We will use Theorem 7.4.1 to obtain the set of stabilizing gains. As a first step, we compute $z_1 \epsilon (0, \frac{\pi}{2})$ satisfying (7.4.5), *i.e.*,

$$\frac{1}{tan(z)} = 2z .$$

The root obtained is $z_1 = 0.6533$. Thus, from (7.4.2) the set of stabilizing gains is given by

$$0 < k_i < 1.0748 .$$

If we now set the controller parameter k_i to $\frac{1}{2}$ then the characteristic quasipolynomial $\delta^*(s)$ of the system is given by

$$\delta^*(s) = \frac{1}{2} + (s + 2s^2)e^s \ .$$

Substituting $s = jw$ we obtain

$$\delta^*(jw) = \left[\frac{1}{2} - wsin(w) - 2w^2cos(w)\right] + j\left[wcos(w) - 2w^2sin(w)\right] \ .$$

Fig. 7.10 shows the plot of the real and imaginary parts of $\delta^*(jw)$. As we can see the roots of the real and imaginary parts interlace. Fig. 7.11 shows the time response of the closed-loop system to a unit step input r. Clearly the closed-loop system is stable and the output tracks the step input.

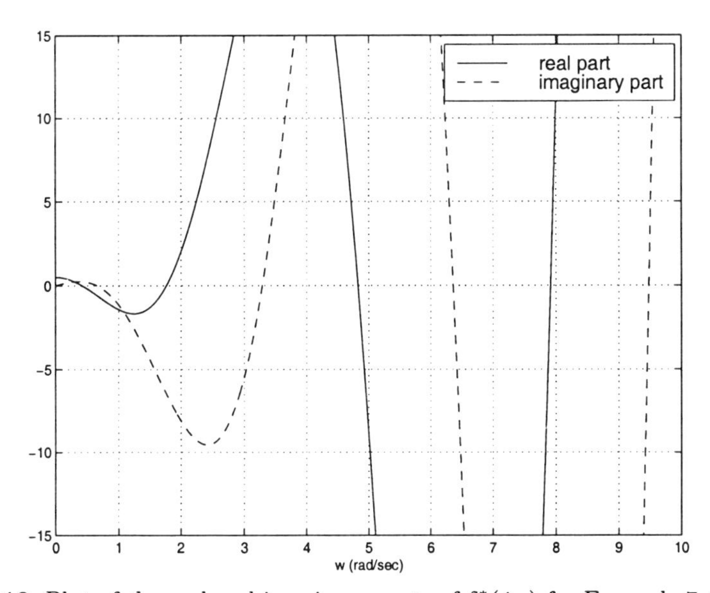

Fig. 7.10. Plot of the real and imaginary parts of $\delta^*(jw)$ for Example 7.4.1.

7.5 Stabilization using a PI Controller

We consider in this section the stabilization problem shown in Fig. 7.2 where the controller now has a proportional term and an integral term, *i.e.*,

$$C(s) = k_p + \frac{k_i}{s} \ .$$

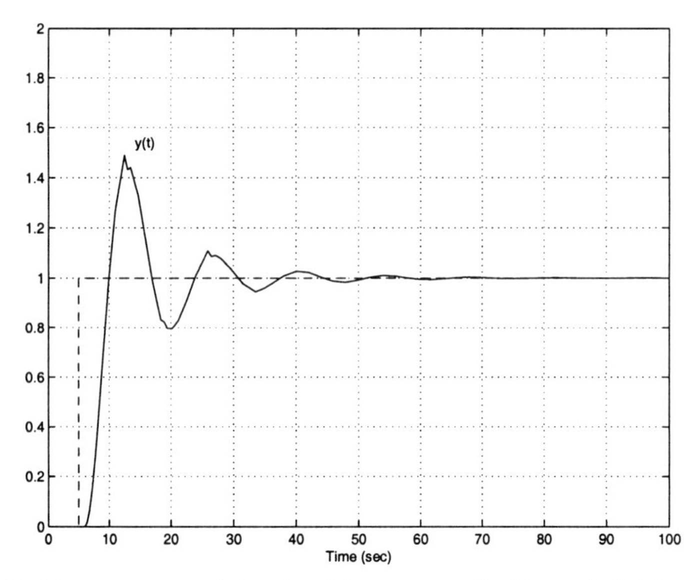

Fig. 7.11. Time response of the closed-loop system for Example 7.4.1.

Our objective is to determine analytically the region in the k_i-k_p parameter space for which the closed-loop system is stable.

When the time delay L of the plant model is zero, the characteristic equation of the closed-loop system is given by

$$\delta(s) = Ts^2 + (kk_p + 1)s + kk_i.$$

From the above equation, we conclude that for the closed-loop stability of the delay-free system, we must have either

$$kk_i > 0\,,\ kk_p + 1 > 0\,,\ T > 0 \tag{7.5.1}$$

or

$$kk_i < 0\,,\ kk_p + 1 < 0\,,\ T < 0 \tag{7.5.2}$$

Clearly (7.5.1) must be satisfied for an open loop stable plant while (7.5.2) must be satisfied for an open loop unstable plant. Assuming that the steady-state gain k of the plant is positive we obtain the following conditions for closed-loop stability of the delay-free system:

$$k_p > -\frac{1}{k} \quad , \quad k_i > 0 \ \text{(Open loop stable plant, } i.e.,\ T > 0) \tag{7.5.3}$$

$$k_p < -\frac{1}{k} \quad , \quad k_i < 0 \ \text{(Open loop unstable plant, } i.e.,\ T < 0) \tag{7.5.4}$$

We now consider the case where the time delay of the model is different from zero. In this case the closed-loop characteristic equation of the system is given by

$$\delta(s) = (kk_i + kk_p s)e^{-Ls} + (1 + Ts)s \ .$$

As in previous sections, we will use Theorem 7.2.1 to find the set of stabilizing PI controllers. As before, we construct the quasipolynomial $\delta^*(s)$, *i.e.*,

$$\delta^*(s) = e^{Ls}\delta(s) = kk_i + kk_p s + (1 + Ts)se^{Ls} \ .$$

Substituting $s = jw$, we have

$$\delta^*(jw) = \delta_r(w) + j\delta_i(w)$$

where

$$\begin{aligned}
\delta_r(w) &= kk_i - wsin(Lw) - Tw^2cos(Lw) \\
\delta_i(w) &= w[kk_p + cos(Lw) - Twsin(Lw)] \ .
\end{aligned}$$

We now consider two different cases.

7.5.1 Open-loop Stable Plant

In this case $T > 0$. Furthermore, as before, let us assume that $k > 0$ and $L > 0$. Clearly, the controller parameter k_i only affects the real part of $\delta^*(jw)$ whereas the controller parameter k_p affects the imaginary part of $\delta^*(jw)$. Moreover, we note that k_i, k_p appear affinely in $\delta_r(w)$, $\delta_i(w)$ respectively. Thus, by sweeping over all real k_p and solving a stabilization problem at each stage (as in Section 7.4), we can determine the set of all stabilizing (k_p, k_i) values for the given plant.

Now, the range of k_p values over which the sweeping needs to be carried out can be narrowed down by using the following Theorem.

Theorem 7.5.1. *Under the above assumptions on k and L, the range of k_p values for which a solution exists to the PI stabilization problem of a given open-loop stable plant with transfer function $G(s)$ as in (7.3.1) is given by*

$$-\frac{1}{k} < k_p < \frac{T}{kL}\sqrt{\alpha_1^2 + \frac{L^2}{T^2}} \tag{7.5.5}$$

where α_1 is the solution of the equation

$$tan(\alpha) = -\frac{T}{L}\alpha$$

in the interval $(\frac{\pi}{2}, \pi)$.

Proof. With the change of variables $z = Lw$ the real and imaginary parts of $\delta^*(jw)$ can be rewritten as:

$$\delta_r(z) = k[k_i - a(z)] \tag{7.5.6}$$

$$\delta_i(z) = \frac{z}{L}\left[kk_p + cos(z) - \frac{T}{L}zsin(z)\right] \tag{7.5.7}$$

where

$$a(z) \triangleq \frac{z}{kL}\left[sin(z) + \frac{T}{L}zcos(z)\right] . \tag{7.5.8}$$

As in the proof of Theorem 7.4.1 we need to check two conditions to ensure the stability of the quasipolynomial $\delta^*(s)$.

Step 1. We start by checking Condition 2 of Theorem 7.2.1:

$$E(w_o) = \delta_i'(w_o)\delta_r(w_o) - \delta_i(w_o)\delta_r'(w_o) > 0$$

for some w_o in $(-\infty, \infty)$. Let us take $w_o = z_o = 0$. Thus $\delta_i(z_o) = 0$ and $\delta_r(z_o) = kk_i$. We also have

$$\begin{aligned} \delta_i'(z) &= \frac{kk_p}{L} + \left(\frac{1}{L} - \frac{T}{L^2}z^2\right)cos(z) - \left(\frac{1}{L}z + \frac{2T}{L^2}z\right)sin(z) \\ \Rightarrow E(z_o) &= \left(\frac{kk_p+1}{L}\right)(kk_i) . \end{aligned}$$

By our initial assumption $k > 0$ and $L > 0$. Thus, if we pick $k_i > 0$ and $k_p > -\frac{1}{k}$, we have $E(z_o) > 0$. Notice that the case where $k_i < 0$ and $k_p < -\frac{1}{k}$ is ruled out since from (7.5.3) it is clear that this is not a stabilizing set for the delay-free case.

Step 2. We now check Condition 1 of Theorem 7.2.1: the interlacing of the roots of $\delta_r(z)$ and $\delta_i(z)$. From (7.5.7) we can compute the roots of the imaginary part, *i.e.*, $\delta_i(z) = 0$. This gives us the following equation

$$\frac{z}{L}\left[kk_p + cos(z) - \frac{T}{L}zsin(z)\right] = 0$$

$$\begin{aligned} \Rightarrow \quad & z = 0 \text{ or} \\ & kk_p + cos(z) - \frac{T}{L}zsin(z) = 0 . \end{aligned} \tag{7.5.9}$$

From this we see that one root of the imaginary part is $z_o = 0$. The other roots are difficult to find since we need to solve (7.5.9) analytically. However, we can plot the terms involved in Equation 7.5.9 and graphically examine the nature of the solution. There are three different cases to consider. In each case, the positive real roots of (7.5.9) will be denoted by z_j, $j = 1, 2, ...$, arranged in increasing order of magnitude.

Case 1: $-\frac{1}{k} < k_p < \frac{1}{k}$. In this case, we graph $\frac{kk_p+cos(z)}{sin(z)}$ and $\frac{T}{L}z$ to obtain the plots shown in Fig. 7.12.

Case 2: $k_p = \frac{1}{k}$. In this case, we sketch $kk_p + cos(z)$ and $\frac{T}{L}zsin(z)$ to obtain the plots shown in Fig. 7.13.

Case 3: $\frac{1}{k} < k_p$. In this case, we sketch $\frac{kk_p+cos(z)}{sin(z)}$ and $\frac{T}{L}z$ to obtain the plots shown in Figs. 7.14(a) and 7.14(b). The plot in Fig. 7.14(a) corresponds to the case where $\frac{1}{k} < k_p < k_u$, and k_u is the largest number so that the plot

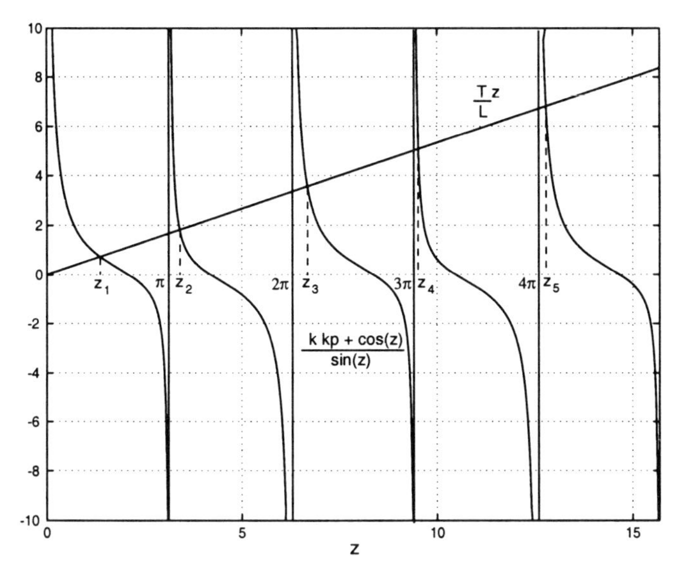

Fig. 7.12. Plot of the terms involved in Equation 7.5.9 for $-\frac{1}{k} < k_p < \frac{1}{k}$.

of $\frac{kk_p+cos(z)}{sin(z)}$ intersects the line $\frac{T}{L}z$ twice in the interval $(0, \pi)$. The plot in Fig. 7.14(b) corresponds to the case where $k_p \geq k_u$ and the plot of $\frac{kk_p+cos(z)}{sin(z)}$ does not intersect the line $\frac{T}{L}z$ twice in the interval $(0, \pi)$.

Let us now use Theorem 7.2.2 to check if $\delta_i(z)$ has only real roots. Substituting $s_1 = Ls$ in the expression for $\delta^*(z)$, we see that for the new quasipolynomial in s_1, $M = 2$ and $N = 1$. Next we choose $\eta = \frac{\pi}{4}$ to satisfy the requirement that $sin(\eta) \neq 0$. Now from Figs. 7.12, 7.13 and 7.14(a), we see that in each of these cases, *i.e.*, for $-\frac{1}{k} < k_p < k_u$, $\delta_i(z)$ has three real roots in the interval $[0, 2\pi - \frac{\pi}{4}] = [0, \frac{7\pi}{4}]$, including a root at the origin. Since $\delta_i(z)$ is an odd function of z, it follows that in the interval $[-\frac{7\pi}{4}, \frac{7\pi}{4}]$, $\delta_i(z)$ will have 5 real roots. Also observe from Figs. 7.12, 7.13 and 7.14(a) that $\delta_i(z)$ has a real root in the interval $(\frac{7\pi}{4}, \frac{9\pi}{4}]$. Thus $\delta_i(z)$ has $4N + M = 6$ real roots in the interval $[-2\pi + \frac{\pi}{4}, 2\pi + \frac{\pi}{4}]$. Moreover, it is clear from Figs. 7.12, 7.13 and 7.14(a) that $\delta_i(z)$ has two real roots in each of the intervals $[2l\pi + \frac{\pi}{4}, 2(l+1)\pi + \frac{\pi}{4}]$ and $[-2(l+1)\pi + \frac{\pi}{4}, -2l\pi + \frac{\pi}{4}]$ for $l = 1, 2, \ldots$. Hence, it follows that $\delta_i(z)$ has exactly $4lN + M$ real roots in $[-2l\pi + \frac{\pi}{4}, 2l\pi + \frac{\pi}{4}]$ for $-\frac{1}{k} < k_p < k_u$. Hence from Theorem 7.2.2, we conclude that for $-\frac{1}{k} < k_p < k_u$, $\delta_i(z)$ has only real roots. Also note that the case $k_p \geq k_u$ corresponding to Fig. 7.14(b) does not merit any further consideration since using Theorem 7.2.2, we can easily argue that in this case, all the roots of $\delta_i(z)$ will not be real, thereby ruling out closed-loop stability.

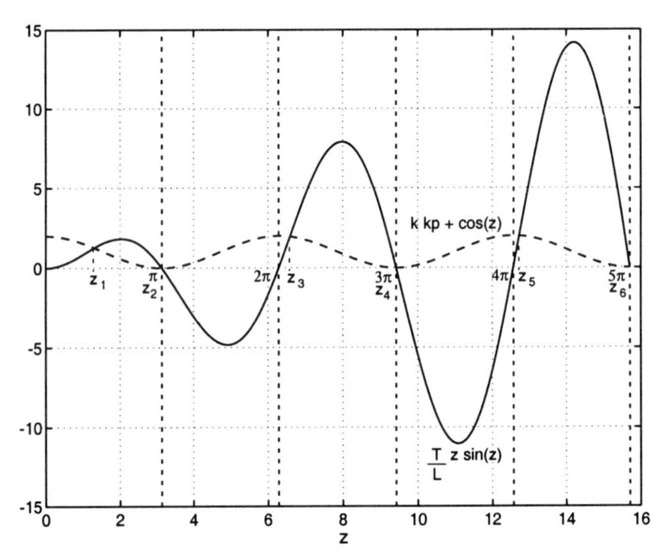

Fig. 7.13. Plot of the terms involved in Equation 7.5.9 for $k_p = \frac{1}{k}$.

We now evaluate $\delta_r(z)$ at the roots of the imaginary part $\delta_i(z)$. For $z_o = 0$, using (7.5.6) we obtain

$$\begin{aligned} \delta_r(z_o) &= k[k_i - a(0)] \\ &= kk_i \,. \end{aligned} \tag{7.5.10}$$

For z_j, where $j = 1, 2, 3, ...$ using (7.5.6) we obtain

$$\delta_r(z_j) = k[k_i - a(z_j)] \tag{7.5.11}$$

Interlacing of the roots of $\delta_r(z)$ and $\delta_i(z)$ is equivalent to $\delta_r(z_o) > 0$ (since $k_i > 0$ as derived in Step 1), $\delta_r(z_1) < 0$, $\delta_r(z_2) > 0$, $\delta_r(z_3) < 0$, and so on . Using this fact and Equations 7.5.10 and 7.5.11 we obtain

$$\begin{aligned} \delta_r(z_o) > 0 \quad &\Rightarrow \quad k_i > 0 \\ \delta_r(z_1) < 0 \quad &\Rightarrow \quad k_i < a_1 \\ \delta_r(z_2) > 0 \quad &\Rightarrow \quad k_i > a_2 \\ \delta_r(z_3) < 0 \quad &\Rightarrow \quad k_i < a_3 \\ \delta_r(z_4) < 0 \quad &\Rightarrow \quad k_i > a_4 \\ &\vdots \end{aligned} \tag{7.5.12}$$

where the bounds a_j for $j = 1, 2, 3, ...$ are given by

$$a_j \triangleq a(z_j) \,. \tag{7.5.13}$$

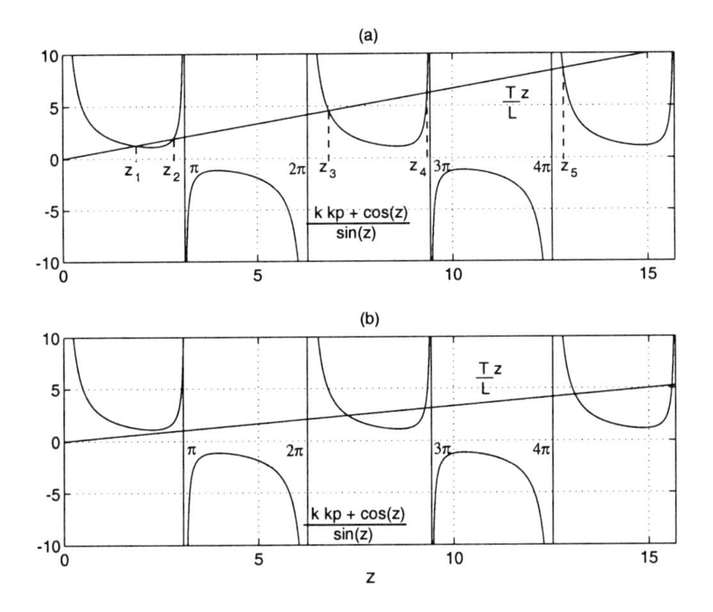

Fig. 7.14. Plot of the terms involved in Equation 7.5.9 for $\frac{1}{k} < k_p$.

Now, from this set of inequalities it is clear that we need the odd bounds (*i.e.*, a_1, a_3, etc.) to be strictly positive in order to obtain a feasible range for the controller parameter k_i. As we will see in the next Lemma, for $k_p > -\frac{1}{k}$, this occurs if and only if

$$k_p < \frac{T}{kL}\sqrt{\alpha_1^2 + \frac{L^2}{T^2}}$$

where α_1 is the solution of the equation

$$tan(\alpha) = -\frac{T}{L}\alpha \; .$$

in the interval $(\frac{\pi}{2}, \pi)$. Moreover, from the same Lemma we will see that the bounds a_j corresponding to even values of j are all negative for $k_p \epsilon$ $(-\frac{1}{k}, \frac{T}{kL}\sqrt{\alpha_1^2 + \frac{L^2}{T^2}})$. Thus, the conditions (7.5.12) reduce to

$$0 < k_i < \min_{j=1,3,5,\ldots} \{a_j\} \; . \tag{7.5.14}$$

As in the earlier sections, one can make use of the interlacing property and the fact that $\delta_i(z)$ has only real roots to establish that for $-\frac{1}{k} < k_p < \frac{T}{kL}\sqrt{\alpha_1^2 + \frac{L^2}{T^2}}$, $\delta_r(z)$ also has only real roots. This completes the proof of the theorem. ♣

Lemma 7.5.1. *For $k_p > -\frac{1}{k}$, a necessary and sufficient condition for a_j defined in (7.5.13) to be positive for odd values of j is that*

$$k_p < \frac{T}{kL}\sqrt{\alpha_1^2 + \frac{L^2}{T^2}}$$

where α_1 is the solution of the equation

$$tan(\alpha) = -\frac{T}{L}\alpha \tag{7.5.15}$$

in the interval $(\frac{\pi}{2}, \pi)$.

Furthermore, for all $k_p \epsilon (-\frac{1}{k}, \frac{T}{kL}\sqrt{\alpha_1^2 + \frac{L^2}{T^2}})$, $a_j < 0$ for even values of j.

Proof. From Figs. 7.12-7.14(a), we see that for $k_p \epsilon (-\frac{1}{k}, \frac{T}{kL}\sqrt{\alpha_1^2 + \frac{L^2}{T^2}})$, the roots of (7.5.9) corresponding to odd values of j satisfy the following properties:

$$z_1 \epsilon (0, \pi),\ z_3 \epsilon (2\pi, 3\pi),\ z_5 \epsilon (4\pi, 5\pi),$$

and so on, *i.e.*, $z_j \epsilon ((j-1)\pi, j\pi)$. Thus, in all these three cases the roots of (7.5.9) corresponding to odd values of j are either in the first quadrant or in the second quadrant. Then,

$$sin(z_j) > 0 \text{ for odd values of } j.$$

Now, recall from (7.5.13) that the parameter a_j was defined as

$$a_j = \frac{z_j}{kL}\left[sin(z_j) + \frac{T}{L} z_j cos(z_j)\right].$$

Thus for $z_j \neq l\pi$, $l = 0, 1, 2, ...$, we can write

$$\begin{aligned} a_j &= \frac{z_j}{kL}\left[sin(z_j) + \frac{kk_p + cos(z_j)}{sin(z_j)} \cdot cos(z_j)\right] \quad \text{[using (7.5.9)]} \\ &= \frac{z_j}{kL}\left[\frac{1 + kk_p cos(z_j)}{sin(z_j)}\right] \end{aligned} \tag{7.5.16}$$

From this expression it is clear that if $z_j \neq l\pi$, then the parameter a_j is positive if and only if

$$\begin{aligned} sin(z_j) > 0 \quad &\text{and} \quad 1 + kk_p cos(z_j) > 0 \text{ or} \\ sin(z_j) < 0 \quad &\text{and} \quad 1 + kk_p cos(z_j) < 0 \,. \end{aligned}$$

and it is negative otherwise. Fig. 7.15 shows the k_p-z plane split into different regions according to the value of parameter a_j. In those regions where $a_j > 0$ a plus sign (+) has been placed and in those regions where $a_j < 0$ a minus sign (-) has been placed. In this figure the dashed line corresponds to the function

$$1 + kk_p cos(z_j) = 0$$

or equivalently

$$k_p = -\frac{1}{kcos(z)} .$$

Although the plot here corresponds to the interval $z\epsilon[0, 2\pi]$, the function being periodic, the plot repeats itself.

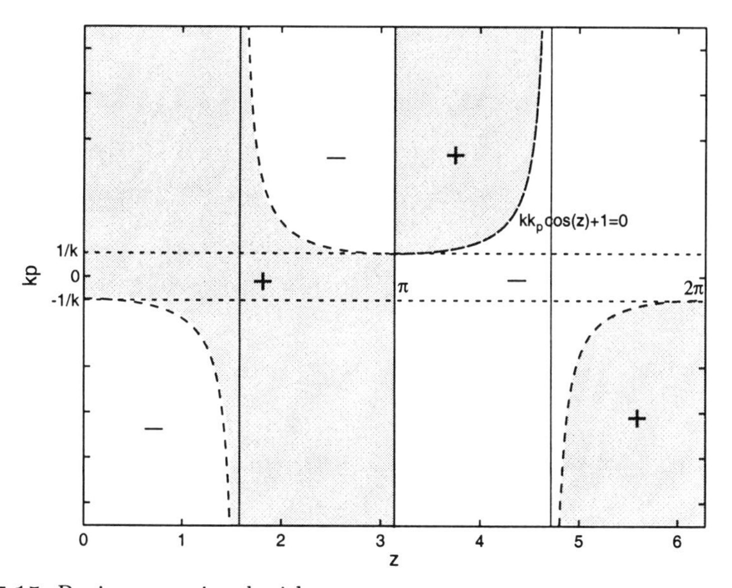

Fig. 7.15. Regions associated with parameter a_j.

We will now graph the solutions of Equation 7.5.9 in the same k_p-z plane. Recall that the solutions of this equation represent the non-zero roots of the imaginary part $\delta_i(z)$. Equation 7.5.9 can be rewritten as

$$k_p = \frac{1}{k}\left[\frac{T}{L} zsin(z) - cos(z)\right] \tag{7.5.17}$$

and Fig. 7.16 shows the graph of this function along with the regions presented in Fig. 7.15. The intersection of Equation 7.5.9 with the curve $1 + kk_p cos(z) = 0$ occurs at five values of the parameter z: 0, α_1, π, α_2 and 2π. Thus each of these values will satisfy the relationship

$$-\frac{1}{cos(z)} = \frac{T}{L} zsin(z) - cos(z) .$$

Furthermore, if $z \neq l\pi$ then simplifying this equation, we obtain

$$tan(z) = -\frac{T}{L} z .$$

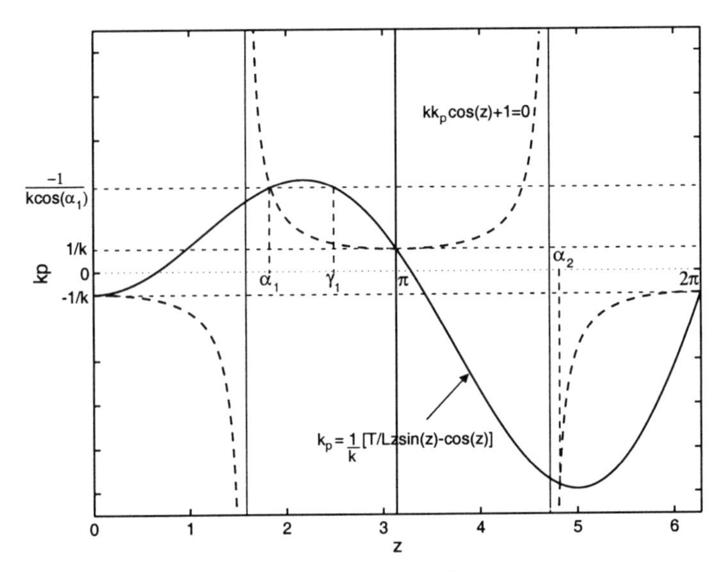

Fig. 7.16. Study of Equation 7.5.9 in the k_p-z plane.

Thus α_1 and α_2 will be solutions of the above equation.
For a given value of k_p, let $z_1(k_p)$ and $z_2(k_p)$ be the positive real roots of (7.5.17) arranged in ascending order of magnitude. Now from Fig. 7.16, it is clear that for $k_p \epsilon (-\frac{1}{k}, -\frac{1}{kcos(\alpha_1)}) - \{\frac{1}{k}\}$,

$$\begin{aligned} a_1 = a(z_1(k_p)) &> 0 \\ \text{and } a_2 = a(z_2(k_p)) &< 0 . \end{aligned}$$

For $k_p = \frac{1}{k}$, from Fig. 7.16, we once again conclude that $a_1 = a(z_1(k_p)) > 0$. Since $z_2 = \pi$, we cannot use Fig. 7.16 or (7.5.16) to determine the sign of $a(z_2(k_p))$. However from the original definition of $a(z)$ in (7.5.8), it follows that $a_2 = a(z_2(k_p)) < 0$. From Fig. 7.16, we also see that for $k_p > -\frac{1}{kcos(\alpha_1)}$,

$$\begin{aligned} a_1 = a(z_1(k_p)) &< 0 \\ \text{and } a_2 = a(z_2(k_p)) &< 0 . \end{aligned}$$

Thus we conclude that $a_1 > 0$ if and only if $k_p < -\frac{1}{kcos(\alpha_1)}$. Since α_1 satisfies the relationship $tan(\alpha_1) = -\frac{T}{L}\alpha_1$ we have that

$$cos(\alpha_1) = -\frac{1}{\sqrt{1 + \frac{T^2}{L^2}\alpha_1^2}}$$

so that

$$-\frac{1}{kcos(\alpha_1)} = \frac{T}{kL}\sqrt{\alpha_1^2 + \frac{L^2}{T^2}} .$$

In view of the above discussion, we conclude that if $-\frac{1}{k} < k_p < \frac{T}{kL}\sqrt{\alpha_1^2 + \frac{L^2}{T^2}}$ then

$$a_1 > 0 \text{ and } a_2 < 0$$

Using a similar approach we can show that

$$\text{if } -\frac{1}{k} < k_p < \frac{T}{kL}\sqrt{\alpha_3^2 + \frac{L^2}{T^2}} \quad \Rightarrow \quad a_3 > 0\,, a_4 < 0$$
$$\text{if } -\frac{1}{k} < k_p < \frac{T}{kL}\sqrt{\alpha_5^2 + \frac{L^2}{T^2}} \quad \Rightarrow \quad a_5 > 0\,, a_6 < 0$$
$$\vdots$$

where α_j is the solution to $tan(\alpha) = -\frac{T}{L}\alpha$ in the interval $((j-\frac{1}{2})\pi, j\pi)$. It is clear that all these upper bounds on k_p increase monotonically with α_j. Thus, it suffices to take the upper bound corresponding to α_1 to guarantee that a_j for odd values of j are strictly positive and a_j for even values of j are strictly negative. This completes the proof. ♣

Remark 7.5.1. In the proof of the above lemma, we have not considered the case where the value of k_p is such that $k_p = \frac{1}{k}[\frac{T}{L}z sin(z) - cos(z)]$ does not have two zeros in the interval $z\epsilon[0, 2\pi]$. This is because we know from the proof of Theorem 7.5.1 that for such a value of k_p, closed loop stability is ruled out.

Remark 7.5.2. As we can see from Figs. 7.12-7.14(a), the odd roots of (7.5.9), *i.e.*, z_j where $j = 1, 3, 5, ...$ are getting closer to $(j-1)\pi$ as j increases. So in the limit for odd values of j we have:

$$\lim_{j\to\infty} cos(z_j) = 1\,.$$

Moreover, since the cosine function is monotonically decreasing between $(j-1)\pi$ and $j\pi$ for odd values of j, and because of the previous observation we have:

$$cos(z_1) < cos(z_3) < cos(z_5) < ...$$

We now present a lemma that will be useful in the development of an algorithm for the PI control problem.

Lemma 7.5.2. *If $cos(z_j) > 0$ then $a_j < a_{j+2}$ for odd values of j.*

Proof. From (7.5.8), (7.5.13) we have

$$\begin{aligned} kLa_j &= z_j sin(z_j) + \frac{T}{L}z_j^2 cos(z_j) \\ \Rightarrow Tka_j &= kk_p + cos(z_j) + \frac{T^2}{L^2}z_j^2 cos(z_j) \;\; \text{[using (7.5.9)]} \\ \Rightarrow Tka_j - kk_p &= cos(z_j)\left(1 + \frac{T^2}{L^2}z_j^2\right)\,. \end{aligned} \tag{7.5.18}$$

We know that the z_j, $j = 1, 3, 5...$ are arranged in increasing order of magnitude, *i.e.*, $z_j < z_{j+2}$, so we have for odd values of j

$$1 + \frac{T^2}{L^2} z_j^2 < 1 + \frac{T^2}{L^2} z_{j+2}^2 \, . \tag{7.5.19}$$

Now, because of Remark 7.5.2 we have

$$0 < cos(z_j) < cos(z_{j+2}) \text{ for odd values of } j. \tag{7.5.20}$$

Since $cos(z_j) > 0$, from (7.5.19) we have

$$\left(1 + \frac{T^2}{L^2} z_j^2\right) cos(z_{j+2}) < \left(1 + \frac{T^2}{L^2} z_{j+2}^2\right) cos(z_{j+2}) \, .$$

Since $1 + \frac{T^2}{L^2} z_j^2 > 0$, from (7.5.20) we have

$$\left(1 + \frac{T^2}{L^2} z_j^2\right) cos(z_j) < \left(1 + \frac{T^2}{L^2} z_j^2\right) cos(z_{j+2}) \, .$$

Combining these two latter inequalities:

$$\begin{aligned} \left(1 + \frac{T^2}{L^2} z_j^2\right) cos(z_j) &< \left(1 + \frac{T^2}{L^2} z_{j+2}^2\right) cos(z_{j+2}) \\ \Rightarrow Tka_j - kk_p &< Tka_{j+2} - kk_p \quad \text{[using (7.5.18)]} \\ \Rightarrow a_j &< a_{j+2} \end{aligned}$$

for odd values of j and this completes the proof. ♣

Remark 7.5.3. Notice that for a fixed value of k_p inside the range proposed by Theorem 7.5.1, we can find the range of k_i such that the closed-loop system is stable. This range is given by (7.5.14) and depends on the bounds a_j corresponding to odd values of j. However, by Lemma 7.5.2, if $cos(z_1) > 0$, then the bound a_1 is the minimum of all the odd bounds, and the range of stabilizing k_i is given by: $0 < k_i < a_1$. If this is not the case, but we have that $cos(z_3) > 0$, then the bound a_3 is less than all the other bounds a_j for $j = 5, 7, 9, ...$ Then, in this case the range of stabilizing k_i is given by: $0 < k_i < \min\{a_1, a_3\}$.

Theorem 7.5.1 and Lemma 7.5.2 together suggest a procedure for determining the set of all stabilizing (k_p, k_i) values for a given plant. This procedure is summarized in the following algorithm.

Algorithm for PI controller.

- **Step 1:** Pick a k_p in the range suggested by Theorem 7.5.1 and initialize $j = 1$.
- **Step 2:** Find the root z_j of Equation 7.5.9.
- **Step 3:** Compute the parameter a_j associated with the z_j previously found by using Equation 7.5.13.

- **Step 4:** If $cos(z_j) > 0$ then go to Step 5. Else, increase $j = j + 2$ and go to Step 2.
- **Step 5:** Determine the lower and upper bounds for k_i as follows:

$$0 < k_i < \min_{l=1,3,5,\ldots,j} \{a_l\} .$$

- **Step 6:** Go to Step 1.

We now present an example to illustrate the procedure involved in determining the stabilizing (k_p, k_i) values for a given plant.

Example 7.5.1. Consider the problem of choosing stabilizing PI gains for the plant given in (7.3.1), where the plant parameters are $k = 1$, $L = 1$ sec, and $T = 4$ sec. From Theorem 7.5.1 we can obtain the range of k_p values over which the sweeping needs to be carried out. First, we compute $\alpha_1 \epsilon (\frac{\pi}{2}, \pi)$ satisfying (7.5.15), *i.e.*,

$$tan(\alpha) = -4\alpha .$$

Solving this equation we obtain $\alpha_1 = 1.7155$. Thus, from (7.5.5) the range of k_p gains is given by

$$-1 < k_p < 6.9345 .$$

We now sweep over this range of k_p gains and use the previous algorithm to determine the range of k_i gains at each stage. Fig. 7.17 shows the stabilizing region obtained in the k_p-k_i plane.

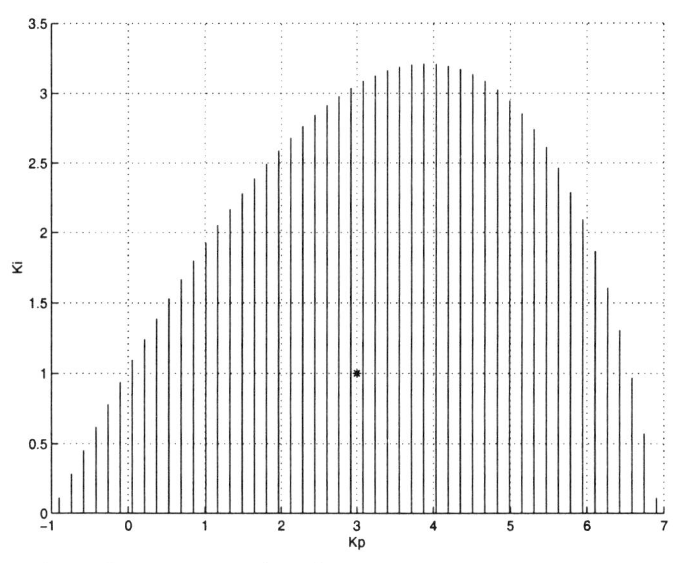

Fig. 7.17. The stabilizing set of (k_p,k_i) values.

We now set the controller parameters k_p and k_i at 3 and 1 respectively. Clearly this point is inside the region sketched in Fig. 7.17. The step response of the closed-loop system with this PI controller is shown in Fig. 7.18. From this figure, we see that the closed-loop system is stable and the output $y(t)$ tracks the step input signal.

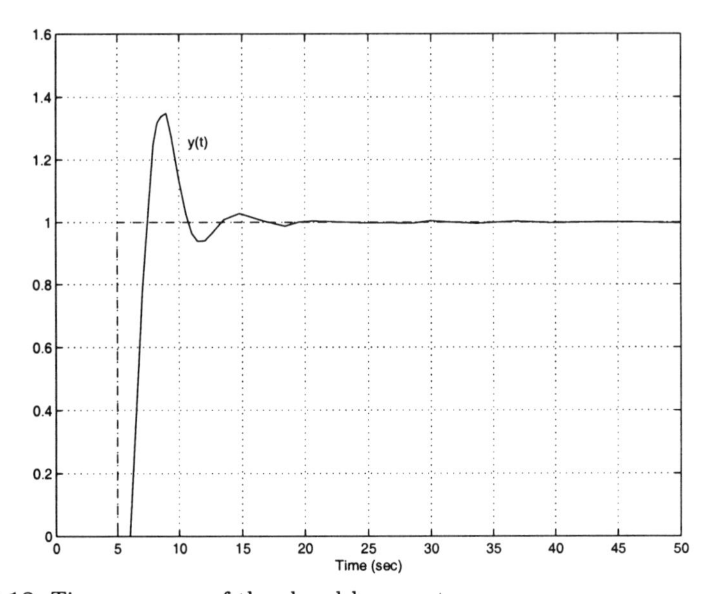

Fig. 7.18. Time response of the closed-loop system.

7.5.2 Open-loop Unstable Plant

In this case, we have $T < 0$. Furthermore, as before, let us assume that $k > 0$ and $L > 0$. Recall from (7.5.4), that for the closed loop stability of the delay-free system, we now require

$$k_p < -\frac{1}{k} \, , \; k_i < 0 \, .$$

The solution to the PI stabilization problem in this case also involves sweeping over all real k_p and solving a constant gain stabilization problem at each stage. The range of k_p values over which the sweeping needs to be carried out can be narrowed down by using the following theorem.

Theorem 7.5.2. *Under the above assumptions on k and L, a necessary condition for a PI controller to simultaneously stabilize the delay-free plant and the plant with delay is $|\frac{T}{L}| > 1$. If this necessary condition is satisfied, then the range of k_p values for which a solution exists to the PI stabilization problem* *of a given open-loop unstable plant with transfer function $G(s)$ as in (7.3.1) is given by*

$$\frac{T}{kL}\sqrt{\alpha_1^2+\frac{L^2}{T^2}} < k_p < -\frac{1}{k} \tag{7.5.21}$$

where α_1 is the solution of the equation

$$tan(\alpha) = -\frac{T}{L}\alpha$$

in the interval $(0, \frac{\pi}{2})$.

Proof. The proof follows along the same lines as that of Theorem 7.5.1. The main differences are that now because of $T < 0$, all the figures such as Figs. 7.12, 7.13, 7.14(a), (b) and the graph of $k_p = \frac{1}{k}[\frac{T}{L}zsin(z) - cos(z)]$ in Fig. 7.16 will have to be appropriately changed. These changes, however, do not pose any difficulty and the reader can verify that the appropriate counterparts of Lemmas 7.5.1 and 7.5.2 can be developed by mimicing the steps used earlier in Section 7.5.1. ♣

Example 7.5.2. Let us now consider the problem of finding the set of stabilizing PI controllers for the plant given in (7.3.1), where the plant parameters are $k = 1$, $L = 1$ sec, and $T = -2$. Since the plant is open-loop unstable we will use Theorem 7.5.2 to find the range of k_p values over which the sweeping needs to be carried out. Since $|\frac{T}{L}| = 2 > 1$ we can proceed to compute $\alpha_1 \epsilon (0, \frac{\pi}{2})$ satisfying the following equation

$$tan(\alpha) = 2\alpha \ .$$

Solving this equation we obtain $\alpha_1 = 1.1656$. Thus, from (7.5.21) the range of k_p values is given by

$$-2.5366 < k_p < -1 \ .$$

By sweeping over the above range of k_p values we can determine the range of k_i values at each stage. Fig. 7.19 shows the stabilizing region obtained in the k_p-k_i plane.

We can now take a point inside the region sketched in Fig. 7.19. For instance, we can take $k_p = -1.8$ and $k_i = -0.03$. Fig. 7.20 shows the step response of the closed-loop system with this PI controller. As we can see from this figure, the closed-loop system is stable and the output $y(t)$ tracks the step input signal.

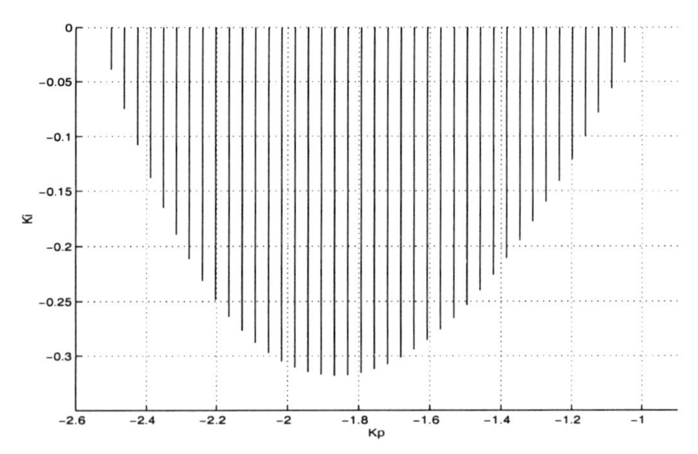

Fig. 7.19. The stabilizing set of (k_p,k_i) values.

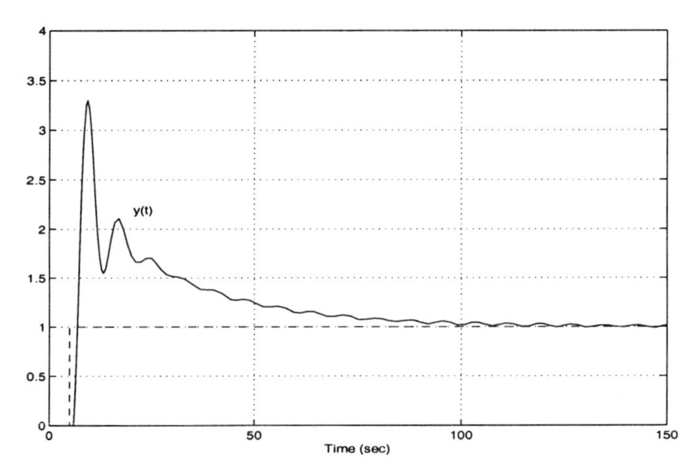

Fig. 7.20. Time response of the closed-loop system.

7.6 Notes and References

The stability and control of time-delay systems has been an active area of research in the controls field. Several time and frequency domain approaches are available for studying the stability of time-delay systems. In [37] the so called Satche's method is presented as an approach for resolving the complexity of the Nyquist diagram for time-delay systems. Also discussed is an extension of the Mikhailov's method due to Sokolov and Miasnikov. A good review of the methods and tools used for the stability analysis and robust stabilization of time-delay systems is available in [18]. The extension of the classical Hermite-Biehler Theorem applicable to quasipolynomials was derived by Pontryagin [42]. An excellent collection of this and other theorems of Pontryagin and their applications can be found in [4], [35]. Even though these theorems cannot be used to easily check stability, they do provide a good route for solving many fixed order/structure stabilization problems as demonstrated in this chapter.

The constant gain and PI results in this chapter are based on Silva, Datta and Bhattacharyya [44] which are part of the doctoral research of Guillermo Silva. Their extension to the case of PID controllers and second order systems with dead time is a topic of current investigation.

CHAPTER 8

CONSTANT GAIN STABILIZATION WITH DESIRED DAMPING

In this chapter, we derive a generalization of the Hermite-Biehler Theorem applicable to polynomials with *complex* coefficients. This result allows us to solve the problem of constant gain stabilization while achieving a desired degree of damping.

8.1 Introduction

A minimum requirement of any control system design is that closed loop stability be guaranteed, *i.e.*, for the continuous-time linear case, the roots of the closed loop characteristic polynomial must lie in the open left half plane. In addition to possessing closed loop stability, a control system must be designed to meet certain performance requirements. In classical control design, some of these performance requirements are stated in terms of the time domain specifications of the maximum overshoot, settling time and rise time of the step response of the closed loop system. These time domain specifications are usually translated by the designer into desired closed loop pole locations using certain rules of thumb. A common description of the desired pole locations in the complex plane usually takes the form of specifications on the *damping ratio* and the *damped natural frequency* of the dominant pair of closed loop complex poles. This is because these parameters play a predominant role in shaping the time domain behaviour of a system. Thus the problem of designing a controller to ensure that the roots of the closed loop characteristic equation satisfy the damping ratio and damped natural frequency requirements arises in practical classical control design.

In Chapter 3, we presented a generalization of the classical Hermite-Biehler Theorem for real polynomials. In Chapter 4, we showed that this generalization could be used for analytically characterizing the set of all stabilizing feedback gains for a given plant as well as for providing a computational characterization of all stabilizing proportional-integral (PI) and proportional-integral-derivative (PID) controllers. In this chapter, we use a similar approach to solve the constant gain stabilization problem subject to a specified damping ratio and damped natural frequency. Our investigation reveals that instead of a generalization of the Hermite-Biehler Theorem for real polynomials as in Chapter 3, we now need a complex version of the generalized Hermite-Biehler Theorem. Accordingly, such a result is derived and used for solving the problem at hand.

The chapter is organized as follows. In Section 8.2, we introduce some notation and formulate the problem to be solved. Specifically, it is shown that the problem of interest to us in this chapter can be decomposed into two sub-problems and that the complex version of the generalized Hermite-Biehler Theorem is needed for solving one of them. Section 8.3 is devoted to the development of such a complex version. First, in Subsection 8.3.1, we state the relationship between the net phase change of the frequency response of a complex polynomial as the frequency ω varies from $-\infty$ to ∞ and the numbers of its roots in the open left-half and open right-half planes. Then, in Subsection 8.3.2, we use this relationship to derive generalizations of the Hermite-Biehler Theorem applicable to complex polynomials. In Section 8.4, we first consider the problem of stabilizing a given plant with respect to a "rotated Hurwitz" stability region using a constant gain compensator. A complete analytical characterization of all stabilizing gain values is provided. These results immediately lead to a solution to the constant gain stabilization problem subject to desired damping ratio and damped natural frequency specifications. In Section 8.5 the detailed calculations involved are illustrated on an example.

8.2 Problem Formulation

We first introduce some notation and terminology to conveniently describe stability regions in the complex plane that are of interest to us in this chapter. Now, given a complex polynomial $\delta(s)$ of degree n, we know that $\delta(s)$ is said to be Hurwitz stable if $\delta(s)$ has all its roots in the open left half plane. Next, let us consider the region S in the left half plane corresponding to a particular damping ratio and damped natural frequency as sketched in Fig. 8.1. Given a real polynomial $\delta(s)$ of degree n, we say that $\delta(s)$ is S Hurwitz stable if $\delta(s)$ has all its roots in the region S. The region S in Fig. 8.1 can be conveniently characterized by using a parallel shift and two rotations of the imaginary axis as follows:

(I) shift the imaginary axis parallel to itself by an appropriate number of units α to the left. We define the *shifted Hurwitz* stability region $S_{-\alpha}$ as illustrated in Fig. 8.2 (a) by

$$S_{-\alpha} := \{s : s \in C, Re[s] < -\alpha\}.$$

Furthermore, given a real polynomial $\delta(s)$ of degree n, we say that $\delta(s)$ is $-\alpha$ Hurwitz stable if $\delta(s)$ has all its roots in $S_{-\alpha}$.

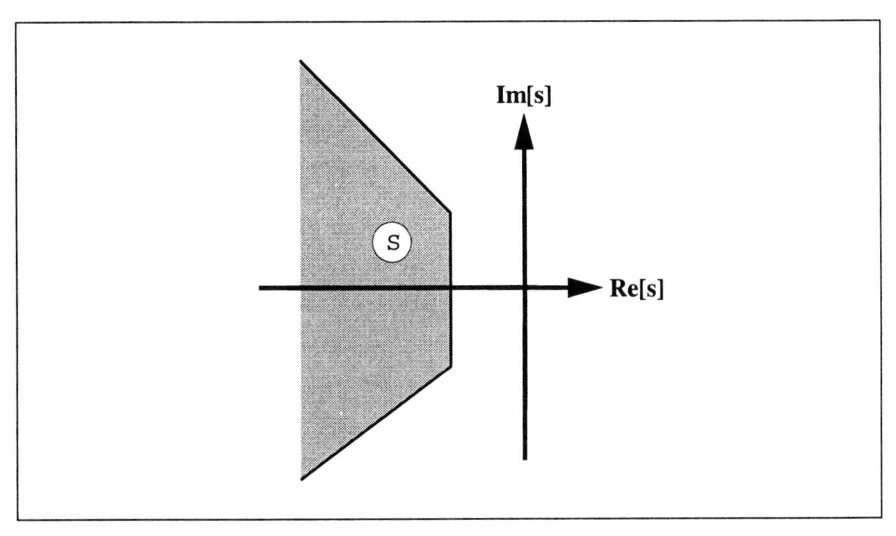

Fig. 8.1. The region S with a specified damping ratio and damped natural frequency.

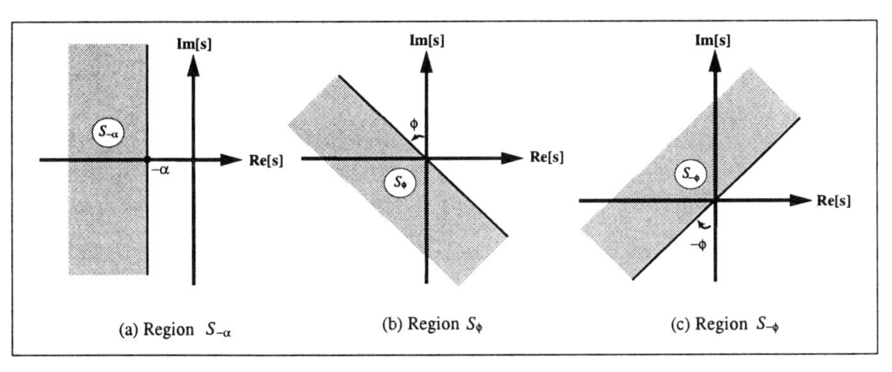

Fig. 8.2. (a) the $-\alpha$ shifted Hurwitz stability region $S_{-\alpha}$, (b) the ϕ rotated Hurwitz stability region S_{ϕ}, and (c) the $-\phi$ rotated Hurwitz stability region $S_{-\phi}$.

(II) rotate the imaginary axis by a fixed positive angle ϕ about the origin. We define the *rotated Hurwitz* stability region S_ϕ as illustrated in Fig. 8.2 (b) by

$$S_\phi := \{s : s \in C, Re[se^{-j\phi}] < 0\}.$$

Furthermore, given a real polynomial $\delta(s)$ of degree n, we say that $\delta(s)$ is ϕ Hurwitz stable if $\delta(s)$ has all its roots in S_ϕ.

(III) rotate the imaginary axis by a fixed negative angle $-\phi$ about the origin. We define the *rotated Hurwitz* stability region $S_{-\phi}$ as illustrated in Fig. 8.2 (c) by

$$S_{-\phi} := \{s : s \in C, Re[se^{j\phi}] < 0\}.$$

Furthermore, given a real polynomial $\delta(s)$ of degree n, we say that $\delta(s)$ is $-\phi$ Hurwitz stable if $\delta(s)$ has all its roots in $S_{-\phi}$.

In view of the preceding definitions, it is clear that

$$S = S_{-\alpha} \cap S_\phi \cap S_{-\phi} \tag{8.2.1}$$

where α and ϕ are appropriately chosen.

Having introduced the relevant terminology, we now turn to the problem that motivates the rest of this chapter. Consider the standard feedback control system shown in Fig. 8.3.

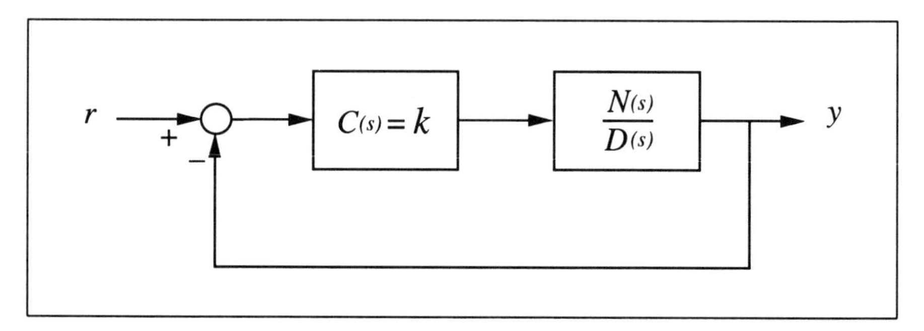

Fig. 8.3. Feedback control system.

Here r is the command signal, y is the output, $G(s) = \frac{N(s)}{D(s)}$ is the plant to be controlled, $N(s)$ and $D(s)$ are coprime polynomials, and $C(s)$ is the controller used for making the closed loop system stable. In this chapter, the controller $C(s)$ is chosen to be a scalar gain k so that the closed loop characteristic polynomial $\delta(s,k)$ is given by

$$\delta(s,k) = D(s) + kN(s).$$

Our objective here is to determine all those values of k, if any, for which the closed loop system is S Hurwitz stable. In view of (8.2.1), this problem can be decomposed into three subproblems: first, determine the set $K_{-\alpha}$ consisting of those values of k, for which the closed loop system is $-\alpha$ Hurwitz stable; next determine the sets K_ϕ and $K_{-\phi}$ consisting of those values of k, for which the closed loop system is ϕ and $-\phi$ Hurwitz stable, respectively. Then the set K of all values of k for which the closed loop characteristic polynomial $\delta(s,k)$ is S Hurwitz stable is given by

$$K = K_{-\alpha} \cap K_\phi \cap K_{-\phi}.$$

We next determine each of the sets $K_{-\alpha}$, K_ϕ and $K_{-\phi}$. For $K_{-\alpha}$, we set $s = s_1 - \alpha$ and write

$$\begin{aligned} \delta(s,k) &= \delta_{-\alpha}(s_1,k) \\ &= D(s_1-\alpha) + kN(s_1-\alpha). \end{aligned}$$

Then the $-\alpha$ Hurwitz stability of $\delta(s,k)$ is equivalent to the Hurwitz stability of $\delta_{-\alpha}(s_1,k)$. Hence, by using the generalized Hermite-Biehler Theorem for real polynomials and the results of Chapter 4, Section 4.2, one can analytically characterize the set $K_{-\alpha}$ of all $-\alpha$ Hurwitz stabilizing gain values.

Next let us consider the determination of K_ϕ and $K_{-\phi}$. For the case of K_ϕ, *i.e.*, the ϕ Hurwitz stabilization problem, we set $s = s_1 e^{j\phi}$ and write

$$\begin{aligned} \delta(s,k) &= \delta_\phi(s_1,k) \\ &= D(s_1 e^{j\phi}) + kN(s_1 e^{j\phi}). \end{aligned}$$

Then the ϕ Hurwitz stability of $\delta(s,k)$ is equivalent to the Hurwitz stability of $\delta_\phi(s_1,k)$. Finally, for the case of $K_{-\phi}$, *i.e.*, the $-\phi$ Hurwitz stabilization problem, we substitute $s = s_1 e^{-j\phi}$ into $\delta(s,k)$ and write

$$\begin{aligned} \delta(s,k) &= \delta_{-\phi}(s_1,k) \\ &= D(s_1 e^{-j\phi}) + kN(s_1 e^{-j\phi}). \end{aligned}$$

Then the $-\phi$ Hurwitz stability of $\delta(s,k)$ is equivalent to the Hurwitz stability of $\delta_{-\phi}(s_1,k)$.

We next show that the two sets K_ϕ and $K_{-\phi}$ are in fact one and the same and hence do not have to be determined separately. To see this, suppose that for a given k', s' is a root of $\delta_\phi(s_1,k')$. Then we have

$$D(s' e^{j\phi}) + k' N(s' e^{j\phi}) = 0$$

Since $D(.)$ and $N(.)$ are real polynomials, it follows that

$$D(s'^{*} e^{-j\phi}) + k' N(s'^{*} e^{-j\phi}) = 0.$$

where s'^{*} is the complex conjugate of s'. This shows that s'^{*} is a root of $\delta_{-\phi}(s_1,k')$. Since s' and s'^{*} have the same real parts, it follows that for a given k', $\delta_\phi(s_1, k')$ is Hurwitz stable if and only if $\delta_{-\phi}(s_1, k')$ is Hurwitz stable. Hence $K_\phi = K_{-\phi}$ so that

$$K = K_{-\alpha} \cap K_\phi.$$

Therefore, our original S Hurwitz stabilization problem reduces to the problem of determining those values of k, if any, for which the $-\alpha$ and ϕ Hurwitz stabilization problems are simultaneously solvable. As already mentioned the $-\alpha$ Hurwitz stabilization problem can be solved using the results of Chapter 4. However, in general, $\delta_\phi(s_1, k)$ is a complex polynomial. Hence, the results of Chapter 4 which are based on a generalized Hermite-Biehler Theorem for *real polynomials* cannot be used for solving the ϕ Hurwitz stabilization problem. This motivates us to derive a generalization of the Hermite-Biehler Theorem for complex polynomials. Such a derivation is presented in the next section.

8.3 Generalized Hermite-Biehler Theorem: Complex Polynomials

We have already presented a generalization of the Hermite-Biehler Theorem in Chapter 3 for polynomials with real coefficients. In this section, we present generalizations of the Hermite-Biehler Theorem applicable to polynomials with complex coefficients. This is carried out in two steps. First, in Subsection 8.3.1, we state the relationship between the net phase change of the "frequency response" of a complex polynomial as the frequency ω varies from $-\infty$ to ∞ and the numbers of its roots in the open left-half and open right-half planes. Thereafter, in Subsection 8.3.2, we derive generalizations of the Hermite Biehler Theorem for complex polynomials that are not necessarily Hurwitz.

The main difference between the development here and that in Chapter 3 is that in the latter case, the roots of the *real* polynomial were symmetrically distributed with respect to the real axis. Consequently in Chapter 3, it was enough to focus on the behaviour of $\delta(j\omega)$ as ω goes from 0 to ∞. In this section, however, it will be necessary to study the behaviour of $\delta(j\omega)$ over the entire frequency range from $\omega = -\infty$ to $\omega = \infty$.

8.3.1 Root Distribution and Net Accumulated Phase

In this subsection we state a fundamental relationship between the net accumulated phase of the frequency response of a complex polynomial and the difference between the numbers of roots of the polynomial in the open left-half and open right-half planes. Let $\mathcal{C}$ denote the complex plane, $\mathcal{C}^-$ the open left half plane and $\mathcal{C}^+$ the open right half plane.

First, we focus on polynomials without zeros on the imaginary axis. Consider a polynomial $\delta(s)$ of degree n:

$$\begin{aligned}\delta(s) &= \delta_0 + \delta_1 s + \delta_2 s^2 + \ldots + \delta_n s^n,\ \delta_i \in \mathcal{C},\ i = 0, 1, \cdots, n,\ \delta_n \neq 0\\ &\qquad \text{such that } \delta(j\omega) \neq 0,\ \forall \omega \in (-\infty,\ \infty).\end{aligned}$$

Let $p(\omega)$ and $q(\omega)$ be two functions defined pointwise by $p(\omega) = Re[\delta(j\omega)]$, $q(\omega) = Im[\delta(j\omega)]$.
With this definition, we have

$$\delta(j\omega) = p(\omega) + jq(\omega)\ \forall \omega.$$

Furthermore $\theta(\omega) \stackrel{\Delta}{=} \angle\delta(j\omega) = \arctan\left[\frac{q(\omega)}{p(\omega)}\right]$. Let $\Delta_{-\infty}^{\infty}\theta$ denote the net change in the argument $\theta(\omega)$ as ω increases from $-\infty$ to ∞ and let $l(\delta)$ and $r(\delta)$ denote the numbers of roots of $\delta(s)$ in $\mathcal{C}^-$ and $\mathcal{C}^+$ respectively.

Lemma 8.3.1. *Let $\delta(s)$ be a complex polynomial with no imaginary axis roots. Then*

$$\Delta_{-\infty}^{\infty}\theta = \pi(l(\delta) - r(\delta)).$$

8.3.2 Generalizations of the Hermite-Biehler Theorem

In this subsection, we derive generalizations of the Hermite-Biehler Theorem by first developing a procedure for systematically determining the net accumulated phase change of the "frequency response" of a polynomial. Proceeding as in Section 3.5.1, we first note that at any given frequency ω,

$$\frac{d\theta(\omega)}{d\omega} = \frac{\dot{q}(\omega)p(\omega) - \dot{p}(\omega)q(\omega)}{p^2(\omega) + q^2(\omega)} \tag{8.3.1}$$

Next, we note that for complex polynomials with *real* leading coefficients, the frequency response plot can be made to approach either the real or imaginary axis as $\omega \to \pm\infty$ by scaling the plot of $\delta(j\omega)$ with $\frac{1}{f(\omega)}$ where $f(\omega) = (1 + \omega^2)^{\frac{n}{2}}$. Accordingly, for complex polynomials with real leading coefficients, we define the normalized frequency response plot by

$$\delta_f(j\omega) = p_f(\omega) + jq_f(\omega) \text{ where } p_f(\omega) := \frac{p(\omega)}{(1+\omega^2)^{\frac{n}{2}}},\ q_f(\omega) := \frac{q(\omega)}{(1+\omega^2)^{\frac{n}{2}}}.$$

The subsequent development in this chapter makes use of this normalized frequency response plot for determining the net accumulated phase change as we move from $\omega = -\infty$ to $\omega = +\infty$. Note that the assumption about real leading coefficients is not restrictive because if the polynomial in question has a complex leading coefficient, one can divide the entire polynomial by its leading coefficient to obtain a new polynomial whose leading coefficient is unity (and hence real) and whose roots are the same as those of the original polynomial.

In view of the preceding discussion, we now consider a complex polynomial $\delta(s)$ of degree n with no zeros on the imaginary axis, and whose leading coefficient is real:

$$\begin{aligned}\delta(s) &= \delta_0 + \delta_1 s + \delta_2 s^2 + \ldots + \delta_n s^n, \ \delta_i \in \mathcal{C}, \ i = 0, 1, \ldots, n-1, \ \delta_n \in \mathcal{R}, \\ &\quad \delta_n \neq 0, \ \text{such that} \ \delta(j\omega) \neq 0, \ \forall \omega \in (-\infty, \ \infty).\end{aligned}$$

Let $p(\omega)$, $q(\omega)$, $p_f(\omega)$, $q_f(\omega)$ be as already defined and let

$$\omega_1 < \omega_2 < \cdots < \omega_{m-1}$$

be the real, distinct finite zeros of $q_f(\omega)$ with odd multiplicities. (Note that the function $q_f(\omega)$ does not change sign while passing through a real zero of even multiplicity; hence such zeros need not be considered while counting the net phase accumulation.) Also let us define $\omega_0 = -\infty$ and $\omega_m = +\infty$.

Then, as in Chapter 3, we can make the following simple observations:

1. If ω_i, ω_{i+1} are both zeros of $q_f(\omega)$ then

$$\Delta_{\omega_i}^{\omega_{i+1}} \theta = \frac{\pi}{2} \left[\operatorname{sgn}[p_f(\omega_i)] - \operatorname{sgn}[p_f(\omega_{i+1})]\right] \cdot \operatorname{sgn}[q_f(\omega_i^+)]. \quad (8.3.2)$$

2. If ω_i is not a zero of $q_f(\omega)$ while ω_{i+1} is a zero of $q_f(\omega)$, a situation possible only when $\omega_i = -\infty$ is a zero of $p_f(\omega)$ and n is odd, then

$$\begin{aligned}\Delta_{\omega_i}^{\omega_{i+1}} \theta &= -\frac{\pi}{2} \operatorname{sgn}[p_f(\omega_{i+1})] \cdot \operatorname{sgn}[q_f(\omega_{i+1}^-)] \\ &= \frac{\pi}{2} \operatorname{sgn}[p_f(\omega_{i+1})] \cdot \operatorname{sgn}[q_f(\omega_{i+1}^+)]. \quad (8.3.3)\end{aligned}$$

3. If ω_i is a zero of $q_f(\omega)$ while ω_{i+1} is not a zero of $q_f(\omega)$, a situation possible only when $\omega_{i+1} = \infty$ is a zero of $p_f(\omega)$ and n is odd, then

$$\Delta_{\omega_i}^{\omega_{i+1}} \theta = \frac{\pi}{2} \operatorname{sgn}[p_f(\omega_i)] \cdot \operatorname{sgn}[q_f(\omega_i^+)]. \quad (8.3.4)$$

4.

$$\operatorname{sgn}[q_f(\omega_{i+1}^+)] = -\operatorname{sgn}[q_f(\omega_i^+)], \ i = 0, \ 1, \ \cdots, \ m-2. \quad (8.3.5)$$

Using (8.3.5) repeatedly, we obtain

$$\operatorname{sgn}[q_f(\omega_i^+)] = (-1)^{m-i-1} \cdot \operatorname{sgn}[q_f(\omega_{m-1}^+)], \ i = 0, 1, \ldots, m-1. \quad (8.3.6)$$

Substituting (8.3.6) into (8.3.2), we see that if ω_i, ω_{i+1} are both zeros of $q_f(\omega)$ then

$$\begin{aligned}\Delta_{\omega_i}^{\omega_{i+1}} \theta = \frac{\pi}{2} &\left[\operatorname{sgn}[p_f(\omega_i)] - \operatorname{sgn}[p_f(\omega_{i+1})]\right] \cdot (-1)^{m-i-1} \\ &\cdot \operatorname{sgn}[q_f(\omega_{m-1}^+)]. \quad (8.3.7)\end{aligned}$$

The above observations enable us to state and prove the following Theorem concerning $l(\delta) - r(\delta)$. However, to simplify the theorem statement, we first define the "imaginary signature" $\sigma_i(\delta)$ of a complex polynomial $\delta(s)$ with real leading coefficient.

Definition 8.3.1. *Let $\delta(s)$ be any complex polynomial of degree n with real leading coefficient. Let $\omega_1 < \omega_2 < \cdots < \omega_{m-1}$ be the real, distinct finite zeros of $q_f(\omega)$ with odd multiplicities. Also define $\omega_0 = -\infty$ and $\omega_m = +\infty$. Then*

$$\sigma_i(\delta) := \begin{cases} \frac{1}{2}\{sgn[p_f(\omega_0)] \cdot (-1)^{m-1} + 2\sum_{i=1}^{m-1} sgn[p_f(\omega_i)] \cdot (-1)^{m-1-i} \\ -sgn[p_f(\omega_m)]\} \cdot sgn[q(\infty)] \\ \qquad \text{if } n \text{ is even} \\ \\ \frac{1}{2}\{2\sum_{i=1}^{m-1} sgn[p_f(\omega_i)] \cdot (-1)^{m-1-i}\} \cdot sgn[q(\infty)] \\ \qquad \text{if } n \text{ is odd} \end{cases} \tag{8.3.8}$$

defines the imaginary signature $\sigma_i(\delta)$ of $\delta(s)$.

Theorem 8.3.1. *Let $\delta(s)$ be a given complex polynomial of degree n with real leading coefficient and no roots on the $j\omega$-axis,* i.e., *the normalized plot $\delta_f(j\omega)$ does not pass through the origin. Then*

$$l(\delta) - r(\delta) = \sigma_i(\delta). \tag{8.3.9}$$

Proof. First, let us suppose that n is even. Then $\omega_0 = -\infty$ and $\omega_m = \infty$ are zeros of $q_f(\omega)$. By repeatedly using (8.3.7) to determine $\Delta_{-\infty}^{\infty}\theta$, applying Lemma 8.3.1, and then using the fact that $\text{sgn}[q_f(\omega_{m-1}^+)] = \text{sgn}[q(\infty)]$, it follows that $l(\delta) - r(\delta)$ is equal to the first expression in (8.3.8). Hence (8.3.9) holds for n even.

Next let us consider the case that n is odd. Then $\omega_0 = -\infty$ and $\omega_m = \infty$ are not zeros of $q_f(\omega)$. Hence,

$$\begin{aligned} \Delta_{-\infty}^{\infty}\theta &= \Delta_{-\infty}^{\omega_1}\theta + \sum_{i=1}^{m-2} \Delta_{\omega_i}^{\omega_{i+1}}\theta + \Delta_{\omega_{m-1}}^{\infty}\theta \\ &= \frac{\pi}{2}\text{sgn}[p_f(\omega_1)] \cdot \text{sgn}[q_f(\omega_1^+)] \\ &\quad + \sum_{i=1}^{m-2} \frac{\pi}{2}\,[\text{sgn}[p_f(\omega_i)] - \text{sgn}[p_f(\omega_{i+1})]] \cdot (-1)^{m-1-i}\text{sgn}[q_f(\omega_{m-1}^+)] \\ &\quad + \frac{\pi}{2}\text{sgn}[p_f(\omega_{m-1})] \cdot \text{sgn}[q_f(\omega_{m-1}^+)] \\ &\quad \text{(using (8.3.3), (8.3.7) and (8.3.4))}. \end{aligned} \tag{8.3.10}$$

From (8.3.6) it follows that

$$\text{sgn}[q_f(\omega_1^+)] = (-1)^{m-2} \cdot \text{sgn}[q_f(\omega_{m-1}^+)]. \tag{8.3.11}$$

Substituting (8.3.11) into (8.3.10), applying Lemma 8.3.1, and then using the fact that $\text{sgn}[q_f(\omega_{m-1}^+)] = \text{sgn}[q(\infty)]$, it follows that $l(\delta) - r(\delta)$ is equal to the second expression in (8.3.8). Hence (8.3.9) also holds for n odd. ♣

We now state the result analogous to Theorem 8.3.1 where $l(\delta) - r(\delta)$ of a complex polynomial $\delta(s)$ is to be determined using the values of the frequencies where $\delta_f(j\omega)$ crosses the imaginary axis. The proof is omitted since it follows along essentially the same lines as that of Theorem 8.3.1. Once again, to simplify the theorem statement, we first define the "real signature" $\sigma_r(\delta)$ of a complex polynomial $\delta(s)$ with real leading coefficient.

Definition 8.3.2. *Let $\delta(s)$ be any given complex polynomial of degree n with real leading coefficient. Let $\omega_1 < \omega_2 < \cdots < \omega_{m-1}$ be the real, distinct finite zeros of $p_f(\omega)$ with odd multiplicities. Also define $\omega_0 = -\infty$ and $\omega_m = +\infty$.*

Then

$$\sigma_r(\delta) := \begin{cases} -\frac{1}{2}\{2\sum_{i=1}^{m-1} sgn[q_f(\omega_i)] \cdot (-1)^{m-1-i}\} \cdot sgn[p(\infty)] \\ \qquad \text{if } n \text{ is even} \\ \\ -\frac{1}{2}\{sgn[q_f(\omega_0)] \cdot (-1)^{m-1} + 2\sum_{i=1}^{m-1} sgn[q_f(\omega_i)] \\ \cdot(-1)^{m-1-i} - sgn[q_f(\omega_m)]\} \cdot sgn[p(\infty)] \\ \qquad \text{if } n \text{ is odd} \end{cases} \tag{8.3.12}$$

defines the real signature $\sigma_r(\delta)$ of $\delta(s)$.

Theorem 8.3.2. *Let $\delta(s)$ be a given complex polynomial of degree n with real leading coefficient and no roots on the $j\omega$-axis,* i.e., *the normalized plot $\delta_f(j\omega)$ does not pass through the origin. Then*

$$l(\delta) - r(\delta) = \sigma_r(\delta). \tag{8.3.13}$$

Remark 8.3.1. Theorems 8.3.1 and 8.3.2 hold for both real as well as complex polynomials provided the latter have real leading coefficients. If we focus only on real polynomials $\delta(s)$, then the real zeros of $q_f(\omega)$ and $p_f(\omega)$ will be symmetrically distributed about the origin and $q_f(\omega)$ will have a zero at the origin of odd multiplicity. Thus in this case, it is enough to consider only the non-negative zeros of $q_f(\omega)$ as was done in Theorems 3.5.1 and 3.5.2. Furthermore, for a Hurwitz polynomial $l(\delta) - r(\delta) = n$. Using these facts, it is easy to show that (8.3.9), (8.3.13) effectively generalize (3.2), (3.3) respectively. Thus Theorems 8.3.1 and 8.3.2 are indeed generalizations of the Hermite-Biehler Theorem.

Theorems 8.3.1 and Theorem 8.3.2 require that the polynomial $\delta(s)$ have no roots on the imaginary axis. We now present the following refinement of Theorem 8.3.1, which states that the conclusion of Theorem 8.3.1 is valid even if $\delta(s)$ has $j\omega$-axis roots.

Theorem 8.3.3. *Let $\delta(s)$ be a given complex polynomial of degree n with real leading coefficient. Then*

$$l(\delta) - r(\delta) = \sigma_i(\delta). \tag{8.3.14}$$

Proof. Now, $\delta(s)$ can be factored as

$$\delta(s) = \delta^*(s)\delta^{'}(s)$$

where $\delta^*(s)$ contains all the $j\omega$ axis roots of $\delta(s)$, while $\delta^{'}(s)$ of degree $n^{'}$ has no $j\omega$ axis roots. Also let $\delta^*(s)$ be of the form

$$\prod_{i_m}(s - j\alpha_{i_m})^{n_{i_m}}, i_m = 1, 2, \cdots; \ \alpha_{i_m} \in \mathcal{R}, \text{ and } n_{i_m} \geq 0.$$

We use an inductive argument to show that multiplying $\delta^{'}(s)$ by $\delta^*(s)$ does not affect (8.3.14).

Let the induction index u be equal to one and consider[1]

$$\delta_1(s) = (s - j\alpha_1)^{n_1}\delta^{'}(s).$$

Define

$$\begin{aligned} \delta_1(j\omega) &:= p_1(\omega) + jq_1(\omega) \\ \text{and } \delta^{'}(j\omega) &:= p^{'}(\omega) + jq^{'}(\omega). \end{aligned}$$

We now consider four different cases, namely $n_1 = 4l_1$, $n_1 = 4l_1 + 1$, $n_1 = 4l_1 + 2$ and $n_1 = 4l_1 + 3$, where l_1 is some positive integer. These four cases correspond to the four different ways in which multiplication by $[j(\omega - \alpha_1)]^{n_1}$ affects the real and imaginary parts of $\delta^{'}(j\omega)$.

Case (I): $n_1 = 4l_1$

For $n_1 = 4l_1$, we have

$$p_1(\omega) = (\omega - \alpha_1)^{4l_1}p^{'}(\omega) \tag{8.3.15}$$

$$q_1(\omega) = (\omega - \alpha_1)^{4l_1}q^{'}(\omega). \tag{8.3.16}$$

Let $\omega_1 < \omega_2 < \cdots < \omega_{m-1}$ be the real, distinct finite zeros of $q^{'}_f(\omega)$ with odd multiplicities. Also define $\omega_0 = -\infty$ and $\omega_m = +\infty$. First let us assume that $\delta^{'}(s)$ has odd degree. Then, from Theorem 8.3.1, we have

[1] Note that in this proof, $\delta_i(s)$ is the polynomial being considered in the ith inductive step. This should not be confused with the coefficients δ_i used earlier to define the polynomial $\delta(s)$.

$$\begin{aligned} l(\delta') - r(\delta') &= \sigma_i(\delta') \\ &= \{\sum_{i=1}^{m-1} \text{sgn}[p'_f(\omega_i)] \cdot (-1)^{m-1-i}\} \cdot \text{sgn}[q'(\infty)]. \end{aligned} \quad (8.3.17)$$

Now, from (8.3.16), it follows that ω_i, $i = 1, 2, \cdots, m-1$ are also the real, distinct finite zeros of $q_{1_f}(\omega)$ with odd multiplicities. Furthermore, from (8.3.15) and (8.3.16), we have

$$\begin{aligned} \text{sgn}[p'_f(\omega_i)] &= \text{sgn}[p_{1_f}(\omega_i)], \ i = 1, 2, \cdots, m-1 \\ \text{sgn}[q'(\infty)] &= \text{sgn}[q_1(\infty)]. \end{aligned}$$

Since $l(\delta_1) - r(\delta_1) = l(\delta') - r(\delta')$, it follows that (8.3.14) is true for $\delta_1(s)$ of odd degree. The fact that (8.3.14) is also true for $\delta_1(s)$ of even degree, can be verified by proceeding along exactly the same lines.

Case (II): $n_1 = 4l_1 + 1$

For $n_1 = 4l_1 + 1$, we have

$$p_1(\omega) = -(\omega - \alpha_1)^{4l_1+1} q'(\omega) \quad (8.3.18)$$

$$q_1(\omega) = (\omega - \alpha_1)^{4l_1+1} p'(\omega). \quad (8.3.19)$$

Once again let us assume that $\delta'(s)$ has odd degree. Let $\omega_1 < \omega_2 < \cdots < \omega_{m-1}$ be the real, distinct finite zeros of $p'_f(\omega)$ with odd multiplicities. Also define $\omega_0 = -\infty$ and $\omega_m = +\infty$. Then, from Theorem 8.3.2, we have

$$\begin{aligned} l(\delta') - r(\delta') &= \sigma_r(\delta') \\ &= -\frac{1}{2}\{\text{sgn}[q'_f(\omega_0)] \cdot (-1)^{m-1} + 2\sum_{i=1}^{m-1} \text{sgn}[q'_f(\omega_i)] \cdot (-1)^{m-1-i} \\ &\quad -\text{sgn}[q'_f(\omega_m)]\} \cdot \text{sgn}[p'(\infty)] \end{aligned} \quad (8.3.20)$$

Let us assume that $\alpha_1 \in (\omega_c, \omega_{c+1})$. Then, from (8.3.18) and (8.3.19), we have

$$\left.\begin{aligned} \text{sgn}[q'_f(\omega_i)] &= \text{sgn}[p_{1_f}(\omega_i)], \ i = 0, 1, \ldots, c \\ \text{sgn}[p_{1_f}(\alpha_1)] &= 0 \\ \text{sgn}[q'_f(\omega_i)] &= -\text{sgn}[p_{1_f}(\omega_i)], \ i = c+1, c+2, \ldots, m \\ \text{sgn}[p'(\infty)] &= \text{sgn}[q_1(\infty)]. \end{aligned}\right\} \quad (8.3.21)$$

Since n' is odd and $n_1 = 4l_1 + 1$, it follows that $\delta_1(s)$ has even degree. Since $l(\delta_1) - r(\delta_1) = l(\delta') - r(\delta')$, from (8.3.20), we have

$$\begin{aligned} l(\delta_1) - r(\delta_1) &= \sigma_r(\delta') \\ &= -\frac{1}{2}\{\text{sgn}[q'_f(\omega_0)] \cdot (-1)^{m-1} + 2\text{sgn}[q'_f(\omega_1)] \cdot (-1)^{m-2} \\ &\quad \vdots \end{aligned}$$

$$
\begin{aligned}
&+2\text{sgn}[q_f^{'}(\omega_c)]\cdot(-1)^{m-(1+c)}+2\text{sgn}[p_{1_f}(\alpha_1)]\cdot(-1)^{m-(c+2)}\\
&+2\text{sgn}[q_f^{'}(\omega_{c+1})]\cdot(-1)^{m-(c+2)}\\
&\vdots\\
&+2\text{sgn}[q_f^{'}(\omega_{m-1})]-\text{sgn}[q_f^{'}(\omega_m)]\}\cdot\text{sgn}[p^{'}(\infty)] \qquad (8.3.22)\\
&(\text{ since } p_{1_f}(\alpha_1)=0).
\end{aligned}
$$

Now, from (8.3.19), it follows that $\omega_1, \omega_2, \ldots, \omega_c, \alpha_1, \omega_{c+1}, \ldots, \omega_{m-1}$ are the real, distinct finite zeros of $q_{1_f}(\omega)$ with odd multiplicities. From (8.3.22) and using (8.3.21), we have

$$
\begin{aligned}
l(\delta_1)-r(\delta_1) &= \frac{1}{2}\{\text{sgn}[p_{1_f}(\omega_0)]\cdot(-1)^m+2\sum_{i=1}^{c}\text{sgn}[p_{1_f}(\omega_i)]\cdot(-1)^{(m-i)}\\
&\quad +2\text{sgn}[p_{1_f}(\alpha_1)]\cdot(-1)^{m-(c+1)}+2\sum_{i=c+1}^{m-1}\text{sgn}[p_{1_f}(\omega_i)]\\
&\quad \cdot(-1)^{m-(1+i)}-\text{sgn}[p_{1_f}(\omega_m)]\}\cdot\text{sgn}[q_1(\infty)]\\
&= \sigma_i(\delta_1).
\end{aligned}
$$

which shows that (8.3.14) is true for $\delta_1(s)$ of even degree. The fact that (8.3.14) holds for $\delta_1(s)$ of odd degree, can be verified by proceeding along exactly the same lines.

Case (III): $n_1 = 4l_1 + 2$

For $n_1 = 4l_1 + 2$, we have

$$
\begin{aligned}
p_1(\omega) &= -(\omega-\alpha_1)^{4l_1+2}p^{'}(\omega) \qquad (8.3.23)\\
q_1(\omega) &= -(\omega-\alpha_1)^{4l_1+2}q^{'}(\omega). \qquad (8.3.24)
\end{aligned}
$$

Let ω_i, $i = 0, 1, \cdots, m$ be as already defined in Case (I). Once again, we assume that $\delta^{'}(s)$ has odd degree. Then, from Theorem 8.3.1, we have

$$
\begin{aligned}
l(\delta^{'})-r(\delta^{'}) &= \sigma_i(\delta^{'})\\
&= \frac{1}{2}\{2\sum_{i=1}^{m-1}\text{sgn}[p_f^{'}(\omega_i)]\cdot(-1)^{m-1-i}\}\cdot\text{sgn}[q^{'}(\infty)].
\end{aligned}
$$

Now, from (8.3.24), it follows that ω_i, $i = 1, 2, \cdots, m-1$ are also the real, distinct finite zeros of $q_{1_f}(\omega)$ with odd multiplicities. Furthermore, from (8.3.23) and (8.3.24), we have

$$
\begin{aligned}
\text{sgn}[p_f^{'}(\omega_i)] &= -\text{sgn}[p_{1_f}(\omega_i)],\ i=1,2,\cdots,m-1\\
\text{sgn}[q^{'}(\infty)] &= -\text{sgn}[q_1(\infty)].
\end{aligned}
$$

Since $l(\delta_1) - r(\delta_1) = l(\delta^{'}) - r(\delta^{'})$, it follows that (8.3.14) is true for $\delta_1(s)$ of odd degree. The fact that (8.3.14) is also true for $\delta_1(s)$ of even degree, can be verified by proceeding along exactly the same lines.

Case (IV): $n_1 = 4l_1 + 3$

For $n_1 = 4l_1 + 3$, we have

$$\begin{aligned} p_1(\omega) &= (\omega - \alpha_1)^{4l_1+3} q'(\omega) & (8.3.25) \\ q_1(\omega) &= -(\omega - \alpha_1)^{4l_1+3} p'(\omega). & (8.3.26) \end{aligned}$$

Once again let us assume that $\delta'(s)$ has odd degree. Then, from Theorem 8.3.2, we have

$$\begin{aligned} l(\delta') - r(\delta') &= \sigma_r(\delta') \\ &= -\frac{1}{2}\{\text{sgn}[q'_f(\omega_0)] \cdot (-1)^{m-1} + 2\sum_{i=1}^{m-1} \text{sgn}[q'_f(\omega_i)] \cdot (-1)^{m-1-i} \\ &\quad -\text{sgn}[q'_f(\omega_m)]\} \cdot \text{sgn}[p'(\infty)] \qquad (8.3.27) \end{aligned}$$

where ω_i, $i = 0, 1, \cdots, m$ are as already defined in Case (II).

Let us assume that $\alpha_1 \in (\omega_c, \omega_{c+1})$. Then, from (8.3.25) and (8.3.26), we have

$$\left.\begin{aligned} \text{sgn}[q'_f(\omega_i)] &= -\text{sgn}[p_{1_f}(\omega_i)],\ i = 0, 1, \ldots, c \\ \text{sgn}[p_{1_f}(\alpha_1)] &= 0 \\ \text{sgn}[q'_f(\omega_i)] &= \text{sgn}[p_{1_f}(\omega_i)],\ i = c+1, c+2, \ldots, m \\ \text{sgn}[p'(\infty)] &= -\text{sgn}[q_1(\infty)]. \end{aligned}\right\} \qquad (8.3.28)$$

Since n' is odd and $n_1 = 4l_1 + 3$, it follows that $\delta_1(s)$ has even degree. Since $l(\delta_1) - r(\delta_1) = l(\delta') - r(\delta')$, from (8.3.27), we have

$$\begin{aligned} l(\delta_1) - r(\delta_1) &= \sigma_r(\delta') \\ &= -\frac{1}{2}\{\text{sgn}[q'_f(\omega_0)] \cdot (-1)^{m-1} + 2\text{sgn}[q'_f(\omega_1)] \cdot (-1)^{m-2} \\ &\quad \vdots \\ &\quad +2\text{sgn}[q'_f(\omega_c)] \cdot (-1)^{m-(1+c)} + 2\text{sgn}[p_{1_f}(\alpha_1)] \cdot (-1)^{m-(c+2)} \\ &\quad +2\text{sgn}[q'_f(\omega_{c+1})] \cdot (-1)^{m-(c+2)} \\ &\quad \vdots \\ &\quad +2\text{sgn}[q'_f(\omega_{m-1})] - \text{sgn}[q'_f(\omega_m)]\} \cdot \text{sgn}[p'(\infty)] \qquad (8.3.29) \\ &\quad (\text{ since } p_{1_f}(\alpha_1) = 0). \end{aligned}$$

Now, from (8.3.26), it follows that $\omega_1, \omega_2, \ldots, \omega_c, \alpha_1, \omega_{c+1}, \cdots, \omega_{m-1}$ are the real, distinct finite zeros of $q_{1_f}(\omega)$ with odd multiplicities.

From (8.3.29) and using (8.3.28), we have

$$\begin{aligned} l(\delta_1) - r(\delta_1) &= \frac{1}{2}\{\text{sgn}[p_{1_f}(\omega_0)] \cdot (-1)^m + 2\sum_{i=1}^{c} \text{sgn}[p_{1_f}(\omega_i)] \cdot (-1)^{(m-i)} \\ &\quad +2\text{sgn}[p_{1_f}(\alpha_1)] \cdot (-1)^{m-(c+1)} + 2\sum_{i=c+1}^{m-1} \text{sgn}[p_{1_f}(\omega_i)] \\ &\quad \cdot(-1)^{m-(i+1)} - \text{sgn}[p_{1_f}(\omega_m)]\} \cdot \text{sgn}[q_1(\infty)] \\ &= \sigma_i(\delta_1) \end{aligned}$$

which shows that (8.3.14) is true for $\delta_1(s)$ of even degree. The fact that (8.3.14) holds for $\delta_1(s)$ of odd degree, can be verified by proceeding along exactly the same lines. This completes the first step of the induction argument. Note that by using similar arguments, it is possible to show that $\delta_1(s)$ also satisfies (8.3.13).

Now let $u = k$ and consider

$$\delta_k(s) = \prod_{i_m=1}^{k} (s - j\alpha_{i_m})^{n_{i_m}} \delta'(s).$$

Assume that (8.3.14), (8.3.13) are true for $\delta_k(s)$ (inductive assumption). Now

$$\delta_{k+1}(s) = \prod_{i_m=1}^{k+1} (s - j\alpha_{i_m})^{n_{i_m}} \delta'(s) \tag{8.3.30}$$

$$= (s - j\alpha_{k+1})^{n_{k+1}} \delta_k(s). \tag{8.3.31}$$

Define

$$\begin{aligned} \delta_k(j\omega) &= p_k(\omega) + jq_k(\omega) \\ \delta_{k+1}(j\omega) &= p_{k+1}(\omega) + jq_{k+1}(\omega). \end{aligned}$$

Once again the proof can be completed by considering four different cases, namely $n_{k+1} = 4l_{k+1}$, $n_{k+1} = 4l_{k+1} + 1$, $n_{k+1} = 4l_{k+1} + 2$ and $n_{k+1} = 4l_{k+1} + 3$, where l_{k+1} is some positive integer. Since each of these cases can be handled by proceeding along similar lines, we do not treat all of the cases here. Instead, we focus on a representative case, say $n_{k+1} = 4l_{k+1} + 1$, and provide a detailed treatment for it.

Now, for $n_{k+1} = 4l_{k+1} + 1$, we have

$$p_{k+1}(\omega) = -(\omega - \alpha_{k+1})^{4l_{k+1}+1} q_k(\omega) \tag{8.3.32}$$

$$q_{k+1}(\omega) = (\omega - \alpha_{k+1})^{4l_{k+1}+1} p_k(\omega). \tag{8.3.33}$$

First let us assume that $\delta_k(s)$ has odd degree. Let $\omega_1 < \omega_2 < \cdots < \omega_{m-1}$ be the real, distinct finite zeros of $p_{k_f}(\omega)$ with odd multiplicities. Also define $\omega_0 = -\infty$ and $\omega_m = +\infty$. Then, since (8.3.13) holds for $\delta_k(s)$, we have

$$\begin{aligned}
l(\delta_k) - r(\delta_k) &= \sigma_r(\delta_k) \\
&= -\frac{1}{2}\{\text{sgn}[q_{k_f}(\omega_0)] \cdot (-1)^{m-1} + 2\sum_{i=1}^{m-1} \text{sgn}[q_{k_f}(\omega_i)] \\
&\quad \cdot (-1)^{m-1-i} - \text{sgn}[q_{k_f}(\omega_m)]\} \cdot \text{sgn}[p_k(\infty)] \qquad (8.3.34)
\end{aligned}$$

Let us now assume that $\alpha_{k+1} \in (\omega_c, \ \omega_{c+1})$. Then, from (8.3.32) and (8.3.33), we have

$$\begin{aligned}
\text{sgn}[q_{k_f}(\omega_i)] &= \text{sgn}[p_{k+1_f}(\omega_i)], \ i = 0, \ 1, \ \ldots, \ c \\
\text{sgn}[p_{k+1_f}(\alpha_{k+1})] &= 0 \\
\text{sgn}[q_{k_f}(\omega_i)] &= -\text{sgn}[p_{k+1_f}(\omega_i)], \ i = c+1, \ c+2, \ \ldots, \ m \\
\text{sgn}[p_k(\infty)] &= \text{sgn}[q_{k+1}(\infty)].
\end{aligned} \qquad (8.3.35)$$

Since $\delta_k(s)$ has odd degree and $n_{k+1} = 4l_{k+1} + 1$, it follows that $\delta_{k+1}(s)$ has even degree. Since $l(\delta_{k+1}) - r(\delta_{k+1}) = l(\delta_k) - r(\delta_k)$, from (8.3.34), we have

$$\begin{aligned}
l(\delta_{k+1}) - r(\delta_{k+1}) &= \sigma_r(\delta_k) \\
&= -\frac{1}{2}\{\text{sgn}[q_{k_f}(\omega_0)] \cdot (-1)^{m-1} + 2\text{sgn}[q_{k_f}(\omega_1)] \cdot (-1)^{m-2} \\
&\quad \vdots \\
&\quad +2\text{sgn}[q_{k_f}(\omega_c)] \cdot (-1)^{m-(c+1)} + 2\text{sgn}[p_{k+1_f}(\alpha_{k+1})] \\
&\quad \cdot (-1)^{m-(c+2)} + 2\text{sgn}[q_{k_f}(\omega_{c+1})] \cdot (-1)^{m-(c+2)} \\
&\quad \vdots \\
&\quad +2\text{sgn}[q_{k_f}(\omega_{m-1})] - \text{sgn}[q_{k_f}(\omega_m)]\} \\
&\quad \cdot \text{sgn}[p_k(\infty)] \ (\text{ since } p_{k+1_f}(\alpha_{k+1}) = 0). \qquad (8.3.36)
\end{aligned}$$

Now, from (8.3.33), it follows that $\omega_1, \omega_2, \cdots, \omega_c, \alpha_{k+1}, \omega_{c+1}, \cdots, \omega_{m-1}$ are the real, distinct finite zeros of $q_{k+1_f}(\omega)$ with odd multiplicities.

From (8.3.36) and using (8.3.35), we have

$$\begin{aligned}
l(\delta_{k+1}) - r(\delta_{k+1}) &= \frac{1}{2}\{\text{sgn}[p_{k+1_f}(\omega_0)] \cdot (-1)^m + 2\sum_{i=1}^{c} \text{sgn}[p_{k+1_f}(\omega_i)] \\
&\quad \cdot (-1)^{m-i} + 2\text{sgn}[p_{k+1_f}(\alpha_{k+1})] \cdot (-1)^{m-(c+1)} \\
&\quad +2\sum_{i=c+1}^{m-1} \text{sgn}[p_{k+1_f}(\omega_i)] \cdot (-1)^{m-(i+1)} \\
&\quad -\text{sgn}[p_{k+1_f}(\omega_m)]\} \cdot \text{sgn}[q_{k+1}(\infty)] \\
&= \sigma_i(\delta_{k+1})
\end{aligned}$$

which shows that (8.3.14) is true for $\delta_{k+1}(s)$ of even degree. The fact that (8.3.14) holds for $\delta_{k+1}(s)$ of odd degree, can be verified by proceeding along exactly the same lines. This completes the induction argument and hence the proof. ♣

Remark 8.3.2. By using similar arguments as in the proof of Theorem 8.3.3, it can be shown that the conclusion of Theorem 8.3.2 is valid even when $\delta(s)$ has $j\omega$ axis roots.

8.4 Stabilization with Damping Margin Using a Constant Gain

We now return to the feedback control system shown in Fig. 8.3. where the plant $G(s) = \frac{N(s)}{D(s)}$ is to be stabilized using a constant feedback gain. The closed loop characteristic polynomial $\delta(s,k)$ is given by

$$\delta(s,k) = D(s) + kN(s).$$

In this section, we will focus on the constant gain ϕ Hurwitz stabilization problem. Accordingly, we set $s = s_1 e^{j\phi}$ and write

$$\begin{aligned} \delta(s,k) &= \delta_\phi(s_1,k) \\ &= D(s_1 e^{j\phi}) + kN(s_1 e^{j\phi}). \end{aligned}$$

As discussed in section 8.2, the ϕ Hurwitz stability of $\delta(s,k)$ is equivalent to the Hurwitz stability of $\delta_\phi(s_1,k)$. Hence, our objective now becomes to determine those values of k, if any, for which $\delta_\phi(s_1,k)$ is Hurwitz stable.

To this end, we consider

$$\begin{aligned} N(s_1 e^{j\phi}) &= (a_0 + jb_0) + (a_1 + jb_1)s_1 + \cdots + (a_{m-1} + jb_{m-1})s_1^{m-1} \\ &\quad +(a_m + jb_m)s_1^m, \ a_m + jb_m \neq 0 \\ D(s_1 e^{j\phi}) &= (c_0 + jd_0) + (c_1 + jd_1)s_1 + \cdots + (c_{n-1} + jd_{n-1})s_1^{n-1} \\ &\quad +(c_n + jd_n)s_1^n, \ c_n + jd_n \neq 0 \end{aligned}$$

where n, the degree of $D(s_1 e^{j\phi})$, is greater than or equal to m, the degree of $N(s_1 e^{j\phi})$. Next, the polynomials $N(s_1 e^{j\phi})$ and $D(s_1 e^{j\phi})$ are scaled to make the leading coefficient of $D(s_1 e^{j\phi})$ equal to unity, *i.e.*, real. This results in two new polynomials $N'(s_1)$ and $D'(s_1)$ defined as

$$\begin{aligned} N'(s_1) &\triangleq \frac{1}{c_n + jd_n} N(s_1 e^{j\phi}) \\ &= (a_0' + jb_0') + (a_1' + jb_1')s_1 + \cdots + (a_{m-1}' + jb_{m-1}')s_1^{m-1} \\ &\quad +(a_m' + jb_m')s_1^m \end{aligned}$$

where

$$a'_i = \frac{a_i c_n + b_i d_n}{c_n^2 + d_n^2}$$

and

$$b'_i = \frac{-a_i d_n + b_i c_n}{c_n^2 + d_n^2}, \ i = 0, 1, \cdots, m$$

and

$$\begin{aligned} D'(s_1) &\triangleq \frac{1}{c_n + jd_n} D(s_1 e^{j\phi}) \\ &= (c'_0 + jd'_0) + (c'_1 + jd'_1)s_1 + \cdots + (c'_{n-1} + jd'_{n-1})s_1^{n-1} + s_1^n \end{aligned}$$

where

$$c'_i = \frac{c_i c_n + d_i d_n}{c_n^2 + d_n^2}$$

and

$$d'_i = \frac{-c_i d_n + d_i c_n}{c_n^2 + d_n^2}, \ i = 0, 1, \cdots, n-1.$$

We now define

$$\begin{aligned} \delta'(s_1, k) &\triangleq \frac{1}{(c_n + jd_n)} \delta_\phi(s_1, k) \\ \text{so that } \delta'(s_1, k) &= D'(s_1) + kN'(s_1). \end{aligned} \tag{8.4.1}$$

Clearly, dividing $\delta_\phi(s_1, k)$ by $(c_n + jd_n)$ results in a new polynomial $\delta'(s_1, k)$ with a real leading coefficient and having the same roots as $\delta_\phi(s_1, k)$.

It is clear from (8.4.1) that, in general, both the real and imaginary parts of $\delta'(j\omega, k)$ will depend on k, thereby making it very difficult, if not impossible, to use Theorem 8.3.1 or Theorem 8.3.2 to determine the ranges of k that make $\delta'(s_1, k)$ Hurwitz stable. The situation can be remedied by proceeding as in Chapter 4 to first construct a polynomial, say $D^*(s_1)$, such that the roots of the imaginary part of $\delta'(j\omega, k)D^*(j\omega)$ are independent of k, and then applying Theorem 8.3.3 to determine the values of k, if any, for which $\delta'(s_1, k)$ is Hurwitz stable. To do so, we consider the following "real-imaginary" decompositions[2] of $N'(s_1)$ and $D'(s_1)$:

$$\begin{aligned} N'(s_1) &= N'_R(s_1) + N'_I(s_1) \\ D'(s_1) &= D'_R(s_1) + D'_I(s_1) \end{aligned}$$

where

$$\begin{aligned} N'_R(s_1) &= a'_0 + jb'_1 s_1 + a'_2 s_1^2 + jb'_3 s_1^3 + \cdots \\ N'_I(s_1) &= jb'_0 + a'_1 s_1 + jb'_2 s_1^2 + a'_3 s_1^3 + \cdots \\ D'_R(s_1) &= c'_0 + jd'_1 s_1 + c'_2 s_1^2 + jd'_3 s_1^3 + \cdots \\ D'_I(s_1) &= jd'_0 + c'_1 s_1 + jd'_2 s_1^2 + c'_3 s_1^3 + \cdots. \end{aligned}$$

[2] This terminology refers to the fact that if the polynomial being decomposed is evaluated at $s_1 = j\omega$, then one of the components evaluates out to the real part while the other component evaluates out to the imaginary part.

Define

$$D^*(s_1) = D_R'(s_1) - D_I'(s_1).$$

Multiplying $\delta'(s_1,k)$ by $D^*(s_1)$ and considering the resulting polynomial, it is clear that

$$\begin{aligned} l(\delta'(s_1,k)D^*(s_1)) - r(\delta'(s_1,k)D^*(s_1)) &= l(\delta'(s_1,k)) - r(\delta'(s_1,k)) \\ &\quad + l(D^*(s_1)) - r(D^*(s_1)). \end{aligned}$$

Now, $\delta'(s_1,k)$ of degree n is Hurwitz if and only if $l(\delta'(s_1,k)) = n$ and $r(\delta'(s_1,k)) = 0$. Furthermore, from Theorem 8.3.3

$$\sigma_i(\delta'(s_1,k)D^*(s_1)) = l(\delta'(s_1,k)D^*(s_1)) - r(\delta'(s_1,k)D^*(s_1)).$$

Therefore we have the following.

Lemma 8.4.1. *$\delta'(s_1,k)$ is Hurwitz if and only if*

$$\sigma_i(\delta'(s_1,k)D^*(s_1)) = n + l(D^*(s_1)) - r(D^*(s_1)). \tag{8.4.2}$$

Our task now is to determine those values of k, if any, for which (8.4.2) holds. Evaluating $\delta'(s_1,k)$ and $D^*(s_1)$ at $s_1 = j\omega$, and making use of the real-imaginary decompositions introduced earlier, we obtain

$$\begin{aligned} \delta'(j\omega,k) &= [D_R'(j\omega) + D_I'(j\omega)] + k[N_R'(j\omega) + N_I'(j\omega)] \\ D^*(j\omega) &= D_R'(j\omega) - D_I'(j\omega). \end{aligned}$$

Since $D_R'(j\omega)$, $N_R'(j\omega)$ are purely real, while $D_I'(j\omega)$, $N_I'(j\omega)$ are purely imaginary, we can write

$$\delta'(j\omega,k)D^*(j\omega) = p(\omega,k) + jkq(\omega)$$

where

$$\begin{aligned} p(\omega,k) &= p_1(\omega) + kp_2(\omega) \\ p_1(\omega) &= D_R'^2(j\omega) - D_I'^2(j\omega) \\ p_2(\omega) &= D_R'(j\omega)N_R'(j\omega) - D_I'(j\omega)N_I'(j\omega) \\ q(\omega) &= \frac{1}{j}[D_R'(j\omega)N_I'(j\omega) - D_I'(j\omega)N_R'(j\omega)]. \end{aligned}$$

Since $\delta'(s_1,k)D^*(s_1)$ has degree $2n$, the normalized functions $p_f(\omega,k)$, $q_f(\omega)$ that appear in the definition of $\sigma_i(\delta'(s_1,k)D^*(s_1))$ become:

$$\begin{aligned} p_f(\omega,k) &= \frac{p(\omega,k)}{(1+\omega^2)^n} \\ q_f(\omega) &= \frac{q(\omega)}{(1+\omega^2)^n}. \end{aligned}$$

Since the case of $k = 0$ can be handled trivially by checking whether the open loop plant is ϕ Hurwitz stable or not, let us assume that $k \neq 0$ so that the zeros of the imaginary part of $\delta'(j\omega, k)D^*(j\omega)$ are independent of k. We now proceed in the spirit of Section 4.2 and introduce some definitions before stating our main result.

Definition 8.4.1. *Let $q_f(\omega)$ be as already defined. Let $\omega_1 < \omega_2 < \cdots < \omega_{m-1}$ be the real, distinct finite zeros of $q_f(\omega)$ with odd multiplicities*[3]. *Also define $\omega_0 = -\infty$ and $\omega_m = +\infty$. Define*

$$A \;:=\; \{\{i_0, i_1, \cdots, i_m, i_{m+1}\}\}$$

where

$$\text{for } t = 0, 1, \cdots, m,\ i_t \;\in\; \begin{cases} \{0\} & \text{if } j\omega_t \text{ is a } j\omega\text{-axis root of } D^*(s_1) \\ \{sgn[p_1(\omega_t)]\} & \text{if } p_2(\omega_t) = 0 \text{ but } p_1(\omega_t) \neq 0 \\ \{-1, 1\} & \text{otherwise} \end{cases}$$

and

$$i_{m+1} \;\in\; \{-1, 1\}.$$

In other words, A is the set of all possible strings of 1's, 0's and -1's of length $m+2$ subject to the following restrictions. For every $\mathcal{I} = \{i_0, i_i, \cdots, i_m, i_{m+1}\} \in A$, i_{m+1} is either -1 or $+1$. For all other $t = 0, \cdots, m$, i_t is equal to 0 if the corresponding $j\omega_t$ is a $j\omega$-axis root of $D^*(s_1)$[4] *; if $p_2(\omega_t) = 0$ but $p_1(\omega_t) \neq 0$ for some $t = 0, 1, 2, \cdots, m$, then $i_t = sgn[p_1(\omega_t)]$; and finally i_t is equal to -1 or $+1$ if neither of these two conditions is satisfied.*

Definition 8.4.2. *Let $q(\omega)$ and $q_f(\omega)$ be as already defined. Let $\omega_1 < \omega_2 < \cdots < \omega_{m-1}$ be the real, distinct finite zeros of $q_f(\omega)$ with odd multiplicities. Also define $\omega_0 = -\infty$ and $\omega_m = +\infty$. For each string $\mathcal{I} = \{i_0, i_1, \cdots, i_m, i_{m+1}\}$ in A, let $\gamma(\mathcal{I})$ denote the "imaginary signature" associated with the string $\mathcal{I}$ defined by*

$$\begin{aligned} \gamma(\mathcal{I}) : \;&=\; \frac{1}{2}\{i_0 \cdot (-1)^{m-1} + 2\sum_{r=1}^{m-1} i_r \cdot (-1)^{m-1-r} - i_m\} \\ &\quad \cdot sgn[i_{m+1} \cdot q(\infty)]. \end{aligned} \tag{8.4.3}$$

Definition 8.4.3. *The set of strings in A with a prescribed imaginary signature $\gamma = \psi$ is denoted by $A(\psi)$. We also define the set of feasible strings for the constant gain ϕ Hurwitz stabilization problem as*

---

[3] Note that these zeros are independent of k.

[4] Note that we are only interested in determining those values of k for which $\delta'(s_1, k)$ is *Hurwitz stable*. For such values of k, $\delta'(j\omega, k)D^*(j\omega)$ will be zero if and only if $D^*(j\omega) = 0$. This fact has been explicitly incorporated in the definition of A.

$$F^* = \{\mathcal{I}|\mathcal{I} = \{i_0, i_1, \cdots, i_m, i_{m+1}\} \in A(n + l(D^*(s_1)) - r(D^*(s_1)))\}.$$

In other words F^* *is the set of strings in* $A(n + l(D^*(s_1)) - r(D^*(s_1)))$.

We are now ready to state the main result of this section.

Theorem 8.4.1. *The constant gain feedback* ϕ *Hurwitz stabilization problem is solvable for a given plant with transfer function* $G(s)$ *if and only if the following conditions hold:*

(I) $G(s_1 e^{j\phi}) = \frac{N(s_1 e^{j\phi})}{D(s_1 e^{j\phi})}$ *is a Hurwitz stable plant where* $N(.)$ *and* $D(.)$ *are as already defined;*

or

(II)

(1) F^* *is not empty where* F^* *is as already defined,* i.e., *at least one feasible string exists*

and

(2) There exists a string $\mathcal{I} = \{i_0, i_1, \cdots, i_m, i_{m+1}\} \in F^*$ *such that*

(a) if $i_{m+1} = -1$ *then for* $t = 1, 2, \cdots, m$

$$\max_{i_t \in \mathcal{I}, i_t \cdot sgn[p_2(\omega_t)] = 1} \left[-\frac{1}{G(j\omega_t e^{j\phi})} \right] < \min \left[0, \min_{i_t \in \mathcal{I}, i_t \cdot sgn[p_2(\omega_t)] = -1} \left[-\frac{1}{G(j\omega_t e^{j\phi})} \right] \right].$$

In this case, the set of the corresponding stabilizing gains is given by

$$K_- = \left(\max_{i_t \in \mathcal{I}, i_t \cdot sgn[p_2(\omega_t)] = 1} \left[-\frac{1}{G(j\omega_t e^{j\phi})} \right], \min \left[0, \min_{i_t \in \mathcal{I}, i_t \cdot sgn[p_2(\omega_t)] = -1} \left[-\frac{1}{G(j\omega_t e^{j\phi})} \right] \right] \right)$$

(b) if $i_{m+1} = 1$ *then for* $t = 1, 2, \cdots, m$

$$\max \left[0, \max_{i_t \in \mathcal{I}, i_t \cdot sgn[p_2(\omega_t)] = 1} \left[-\frac{1}{G(j\omega_t e^{j\phi})} \right] \right] < \min_{i_t \in \mathcal{I}, i_t \cdot sgn[p_2(\omega_t)] = -1} \left[-\frac{1}{G(j\omega_t e^{j\phi})} \right].$$

In this case, the set of the corresponding stabilizing gains is given by

$$K_+ = \left(\max \left[0, \max_{i_t \in \mathcal{I}, i_t \cdot sgn[p_2(\omega_t)] = 1} \left[-\frac{1}{G(j\omega_t e^{j\phi})} \right] \right], \min_{i_t \in \mathcal{I}, i_t \cdot sgn[p_2(\omega_t)] = -1} \left[-\frac{1}{G(j\omega_t e^{j\phi})} \right] \right)$$

where $p_2(\omega)$ and $\omega_0, \omega_1, \omega_2, \cdots, \omega_m$ are as already defined.

Furthermore, if Condition (I) holds or if the Condition (II)(2)(a) or (II)(2)(b) is satisfied by the feasible strings $\mathcal{I}_1, \mathcal{I}_2, \cdots, \mathcal{I}_s \in F^*$, then the set of all ϕ Hurwitz stabilizing gains is given by $K_\phi = \cup_{r=0}^{s} K_r$ where $K_0 = \{0\}$ if (I) holds and K_0 is the empty set Φ otherwise, and K_r, $r \neq 0$ is the set of ϕ Hurwitz stabilizing gains corresponding to the feasible string $\mathcal{I}_r$.

Proof. We now consider three different cases, namely $k = 0$, $k < 0$ and $k > 0$.

Case (I) $k = 0$:
It is a trivial observation that $G(s_1 e^{j\phi})$ is stabilizable by $k = 0$ if and only if $G(s_1 e^{j\phi})$ is a Hurwitz stable plant. This immediately leads to Condition (I) in the theorem statement.

Case (II) $k < 0$:
Now from (8.4.2), we know that $\delta'(s_1, k)$ is Hurwitz if and only if

$$\sigma_i(\delta'(s_1, k) D^*(s_1)) = n + l(D^*(s_1)) - r(D^*(s_1)).$$

Thus $\delta'(s_1, k)$ is Hurwitz if and only if $\mathcal{I} \in A(n + l(D^*(s_1)) - r(D^*(s_1)))$ where $\mathcal{I} = \{i_0, i_1, \cdots, i_m, i_{m+1}\}$, $i_l = sgn[p_f(\omega_l, k)]$ for $l = 0, 1, \cdots, m$ and $i_{m+1} = sgn[k]$. In this case, we have assumed $k < 0$ so that $i_{m+1} = -1$. Now there are two different possibilities:

(1) $p_2(\omega_l) = 0$ for some $l \in \{0, 1, \cdots, m\}$: If $p_2(\omega_l) = 0$ then $\text{sgn}[p_f(\omega_l, k)] = \text{sgn}[p_1(\omega_l)]$ so that the requirement $i_l = \text{sgn}[p_f(\omega_l, k)]$ translates to $i_l = sgn[p_1(\omega_l)]$, something which was explicitly accounted for in the definition of A. We note that this case does not lead to any constraints on k.

(2) $p_2(\omega_l) \neq 0$: In this case, there are again three distinct possibilities:

(a) $i_l > 0$: If $i_l > 0$, then the stability requirement is $p_1(\omega_l) + k p_2(\omega_l) > 0$. Thus if $p_2(\omega_l) > 0$ then

$$k > -\frac{p_1(\omega_l)}{p_2(\omega_l)}. \tag{8.4.4}$$

If, on the other hand, $p_2(\omega_l) < 0$, then

$$k < -\frac{p_1(\omega_l)}{p_2(\omega_l)}. \tag{8.4.5}$$

(b) $i_l < 0$: If $i_l < 0$, then the stability requirement is $p_1(\omega_l) + k p_2(\omega_l) < 0$. Thus if $p_2(\omega_l) > 0$ then

$$k < -\frac{p_1(\omega_l)}{p_2(\omega_l)}. \tag{8.4.6}$$

If, on the other hand, $p_2(\omega_l) < 0$, then

$$k > -\frac{p_1(\omega_l)}{p_2(\omega_l)}. \tag{8.4.7}$$

(c) $i_l = 0$: This case can only happen when $j\omega_l$ is a $j\omega$-axis root of $D^*(s_1)$ (see definition of A). Hence it follows that $D^*(j\omega_l) = D_R^{'}(j\omega_l) = D_I^{'}(j\omega_l) = 0$ so that $p_1(\omega_l) = p_2(\omega_l) = p(\omega_l, k) = 0$ for all k. This contradicts the underlying assumption $p_2(\omega_l) \neq 0$ in possibility (2). Thus this case would really occur under possibility (1) and, as expected, would not impose any constraint on k.

Combining (8.4.4)-(8.4.7) we see that

$$\begin{cases} -\frac{p_1(\omega_l)}{p_2(\omega_l)} < k & \text{if } i_l \cdot p_2(\omega_l) > 0 \\ k < -\frac{p_1(\omega_l)}{p_2(\omega_l)} & \text{if } i_l \cdot p_2(\omega_l) < 0 \end{cases}$$

Thus each i_l in the string $\mathcal{I} \in A(n + l(D^*(s_1)) - r(D^*(s_1)))$ for which $i_l p_2(\omega_l) > 0$ contributes a lower bound on k while each i_l for which $i_l p_2(\omega_l) < 0$ contributes an upper bound on k. Furthermore, let us recall that we are currently focussing on the case where $i_{m+1} = -1$ or equivalently k is negative so that 0 is also an upper bound on k. Thus, in this case if the string $\mathcal{I} \in A(n + l(D^*(s_1)) - r(D^*(s_1)))$ is to correspond to a stabilizing k then we must have that the maximum of all the lower bounds must be smaller than the minimum of all the upper bounds, *i.e.*,

$$\max_{i_t \in \mathcal{I}, i_t \cdot \text{sgn}[p_2(\omega_t)]=1} \left[-\frac{p_1(\omega_l)}{p_2(\omega_l)} \right]$$
$$< \min \left[0, \min_{i_t \in \mathcal{I}, i_t \cdot \text{sgn}[p_2(\omega_t)]=-1} \left[-\frac{p_1(\omega_l)}{p_2(\omega_l)} \right] \right]. \quad (8.4.8)$$

We next proceed to express Condition 8.4.8 in terms of $G(s)$. Now

$$G(s) = \frac{N(s)}{D(s)}$$

so that

$$\begin{aligned} \frac{1}{G(j\omega e^{j\phi})} &= \frac{D(j\omega e^{j\phi})}{N(j\omega e^{j\phi})} \\ &= \frac{D^{'}(j\omega)}{N^{'}(j\omega)} \\ &= \frac{D_R^{'}(j\omega) + D_I^{'}(j\omega)}{N_R^{'}(j\omega) + N_I^{'}(j\omega)} \\ &= \frac{[D_R^{'}(j\omega) + D_I^{'}(j\omega)][D_R^{'}(j\omega) - D_I^{'}(j\omega)]}{[N_R^{'}(j\omega) + N_I^{'}(j\omega)][D_R^{'}(j\omega) - D_I^{'}(j\omega)]} \\ &= \frac{p_1(\omega)}{p_2(\omega) + jq(\omega)}. \end{aligned} \quad (8.4.9)$$

Since $q(\omega_t) = 0$, it follows that $-\frac{p_1(\omega_t)}{p_2(\omega_t)} = -\frac{1}{G(j\omega_t e^{j\phi})}$. Thus, from (8.4.8), we must have

$$\max_{i_t \in \mathcal{I}, i_t \cdot \mathrm{sgn}[p_2(\omega_t)]=1} \left[-\frac{1}{G(j\omega_t e^{j\phi})} \right] < \min \left[0, \min_{i_t \in \mathcal{I}, i_t \cdot \mathrm{sgn}[p_2(\omega_t)]=-1} \left[-\frac{1}{G(j\omega_t e^{j\phi})} \right] \right].$$

which is Condition (II)(2)(a) in the theorem statement.

Case (III) $k > 0$: This case leads to Condition (II)(2)(b) in the theorem statement. The derivation of this condition can be carried out using arguments similar to the ones that we just used for deriving (II)(2)(a).

This completes the proof of the necessary and sufficient conditions for the existence of a ϕ Hurwitz stabilizing k. The set of all ϕ Hurwitz stabilizing k's is now obtained by taking the union of all k's that are determined from the feasible strings that satisfy Condition (II) and then tagging on the value $k = \{0\}$ if Condition (I) is also satisfied. ♣

Note that when $\phi = 0$, Theorem 8.4.1 becomes the analytical solution of the constant gain Hurwitz stabilization problem. Theorem 8.4.1 therefore encompasses the constant gain stabilization results of Chapter 4 as a special case.

8.5 Example

Example 8.5.1. Consider the plant $G(s) = \frac{N(s)}{D(s)}$ where

$$\begin{aligned} N(s) &= s^2 + 2s - 2 \\ D(s) &= s^3 + 3s^2 + 4s. \end{aligned}$$

Suppose that we want to use a constant gain k to place the closed loop poles in a region S described as in Section 8.2 by the parameters $\alpha = 0.5$ and $\phi = \frac{\pi}{6}$. The closed loop characteristic polynomial is

$$\delta(s,k) = D(s) + kN(s).$$

We first consider the ϕ Hurwitz stabilization problem. Then

$$\begin{aligned} N(s_1 e^{j\frac{\pi}{6}}) &= (0.5 + 0.866j)s_1^2 + (1.7321 + j)s_1 - 2 \\ D(s_1 e^{j\frac{\pi}{6}}) &= js_1^3 + (1.5 + 2.5981j)s_1^2 + (3.4641 + 2j)s_1. \end{aligned}$$

Dividing $N(s_1 e^{j\frac{\pi}{6}})$ and $D(s_1 e^{j\frac{\pi}{6}})$ by j, we have

$$\begin{aligned} N'(s_1) &= (0.866-0.5j)s_1^2+(1-1.7321j)s_1+2j \\ D'(s_1) &= s_1^3+(2.5981-1.5j)s_1^2+(2-3.4641j)s_1. \end{aligned}$$

Carrying out the real-imaginary decomposition of $D'(s_1)$, we obtain:

$$\begin{aligned} D'(s_1) &= D'_R(s_1)+D'_I(s_1) \\ &\text{with} \\ D'_R(s_1) &= 2.5981s_1^2-3.4641js_1 \\ D'_I(s_1) &= s_1^3-1.5js_1^2+2s_1. \end{aligned}$$

Thus,

$$\begin{aligned} D^*(s_1) &= D'_R(s_1)-D'_I(s_1) \\ &= -s_1^3+(2.5981+1.5j)s_1^2+(-2-3.4641j)s_1. \end{aligned}$$

Now

$$\delta'(s_1,k)=D'(s_1)+kN'(s_1)$$

so that

$$\delta'(j\omega,k)D^*(j\omega) = p_1(\omega)+kp_2(\omega)+jkq(\omega).$$

where

$$\begin{aligned} p_1(\omega) &:= \omega^6-3\omega^5+5\omega^4-12\omega^3+16\omega^2 \\ p_2(\omega) &:= -0.5\omega^5+2\omega^4-7\omega^3+11\omega^2+4\omega \\ q(\omega) &:= -0.866\omega^5+1.7321\omega^4-1.7321\omega^3-5.1962\omega^2+6.9282\omega. \end{aligned}$$

The real, distinct finite zeros of $q_f(\omega)$ with odd multiplicities are:

$$\omega_1=-1.5249,\ \omega_2=0,\ \omega_3=1.1209.$$

We also define $\omega_0=-\infty$ and $\omega_4=+\infty$. Now, from Definition 8.4.1, the set A becomes

$$A=$$

$$\left\{\begin{array}{ccc}
\{-1,-1,0,-1,-1,-1\} & \{1,-1,0,-1,-1,-1\} & \{-1,1,0,-1,-1,-1\} \\
\{1,1,0,-1,-1,-1\} & \{-1,-1,0,1,-1,-1\} & \{1,-1,0,1,-1,-1\} \\
\{-1,1,0,1,-1,-1\} & \{1,1,0,1,-1,-1\} & \{-1,-1,0,-1,1,-1\} \\
\{1,-1,0,-1,1,-1\} & \{-1,1,0,-1,1,-1\} & \{1,1,0,-1,1,-1\} \\
\{-1,-1,0,1,1,-1\} & \{1,-1,0,1,1,-1\} & \{-1,1,0,1,1,-1\} \\
\{1,1,0,1,1,-1\} & \{-1,-1,0,-1,-1,1\} & \{1,-1,0,-1,-1,1\} \\
\{-1,1,0,-1,-1,1\} & \{1,1,0,-1,-1,1\} & \{-1,-1,0,1,-1,1\} \\
\{1,-1,0,1,-1,1\} & \{-1,1,0,1,-,1\} & \{1,1,0,1,-1,1\} \\
\{-1,-1,0,-1,1,1\} & \{1,-1,0,-1,1,1\} & \{-1,1,0,-1,1,1\} \\
\{1,1,0,-1,1,1\} & \{-1,-1,0,1,1,1\} & \{1,-1,0,1,1,1\} \\
\{-1,1,0,1,1,1\} & \{1,1,0,1,1,1\} &
\end{array}\right\}$$

Since $l(D^*(s)) - r(D^*(s)) = -2$ and $sgn[q(\infty)] = -1$, it follows using Definition 8.4.3 that every string $\mathcal{I} = \{i_0,\ i_1,\ i_2,\ i_3,\ i_4,\ i_5\} \in F^*$ must satisfy

$$\frac{1}{2}(-i_0 + 2i_1 - 2i_2 + 2i_3 - i_4) \cdot sgn[i_5 \cdot q(\infty)] = 1.$$

Hence $F^* = \{\mathcal{I}_1, \mathcal{I}_2, \mathcal{I}_3, \mathcal{I}_4, \mathcal{I}_5, \mathcal{I}_6\}$ where

$$\begin{aligned}
\mathcal{I}_1 &= \{-1, 1, 0, -1, -1, -1\} \\
\mathcal{I}_2 &= \{-1, -1, 0, 1, -1, -1\} \\
\mathcal{I}_3 &= \{1, 1, 0, 1, 1, -1\} \\
\mathcal{I}_4 &= \{-1, -1, 0, -1, -1, 1\} \\
\mathcal{I}_5 &= \{1, 1, 0, -1, 1, 1\} \\
\mathcal{I}_6 &= \{1, -1, 0, 1, 1, 1\}.
\end{aligned}$$

Furthermore,

$$\begin{aligned}
-\frac{1}{G(j\omega_0\, e^{j\frac{\pi}{6}})} &= -\infty,\ p_2(\omega_0) > 0 \\
-\frac{1}{G(j\omega_1\, e^{j\frac{\pi}{6}})} &= -2.4326,\ p_2(\omega_1) > 0 \\
-\frac{1}{G(j\omega_3\, e^{j\frac{\pi}{6}})} &= -0.725,\ p_2(\omega_3) > 0 \\
-\frac{1}{G(j\omega_4\, e^{j\frac{\pi}{6}})} &= \infty,\ p_2(\omega_4) < 0.
\end{aligned}$$

Hence from Theorem 8.4.1, we have

$$\begin{cases}
K_0 = \emptyset \\
K_1 = \emptyset \text{ for } \mathcal{I}_1 \\
K_2 = \emptyset \text{ for } \mathcal{I}_2 \\
K_3 = (-0.725,\ 0) \text{ for } \mathcal{I}_3 \\
K_4 = \emptyset \text{ for } \mathcal{I}_4 \\
K_5 = \emptyset \text{ for } \mathcal{I}_5 \\
K_6 = \emptyset \text{ for } \mathcal{I}_6.
\end{cases}$$

Therefore $K_\phi = (-0.725,\ 0)$.

Next we consider the $-\alpha$ Hurwitz stabilization problem with $\alpha = 0.5$. We set $s = s_1 - 0.5$ and obtain

$$\delta_{-\alpha}(s_1, k) = D(s_1 - 0.5) + kN(s_1 - 0.5)$$

where

$$\begin{aligned}
N(s_1 - 0.5) &= s_1^2 + s_1 - 2.75 \\
D(s_1 - 0.5) &= s_1^3 + 1.5s_1^2 + 1.75s_1 - 1.3750.
\end{aligned}$$

Now setting $\phi = 0$ and using Theorem 8.4.1 to solve the ϕ Hurwitz stabilization problem for $\delta_{-\alpha}(s_1, k)$, we obtain $K_{-\alpha} = (-0.7639,\ -0.5)$. Therefore, the set K of all S Hurwitz stabilizing gain values is given by

$$\begin{aligned} K &= K_{-\alpha} \cap K_{\phi} \\ &= (-0.725,\ -0.5). \end{aligned}$$

The root locus of the closed loop system corresponding to this set of gain values is shown in Fig. 8.4. From this root locus, it is clear that the set of gains determined above does place the closed loop poles in the desired region S.

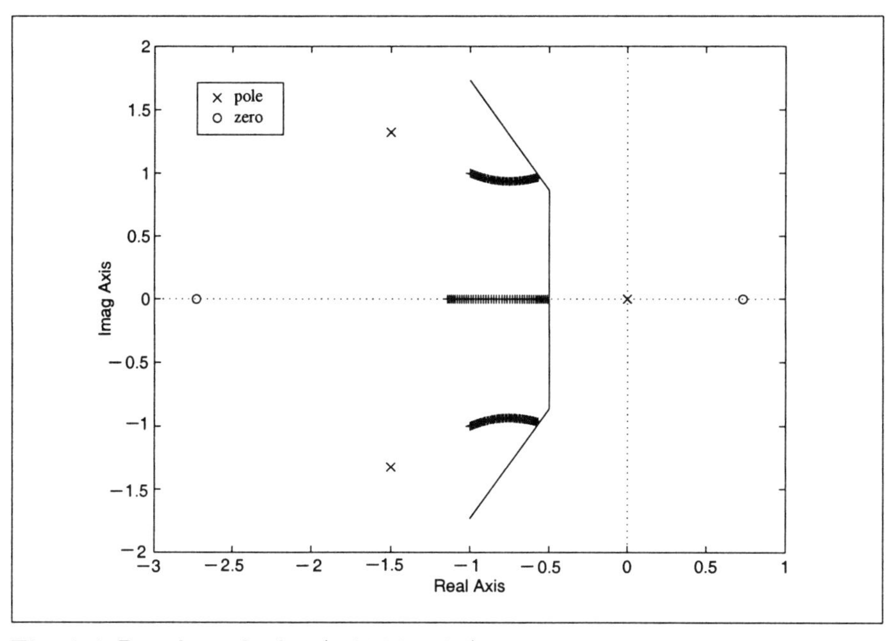

Fig. 8.4. Root locus for $k \in (-0.725,\ -0.5)$.

8.6 Notes and References

The results in this chapter are based on Ho, Datta and Bhattacharyya [25]. The extension of the constant gain results of this chapter to the case of PI and PID controllers is a topic for further research.

CHAPTER 9

CONSTANT GAIN STABILIZATION OF DISCRETE-TIME PLANTS

In this chapter we generalize the Hermite-Biehler Theorem to the case of the unit circle. This result is then used to characterize the set of all stabilizing gains for a discrete-time plant.

9.1 Introduction

In Chapter 4, Section 4.2, we provided a complete analytical solution to the problem of stabilizing a given plant using a constant gain. Our solution was based on appropriate generalizations of the Hermite Biehler Theorem. The constant gain results were subsequently extended in Sections 4.3 and 4.4 to provide an elegant linear programming characterization of all stabilizing PI and PID controllers for a given plant. The aim of this chapter is to focus on the development of analogous results for the discrete-time case.

As a first step, we focus on the constant gain stabilization problem which itself presents unique challenges not encountered in the continuous-time case.

The Chapter is organized as follows. In Section 9.2, we provide a statement of the Hermite-Biehler Theorem for Schur polynomials as well as some equivalent characterizations. In Section 9.3, we study the relationship between the net phase change of the "frequency response" of a real rational function as the frequency ω varies from 0 to π along the unit circle and the difference between the numbers of zeros and poles in the open unit disc. In Section 9.4, we derive two results each of which in effect is a generalization of the Hermite-Biehler Theorem to the case of rational functions and not necessarily Schur real polynomials. In Section 9.5, we consider the problem of stabilizing a given plant using a zeroth order compensator *i.e.*, a constant gain. A complete analytical characterization of all stabilizing gain values is provided under the assumption that the plant in question has no zeros on the unit circle. The relaxation of this assumption, as also the derivation of discrete-time results for PI and PID stabilization are topics of current research.

9.2 The Hermite-Biehler Theorem for Schur Polynomials

In this section, we first state the Hermite-Biehler Theorem for Schur polynomials which provides necessary and sufficient conditions for the Schur stability of a given real polynomial. The proof can be found in [12, 7, 9]

Theorem 9.2.1. *(Hermite-Biehler Theorem: Unit Disc): Let $\delta(z) = \delta_0 + \delta_1 z + \cdots + \delta_n z^n$ be a given real polynomial of degree n. Write*

$$\delta(z) = \delta_s(z) + \delta_a(z)$$

where

$$\delta_s(z) = \frac{\delta(z) + z^n\delta(\frac{1}{z})}{2}$$

and

$$\delta_a(z) = \frac{\delta(z) - z^n\delta(\frac{1}{z})}{2}$$

represent the "symmetric" and "anti-symmetric" parts of $\delta(z)$. Let z_{s_1}, $z_{s_2}, \cdots$ denote the unit circle zeros of $\delta_s(z)$ with corresponding arguments ω_{s_1}, $\omega_{s_2}, \cdots$ and let z_{a_0}, $z_{a_1}, \cdots$ denote the unit circle zeros of $\delta_a(z)$ with corresponding arguments ω_{a_0}, $\omega_{a_1}, \cdots$. Both z_{s_1}, $z_{s_2}, \cdots$ and z_{a_0}, $z_{a_1}, \cdots$ are arranged in ascending order of argument. Then $\delta(z)$ is Schur stable if and only if all the zeros of $\delta_s(z)$, $\delta_a(z)$ lie on the unit circle, and the upper half plane unit circle zeros satisfy the following interlacing property

$$0 = \omega_{a_0} < \omega_{s_1} < \omega_{a_1} < \omega_{s_2} < \cdots = \pi$$

Instead of the "symmetric-antisymmetric" decomposition used above, the conclusion of Theorem 9.2.1 is still valid if we use the following "real-imaginary" decomposition of a real polynomial $\delta(z)$:
Define

$$\delta(z) = \delta_r(z) + \delta_i(z)$$

where

$$\delta_r(z) = \frac{1}{2}[\delta(z) + \delta(\frac{1}{z})]$$
$$\delta_i(z) = \frac{1}{2}[\delta(z) - \delta(\frac{1}{z})]$$

Remark 9.2.1. Note that the unit circle zeros of $\mathrm{Re}[\delta(z)]$, $\mathrm{Im}[\delta(z)]$ are identical to the unit circle zeros of $\delta_r(z)$, $\delta_i(z)$ respectively. This justifies the terminology "real-imaginary" decomposition. Of course, it should be pointed out that $\delta_r(z)$ and $\delta_i(z)$ are not polynomials but rational functions in z.

The discrete-time stabilization results of this chapter are based on appropriate generalizations of the above Hermite-Biehler Theorem. To develop this we start with an alternative statement of this Theorem.

Lemma 9.2.1. *Let $\delta(z) = \delta_0 + \delta_1 z + \cdots + \delta_n z^n$ be a given real polynomial of degree n. Write*

$$\delta(z) = \delta_r(z) + \delta_i(z)$$

where

$$\begin{aligned} \delta_r(z) &= \frac{1}{2}[\delta(z) + \delta(\frac{1}{z})] \\ \delta_i(z) &= \frac{1}{2}[\delta(z) - \delta(\frac{1}{z})]. \end{aligned}$$

Denote $\delta(e^{j\omega}) = p(\omega) + jq(\omega)$, $\omega \in [0, 2\pi]$ where $p(\omega) = \delta_r(e^{j\omega})$, $q(\omega) = \frac{1}{j}\delta_i(e^{j\omega})$. Let $z_{r_1}, z_{r_2}, \cdots$ denote the upper half plane unit circle zeros of $\delta_r(z)$ with corresponding arguments $\omega_{r_1}, \omega_{r_2}, \cdots$ and let $z_{i_0}, z_{i_1}, \cdots$ denote the upper half plane unit circle zeros of $\delta_i(z)$ with corresponding arguments $\omega_{i_0}, \omega_{i_1}, \cdots$. Both $z_{r_1}, z_{r_2}, \cdots$ and $z_{i_0}, z_{i_1}, \cdots$ are arranged in ascending order of argument. Then the following conditions are equivalent:

(i) $\delta(z)$ is Schur stable.
(ii)

$$\begin{aligned} n &= \frac{1}{2} sgn[\delta(1)] \cdot \{sgn[p(\omega_{i_0})] - 2sgn[p(\omega_{i_1})] + 2sgn[p(\omega_{i_2})] + \cdots \\ &\quad +(-1)^{n-1} \cdot 2sgn[p(\omega_{i_{n-1}})] + (-1)^n \cdot sgn[p(\omega_{i_n})]\} \qquad (9.2.1) \end{aligned}$$

(iii)

$$\begin{aligned} n &= sgn[\delta(1)] \cdot \{sgn[q(\omega_{r_1})] - sgn[q(\omega_{r_2})] + sgn[q(\omega_{r_3})] + \cdots \\ &\quad +(-1)^{n-2} \cdot sgn[q(\omega_{r_{n-1}})] + (-1)^{n-1} \cdot sgn[q(\omega_{r_n})]\} \quad (9.2.2) \end{aligned}$$

Proof. (1) $(i) \Leftrightarrow (ii)$

We first show that $(i) \Rightarrow (ii)$.
The phase of a Schur polynomial, evaluated on the unit circle, increases monotonically as we traverse the unit circle in the counterclockwise sense starting from $z = e^{j0} = 1$. Using this monotonic phase property, we can show that the plot of $\delta(e^{j\omega}) = p(\omega) + jq(\omega)$ must move strictly counterclockwise and goes through $2n$ quadrants in turn as ω increases from 0 to π [7]. For Schur $\delta(z)$, the possible plots of $\delta(e^{j\omega})$ are illustrated in Fig. 9.1.
From Fig. 9.1, it is clear that

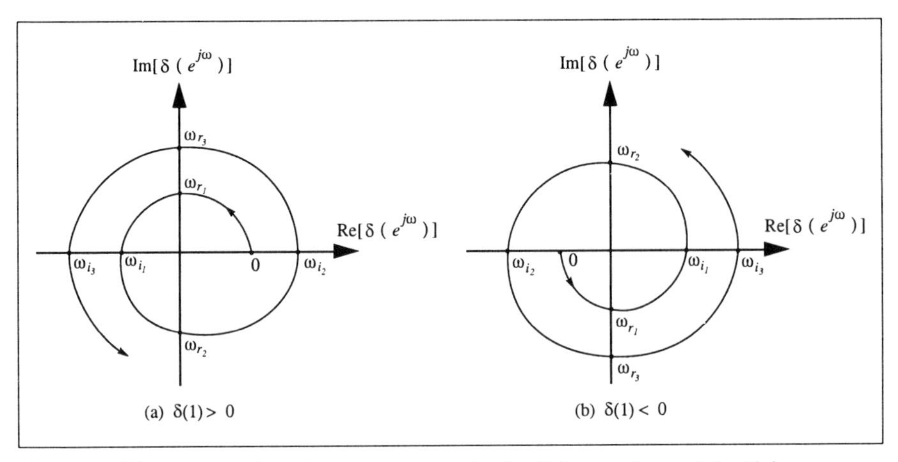

Fig. 9.1. Monotonic phase increase property for Schur polynomials $\delta(z)$.

$$\begin{cases} \operatorname{sgn}[\delta(1)] \cdot \operatorname{sgn}[p(\omega_{i_0})] > 0 \\ -\operatorname{sgn}[\delta(1)] \cdot \operatorname{sgn}[p(\omega_{i_1})] > 0 \\ \vdots \\ (-1)^{n-1}\operatorname{sgn}[\delta(1)] \cdot \operatorname{sgn}[p(\omega_{i_{n-1}})] > 0 \\ (-1)^{n}\operatorname{sgn}[\delta(1)] \cdot \operatorname{sgn}[p(\omega_{i_n})] > 0 \end{cases} \tag{9.2.3}$$

From (9.2.3), it follows that (9.2.1) holds.

$(ii) \Rightarrow (i)$
Equation 9.2.1 holds if and only if $[p(\omega_{i_{l-1}})]$ and $[p(\omega_{i_l})]$ are of opposite signs for $l = 1,\ 2,\ \cdots,\ n$. By the continuity of $p(\omega)$, there exists at least one $\omega_r \in (\omega_{i_{l-1}}\ \omega_{i_l})$ such that $p(\omega_r) = 0$. Moreover, since the maximum possible number of zeros of $p(\cdot)$ in $[0,\ \pi]$ is n, it follows that there exists one and only one $\omega_r \in (\omega_{i_{l-1}},\ \omega_{i_l})$ such that $p(\omega_r) = 0$, thereby leading us to the interlacing property.

(2) $(i) \Leftrightarrow (iii)$
The proof of (2) follows along the same lines as that of (1).

♣

Note that from Lemma 9.2.1 if $\delta(z)$ is Schur stable then all the zeros of $\delta_r(z)$ and $\delta_i(z)$ must be distinct and lie on the unit circle, otherwise (9.2.1) and (9.2.2) will fail.

We now present an example to illustrate the application of Lemma 9.2.1 to verify the interlacing property.

Example 9.2.1. Consider the real polynomial

$$\delta(z) = z^5 + 0.1z^4 + 0.2z^3 + 0.4z^2 + 0.03z + 0.01$$

Then

$$\delta(e^{j\omega}) = p(\omega) + jq(\omega)$$

where

$$\begin{aligned}
p(\omega) &= \delta_r(e^{j\omega}) \\
&= \frac{1}{2}[\delta(e^{j\omega}) + \delta(e^{-j\omega})] \\
&= \cos(5\omega) + 0.1\cos(4\omega) + 0.2\cos(3\omega) + 0.4\cos(2\omega) \\
&\quad +0.03\cos(\omega) + 0.01 \\
q(\omega) &= \frac{\delta_i(e^{j\omega})}{j} \\
&= \frac{1}{2j}[\delta(e^{j\omega}) - \delta(e^{-j\omega})] \\
&= \sin(5\omega) + 0.1\sin(4\omega) + 0.2\sin(3\omega) + 0.4\sin(2\omega) + 0.03\sin(\omega)
\end{aligned}$$

The plots of $p(\omega)$ and $q(\omega)$ are shown in Fig. 9.2. They show that the polynomial $\delta(z)$ satisfies the interlacing property.

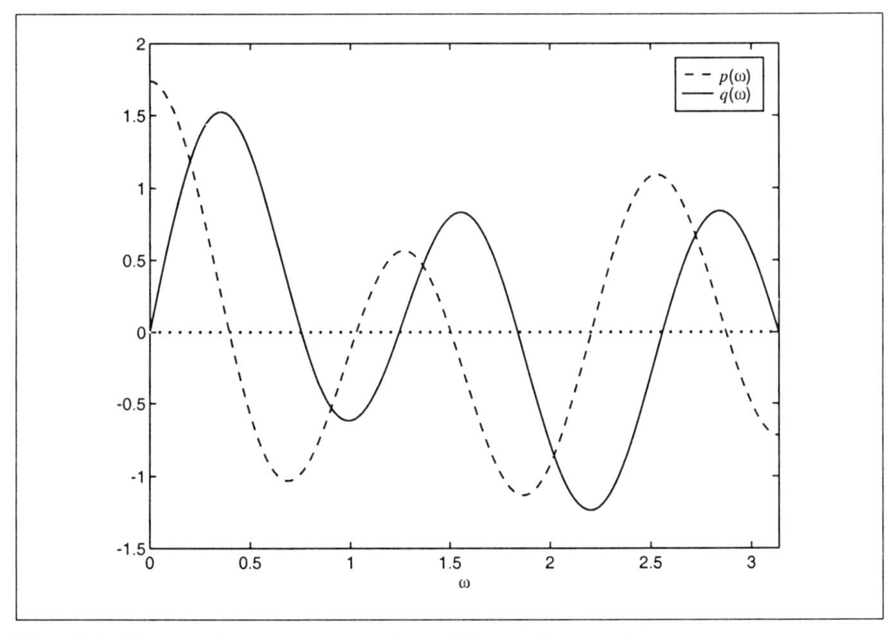

Fig. 9.2. The interlacing property for a Schur polynomial.

Carrying out the real-imaginary decomposition of $\delta(z)$, we obtain

$$\delta_r(z) = \frac{1}{2}[\delta(z) + \delta(\frac{1}{z})]$$

$$= z^{-5}(0.5z^{10} + 0.05z^9 + 0.1z^8 + 0.2z^7 + 0.015z^6 + 0.01z^5 + 0.015z^4 + 0.2z^3 + 0.1z^2 + 0.05z + 0.5)$$

and

$$\begin{aligned}\delta_i(z) &= \frac{1}{2}[\delta(z) - \delta(\frac{1}{z})] \\ &= z^{-5}(0.5z^{10} + 0.05z^9 + 0.1z^8 + 0.2z^7 + 0.015z^6 - 0.015z^4 - 0.2z^3 \\ &\quad -0.1z^2 - 0.05z + 0.5)\end{aligned}$$

Ordering the upper half plane unit circle zeros of $\delta_r(z)$ and $\delta_i(z)$ arranged in ascending order of arguments, we obtain

$$z_{r_1} = 0.9230 + 0.3849j,\ z_{r_2} = 0.5160 + 0.8566j,\ z_{r_3} = 0.0667 + 0.9978j$$
$$z_{r_4} = -0.5911 + 0.8066j,\ z_{r_5} = -0.9646 + 0.2637j$$

and

$$z_{i_0} = 1.0,\ z_{i_1} = 0.7292 + 0.6843j,\ z_{i_2} = 0.3212 + 0.9470j,$$
$$z_{i_3} = -0.2653 + 0.9642j,\ z_{i_4} = -0.8350 + 0.5502j,\ z_{i_5} = -1.0.$$

Then

$$\begin{aligned}&\mathrm{sgn}[p(\omega_{i_0})] = \mathrm{sgn}[\delta_r(z_{i_0})] = 1,\ \mathrm{sgn}[p(\omega_{i_1})] = \mathrm{sgn}[\delta_r(z_{i_1})] = -1,\\ &\mathrm{sgn}[p(\omega_{i_2})] = \mathrm{sgn}[\delta_r(z_{i_2})] = 1,\ \mathrm{sgn}[p(\omega_{i_3})] = \mathrm{sgn}[\delta_r(z_{i_3})] = -1,\\ &\mathrm{sgn}[p(\omega_{i_4})] = \mathrm{sgn}[\delta_r(z_{i_4})] = 1,\ \mathrm{sgn}[p(\omega_{i_5})] = \mathrm{sgn}[\delta_r(z_{i_5})] = -1.\end{aligned}$$

Now $\delta(z)$ is of degree $n = 5$, and

$$\begin{aligned}&\frac{1}{2}\mathrm{sgn}[\delta(1)] \cdot \{\mathrm{sgn}[p(\omega_{i_0})] - 2\mathrm{sgn}[p(\omega_{i_1})] + 2\mathrm{sgn}[p(\omega_{i_2})]\\ &-2\mathrm{sgn}[p(\omega_{i_3})] + 2\mathrm{sgn}[p(\omega_{i_4})] - \mathrm{sgn}[p(\omega_{i_5})]\} = 5\end{aligned}$$

which shows that (9.2.1) holds.

Also, we have

$$\begin{aligned}&\mathrm{sgn}[q(\omega_{r_1})] = \mathrm{sgn}[\frac{\delta_i(z_{r_1})}{j})] = 1,\ \mathrm{sgn}[q(\omega_{r_2})] = \mathrm{sgn}[\frac{\delta_i(z_{r_2})}{j}] = -1,\\ &\mathrm{sgn}[q(\omega_{r_3})] = \mathrm{sgn}[\frac{\delta_i(z_{r_3})}{j}] = 1,\ \mathrm{sgn}[q(\omega_{r_4})] = \mathrm{sgn}[\frac{\delta_i(z_{r_4})}{j}] = -1,\\ &\mathrm{sgn}[q(\omega_{r_5})] = \mathrm{sgn}[\frac{\delta_i(z_{r_5})}{j}] = 1\end{aligned}$$

so that

$$\begin{aligned}&\mathrm{sgn}[\delta(1)] \cdot \{\mathrm{sgn}[q(\omega_{r_1})] - \mathrm{sgn}[q(\omega_{r_2})] + \mathrm{sgn}[q(\omega_{r_3})]\\ &-\mathrm{sgn}[q(\omega_{r_4})] + \mathrm{sgn}[q(\omega_{r_5})]\} = 5\end{aligned}$$

Once again, this verifies (9.2.2).

To verify that $\delta(z)$ is indeed a Schur polynomial, we solve for the roots of $\delta(z)$:

$$0.3135 \pm 0.698j \qquad -0.0330 \pm 0.1572j$$
$$-0.6611$$

We see that all the roots of $\delta(z)$ are in the open unit disk so that $\delta(z)$ is Schur.

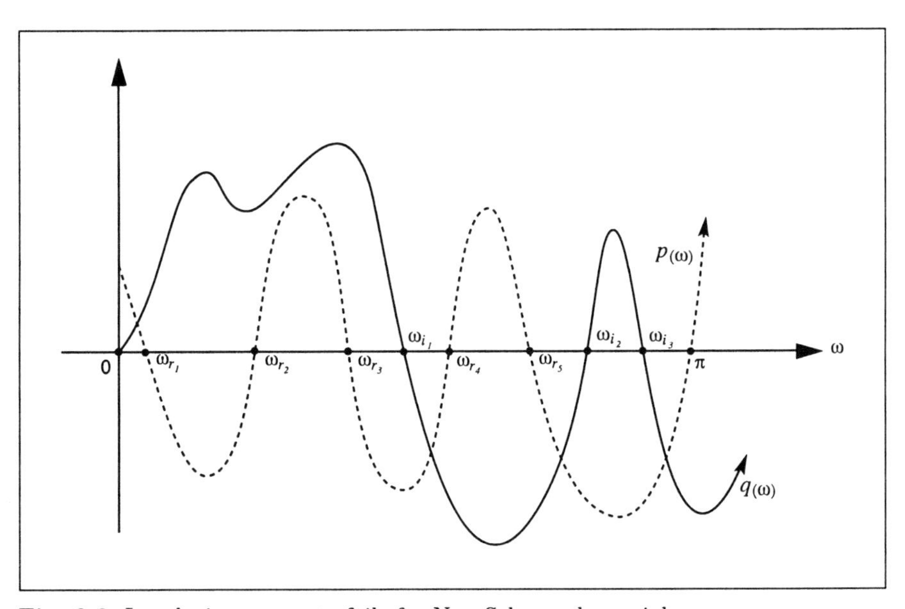

Fig. 9.3. Interlacing property fails for Non-Schur polynomials.

Now consider $\delta(e^{j\omega}) = p(\omega) + jq(\omega)$ as illustrated in Fig. 9.3. From Fig. 9.3, we know that the polynomial $\delta(z)$ is not a Schur polynomial because it fails to satisfy the interlacing property. However, as in Chapter 3, it is logical to ask: does Fig. 9.3 provide us with more information about $\delta(z)$, beyond whether or not it is Schur? Once again, it is possible to know the number of roots of $\delta(z)$ in the open unit disk from the above graph. This motivates us to derive generalized versions of the Hermite-Biehler Theorem for not necessarily Schur polynomials. Such a generalization is also needed for solving the constant gain stabilization problem, which additionally mandates that the generalization also encompass the class of rational functions. This will be clearly borne out by the following discussion.

Consider the discrete-time constant gain stabilization problem shown in Fig. 9.4. Here r is the command signal, y is the output, $G(z) = \frac{N(z)}{D(z)}$ is the

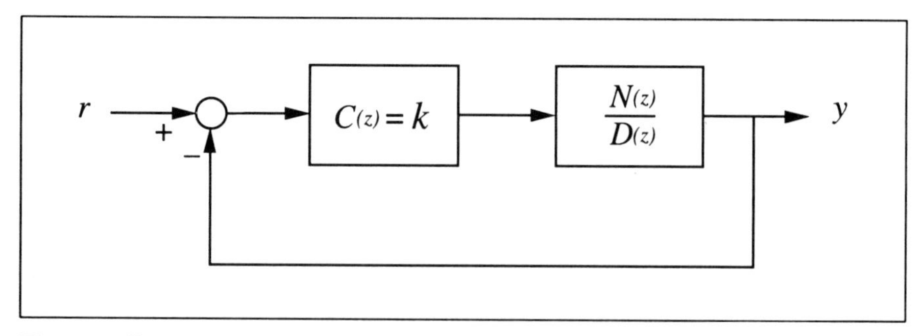

Fig. 9.4. Feedback control system.

plant to be controlled, $N(z)$ and $D(z)$ are coprime polynomials, and $C(z) = k$, *i.e.*, the controller is of zeroth order. Then the closed loop characteristic polynomial is $\delta(z,k) = D(z) + kN(z)$, where k is a scalar. Our objective is to analytically determine k, if any, for which $\delta(z,k)$ is Schur. Now let $D(z)$, $N(z)$ have the following real-imaginary decompositions:

$$\begin{aligned} D(z) &= D_r(z) + D_i(z) \\ N(z) &= N_r(z) + N_i(z) \end{aligned}$$

Then

$$\delta(z,k) = [D_r(z) + kN_r(z)] + [D_i(z) + kN_i(z)]$$

Substituting $z = e^{j\omega}$, we have

$$\delta(e^{j\omega},k) = [D_r(e^{j\omega}) + kN_r(e^{j\omega})] + [D_i(e^{j\omega}) + kN_i(e^{j\omega})]$$

Denote

$$\begin{aligned} \bar{p}(\omega,k) &= D_r(e^{j\omega}) + kN_r(e^{j\omega}) \\ \bar{q}(\omega,k) &= \frac{1}{j}[D_i(e^{j\omega}) + kN_i(e^{j\omega})] \end{aligned}$$

Then

$$\delta(e^{j\omega},k) = \bar{p}(\omega,k) + j\bar{q}(\omega,k)$$

Since $\bar{p}(\omega,k)$, $\bar{q}(\omega,k)$ are both functions of k, it would in general be very difficult to use Lemma 9.2.1 to determine the values of k for which $\delta(z,k)$ is Schur stable. To overcome this difficulty, we proceed as follows.

Let

$$N^*(z) = N(\frac{1}{z}).$$

Then

$$\begin{aligned}\delta(z,k)N^*(z) &= D(z)N(\frac{1}{z}) + kN(z)N(\frac{1}{z})\\ &= [H_r(z) + kN(z)N(\frac{1}{z})] + H_i(z)\end{aligned}$$

where

$$\begin{aligned}H_r(z) &= \frac{1}{2}[D(z)N(\frac{1}{z}) + D(\frac{1}{z})N(z)]\\ H_i(z) &= \frac{1}{2}[D(z)N(\frac{1}{z}) - D(\frac{1}{z})N(z)].\end{aligned}$$

Substituting $z = e^{j\omega}$, we obtain

$$\delta(e^{j\omega},k)N^*(e^{j\omega}) = p(\omega,k) + jq(\omega) \tag{9.2.4}$$

where

$$\begin{aligned}p(\omega,k) &= \frac{1}{2}[D(e^{j\omega})N(e^{-j\omega}) + D(e^{-j\omega})N(e^{j\omega})]\\ &\quad +kN(e^{j\omega})N(e^{-j\omega}) &(9.2.5)\\ q(\omega) &= \frac{1}{2j}[D(e^{j\omega})N(e^{-j\omega}) - D(e^{-j\omega})N(e^{j\omega})]. &(9.2.6)\end{aligned}$$

From (9.2.6), it is clear that $q(\omega)$ is independent of k. However, $\delta(z,k)N^*(z)$ is not a polynomial but a *rational function.* Furthermore, since $N^*(z)$ *may not* have all of its zeros inside the unit disc, it is possible for $\delta(z,k)N^*(z)$ to have zeros outside the unit disc, even when $\delta(z,k)$ itself is Schur stable. These facts motivate us to derive an appropriate generalization of Lemma 9.2.1 for rational functions, some of whose zeros may lie outside the unit disk. This is carried out in the next two sections.

9.3 Poles and Zeros in the Unit Disc and Net Accumulated Phase

In this section we develop, as a preliminary step to the Generalized Hermite-Biehler Theorem, a fundamental relationship between the net accumulated phase of the frequency response of a real rational function and the difference between the numbers of zeros and poles of the rational function in the open unit disk. To this end, let $\mathcal{D}$ denote the open unit disk.

We focus on rational functions without zeros or poles on the unit circle. We consider a real rational function $\delta(z)$:

$$\delta(z) = \frac{h(z)}{g(z)}$$

where $h(z)$ and $g(z)$ are coprime polynomials, and

$$\begin{aligned} h(z) &= h_0 + h_1 z + h_2 z^2 + \ldots + h_m z^m, \ h_i \in \mathcal{R}, \ i = 0, 1, \cdots, m, \ h_m \neq 0 \\ & \quad \text{such that } h(e^{j\omega}) \neq 0, \ \forall \omega \in [0, \ 2\pi] \\ g(z) &= g_0 + g_1 z + g_2 z^2 + \ldots + g_n z^n, \ g_i \in \mathcal{R}, \ i = 0, 1, \cdots, n, \ g_n \neq 0 \\ & \quad \text{such that } g(e^{j\omega}) \neq 0, \ \forall \omega \in [0, \ 2\pi]. \end{aligned}$$

Let $p(\omega)$ and $q(\omega)$ be two functions defined pointwise by $p(\omega) = Re[\delta(e^{j\omega})]$, $q(\omega) = Im[\delta(e^{j\omega})]$.
With this definition, we have

$$\delta(e^{j\omega}) = p(\omega) + jq(\omega) \ \forall \omega \in [0, \ 2\pi]$$

Furthermore $\theta(\omega) \triangleq \angle\delta(e^{j\omega}) = \arctan\left[\frac{q(\omega)}{p(\omega)}\right]$. Let $\Delta_0^\pi \theta$ denote the net change in argument $\theta(\omega)$ as ω increases from 0 to π and let $Z(\delta)$, $P(\delta)$ denote the numbers of zeros and poles of $\delta(z)$ in $\mathcal{D}$. Then we can state the following Lemma:

Lemma 9.3.1. *Let $\delta(z)$ be a real rational function with no unit circle zeros or poles. Then*

$$\Delta_0^\pi \theta = \pi(Z(\delta) - P(\delta))$$

Proof: The proof is an immediate consequence of the principle of the argument.

9.4 Discrete-time Versions of the Generalized Hermite-Biehler Theorem

In this section, we derive two generalizations of the Hermite-Biehler Theorem by first developing a procedure for systematically determining the net accumulated phase change of the "frequency response" of a rational function. The development here is similar to that in Sections 3.5.1 and 8.3.2. The main differences are: (1) here $\delta(z)$ is a rational function and not a polynomial; and (2) no normalization is necessary here since, for $\delta(z)$ having no poles on the unit circle, $\delta(e^{j\omega})$ is finite for all ω.

As in Section 9.3, we consider a rational function $\delta(z)$ without zeros or poles on the unit circle:

$$\delta(z) = \frac{h(z)}{g(z)}$$

where $h(z)$ and $g(z)$ are coprime polynomials.

Let $\delta_r(z)$, $\delta_i(z)$, $p(\omega)$, $q(\omega)$ be as already defined and let

$$0 = \omega_0 < \omega_1 < \omega_2 < \cdots < \omega_m = \pi$$

be the real, non-negative, distinct zeros of $q(\omega)$ with odd multiplicities[1]. Note that ω_i, $i = 0, 1, 2, \cdots, m$ are arguments of the distinct upper half plane unit circle zeros of $\delta_i(z)$ with odd multiplicities.

Then we can make the following simple observations:

1.

$$\Delta_{\omega_i}^{\omega_{i+1}}\theta = \frac{\pi}{2}\left[\mathrm{sgn}[p(\omega_i)] - \mathrm{sgn}[p(\omega_{i+1})]\right] \cdot \mathrm{sgn}[q(\omega_i^+)] \quad (9.4.7)$$

2.

$$\mathrm{sgn}[q(\omega_{i+1}^+)] = -\mathrm{sgn}[q(\omega_i^+)],\ i = 0, 1, 2, \cdots, m-1 \quad (9.4.8)$$

Equation 9.4.7 above is obvious while Equation 9.4.8 simply states that $q(\omega)$ changes sign when it passes through a zero of odd multiplicity.

Using (9.4.8) repeatedly, we obtain

$$\mathrm{sgn}[q(\omega_i^+)] = (-1)^i \cdot \mathrm{sgn}[q(\omega_0^+)],\ i = 0,\ 1, \cdots,\ m \quad (9.4.9)$$

Substituting (9.4.9) into (9.4.7), we have

$$\Delta_{\omega_i}^{\omega_{i+1}}\theta = \frac{\pi}{2}\left[\mathrm{sgn}[p(\omega_i)] - \mathrm{sgn}[p(\omega_{i+1})]\right] \cdot (-1)^i \cdot \mathrm{sgn}[q(\omega_0^+)]. \quad (9.4.10)$$

The above observations enable us to state and prove the following Theorem concerning $Z(\delta) - P(\delta)$. However, to simplify the theorem statement, we first define the "imaginary signature" $\sigma_i(\delta)$ of a real rational function $\delta(z)$.

Definition 9.4.1. *Let $\delta(z)$ be any given real rational function. Let $1 = z_{i_0}$, $z_{i_1}, \cdots, z_{i_m} = -1$ denote the distinct upper half plane unit circle zeros of $\delta_i(z)$ with odd multiplicities and let the corresponding arguments be $0 = \omega_0 < \omega_1 < \omega_2 < \cdots < \omega_m = \pi$. Then*

$$\begin{aligned}\sigma_i(\delta) := \frac{1}{2}\{&sgn[p(\omega_0)] - 2sgn[p(\omega_1)] + 2sgn[p(\omega_2)] + \cdots \\ &+(-1)^{m-1}2sgn[p(\omega_{m-1})] + (-1)^m sgn[p(\omega_m)]\} \cdot sgn[q(\omega_0^+)]\end{aligned} \quad (9.4.11)$$

defines the imaginary signature $\sigma_i(\delta)$ of $\delta(z)$.

Theorem 9.4.1. *Let $\delta(z)$ be a given real rational function with no zeros or poles on the unit circle,* i.e., *the plot $\delta(e^{j\omega})$ does not pass through the origin and is finite. Then*

$$Z(\delta) - P(\delta) = \sigma_i(\delta) \quad (9.4.12)$$

[1] The function $q(\omega)$ does not change sign while passing through a real zero of even multiplicity; hence such zeros can be skipped while counting the net phase accumulation.

Proof.

$$\begin{aligned} \Delta_0^{\pi}\theta &= \sum_{i=0}^{m-1} \Delta_{\omega_i}^{\omega_{i+1}}\theta \\ &= \sum_{i=0}^{m-1} \frac{\pi}{2}\left[\mathrm{sgn}[p(\omega_i)] - \mathrm{sgn}[p(\omega_{i+1})]\right] \cdot (-1)^i \cdot \mathrm{sgn}[q(\omega_0^+)] \\ &\quad \text{(using (9.4.10))} \end{aligned} \tag{9.4.13}$$

Applying Lemma 9.3.1, the desired expression follows. ♣

We now state the counterpart of Theorem 9.4.1 where $Z(\delta) - P(\delta)$ of a real rational function $\delta(z)$ is determined using the values of the frequencies where $\delta(e^{j\omega})$ crosses the imaginary axis. Once again, to simplify the theorem statement, we define the "real signature" $\sigma_r(\delta)$ of a real rational function $\delta(z)$.

Definition 9.4.2. *Let $\delta(z)$ be any given real rational function. Let z_{r_0}, $z_{r_1}, \cdots, z_{r_m}$ denote the distinct upper half plane unit circle zeros of $\delta_r(z)$ with odd multiplicities and let the corresponding arguments be $0 < \omega_0 < \omega_1 < \cdots < \omega_m < \pi$. Then*

$$\begin{aligned} \sigma_r(\delta) \;:=\; & -\{sgn[q(\omega_0)] - sgn[q(\omega_1)] + sgn[q(\omega_2)] + \cdots \\ & +(-1)^{m-1} sgn[q(\omega_{m-1})] + (-1)^m sgn[q(\omega_m)]\} \cdot sgn[p(\omega_0^+)] \end{aligned} \tag{9.4.14}$$

defines the real signature $\sigma_r(\delta)$ of $\delta(z)$.

Theorem 9.4.2. *Let $\delta(z)$ be a given real rational function with no zeros or poles on the unit circle,* i.e., *the plot $\delta(e^{j\omega})$ does not pass through the origin and is finite. Then*

$$Z(\delta) - P(\delta) = \sigma_r(\delta) \tag{9.4.15}$$

Remark 9.4.1. Theorems 9.4.1 and 9.4.2 require that the rational function $\delta(z)$ have no zeros or poles on the unit circle. For the constant gain stabilization problem, this translates to the requirement that the plant to be stabilized *does not* have any zeros on the unit circle. The stabilization results of this chapter will be derived assuming that this requirement is met.

9.5 Stabilization Using a Constant Gain

In this section, we make use of Theorem 9.4.1 to provide a *complete analytical* solution to the constant gain stabilization problem of Fig. 9.4. As already mentioned, we will be assuming that the plant does not have any zeros on the unit circle. The closed loop characteristic polynomial $\delta(z,k)$ is given by

$$\delta(z,k) = D(z) + kN(z) \tag{9.5.16}$$

Our objective is to determine those values of k, if any, for which the closed loop system is stable, *i.e.*, $\delta(z,k)$ is Schur.

There are several classical approaches for solving this problem: the root locus technique, the Nyquist stability criterion, and the Jury criterion. Of these approaches, the root locus technique and the Nyquist stability criterion solve this problem by plotting the root loci of $\delta(z,k)$ and the Nyquist plot of $G(z) = \frac{N(z)}{D(z)}$, respectively. Hence, both of these methods are graphical in nature and fail to provide us with an analytical characterization of all stabilizing k's. The Jury criterion, on the other hand, does provide us with an analytical solution. However, k must be determined by solving a set of polynomial inequalities, a task which is not straight forward especially for higher order plants. This is clearly borne out by the following example.

Example 9.5.1. Consider the plant $G(z) = \frac{N(z)}{D(z)}$ where

$$\begin{aligned} N(z) &= 100z^3 + 2z^2 + 3z + 11 \\ D(z) &= 100z^5 + 2z^4 + 5z^3 - 41z^2 + 52z + 70. \end{aligned}$$

This plant is to be stabilized, if possible, using a constant feedback gain k and we are interested in knowing all the possible ranges of k for which the closed loop system is stable. The closed loop characteristic polynomial is given by

$$\begin{aligned} \delta(z,k) &= D(z) + kN(z) \\ &= 100z^5 + 2z^4 + (5+100k)z^3 + (-41+2k)z^2 + (52+3k)z \\ &\quad +(70+11k) \end{aligned}$$

From the Jury criterion, $\delta(z,k)$ is Schur if and only if the following inequalities hold.

$$\begin{cases} 116k + 188 > 0 \\ 90k + 126 > 0 \\ |\,11k + 70\,| < 100 \\ |\,121k^2 + 1540k - 5100\,| > |\,-278k - 5060\,| \\ |\,14641k^4 + 372680k^3 + 1060116k^2 - 18521360k + 406400\,| \\ > |\,133100k^4 + 2532629k^3 + 5869526k^2 - 23194260k - 5288600\,| \\ |\,-17501251119k^8 - 663273024040k^7 - 7806744773729k^6 \\ -23308525917468k^5 + 71772066123060k^4 + 260100081848800k^3 \\ -131988108006000k^2 - 260384488280000k - 27804129000000\,| \\ > |\,-492493958k^8 - 86862007672k^7 - 3081711813312k^6 \\ -38664130460142k^5 - 136915545679108k^4 + 326497012343760k^3 \\ +979786905624800k^2 + 307435502724000k + 26354990520000\,| \end{cases} \tag{9.5.17}$$

Clearly, the inequalities (9.5.17) are nonlinear and there is really no elegant method for their solution.

Let us now see how the results of the last section can be used to determine the values of k, if any, for which $\delta(z,k)$ in (9.5.16) is Schur stable. As discussed in Section 9.2, both the "real" as well as the "imaginary" parts of $\delta(z,k)$ depend on k and this creates difficulties when trying to use Lemma 9.2.1 to ensure the Schur stability of $\delta(z,k)$. Consequently, starting from $\delta(z,k)$, we will now construct a rational function for which only the "real" part depends on k, and to which Theorem 9.4.1 is applicable.

Suppose that the degree of $D(z)$ is n while the degree of $N(z)$ is m and $m \leq n$ and define

$$N^*(z) = N(\frac{1}{z}).$$

Now, multiplying $\delta(z,k)$ by $N^*(z)$ and examining the resulting rational function, we obtain

$$\begin{aligned} & Z(\delta(z,k)N^*(z)) - P(\delta(z,k)N^*(z)) \\ = \; & Z(\delta(z,k)) - P(\delta(z,k)) + Z(N^*(z)) - P(N^*(z)) \\ = \; & Z(\delta(z,k)) - P(\delta(z,k)) + Z(N(\frac{1}{z})) - P(N(\frac{1}{z})) \\ = \; & Z(\delta(z,k)) - P(\delta(z,k)) - (Z(N(z)) - P(N(z))) \end{aligned}$$

Now, $\delta(z,k)$ of degree n is Schur if and only if $Z(\delta(z,k)) = n$. Also since $\delta(z,k)$ is a polynomial, $P(\delta(z,k))$ is necessarily equal to zero. We further note that, *if $N(z)$ has no zeros on the unit circle*, then for those values of

k for which $\delta(z,k)$ is Schur, the rational function $\delta(z,k)N^*(z)$ satisfies the hypothesis of Theorem 9.4.1 so that

$$\sigma_i(\delta(z,k)N^*(z)) = Z(\delta(z,k)N^*(z)) - P(\delta(z,k)N^*(z)).$$

Combining these facts, we have the following.

Lemma 9.5.1. *$\delta(z,k)$ is Schur if and only if*

$$\sigma_i(\delta(z,k)N^*(z)) = n - (Z(N(z)) - P(N(z))) \tag{9.5.18}$$

Our task now is to determine those values of k for which (9.5.18) holds. Now

$$\delta(z,k)N^*(z) = D(z)N(\frac{1}{z}) + kN(z)N(\frac{1}{z})$$

Denote

$$H(z) = D(z)N(\frac{1}{z})$$

Then the real-imaginary decomposition of $H(z)$ is

$$H(z) = H_r(z) + H_i(z)$$

where

$$\begin{aligned} H_r(z) &= \frac{1}{2}[D(z)N(\frac{1}{z}) + D(\frac{1}{z})N(z)] \\ H_i(z) &= \frac{1}{2}[D(z)N(\frac{1}{z}) - D(\frac{1}{z})N(z)] \end{aligned}$$

Thus, we have

$$\delta(z,k)N^*(z) = [H_r(z) + kN(z)N(\frac{1}{z})] + H_i(z)$$

Substituting $z = e^{j\omega}$, we obtain

$$\delta(e^{j\omega},k)N^*(e^{j\omega}) = p(\omega,k) + jq(\omega) \tag{9.5.19}$$

where

$$\begin{aligned} p(\omega,k) &= p_1(\omega) + kp_2(\omega) \\ p_1(\omega) &= \frac{1}{2}[D(e^{j\omega})N(e^{-j\omega}) + D(e^{-j\omega})N(e^{j\omega})] \\ p_2(\omega) &= N(e^{j\omega})N(e^{-j\omega}) \\ q(\omega) &= \frac{1}{2j}[D(e^{j\omega})N(e^{-j\omega}) - D(e^{-j\omega})N(e^{j\omega})] \end{aligned}$$

Note that $p_1(\omega)$, $p_2(\omega)$, and $q(\omega)$ are all purely real functions.

Now, since $Z(N(z)) - P(N(z))$ is known, the stabilizing values of k can be determined from (9.5.18).

We now proceed in the spirit of Sections 4.2 and 8.4, and introduce some definitions before stating our main result.

Definition 9.5.1. *Let the function* $q(\omega)$ *be as already defined. Let* $0 \leq \omega_0 < \omega_1 < \omega_2 < \cdots < \omega_l \leq \pi$ *be the real distinct zeros of* $q(\omega)$ *with odd multiplicities*[2]. *Define*

$$A = \{\{i_0, i_1, \cdots, i_l\}\}$$

where $i_t \in \{-1, 1\}$ *for* $t = 0, 1, \ldots, l$. *In other words A is the set of all possible strings of 1's and -1's of length* $l+1$.

Definition 9.5.2. *Let the function* $q(\omega)$ *be as already defined. Let* $0 \leq \omega_0 < \omega_1 < \omega_2 < \cdots < \omega_l \leq \pi$ *be the distinct zeros of* $q(\omega)$ *with odd multiplicities. For each string* $\mathcal{I} = \{i_0, i_1, \cdots, i_l\}$ *in A, let* $\gamma(\mathcal{I})$ *denote the "imaginary signature" associated with the string* $\mathcal{I}$ *defined by*

$$\gamma(\mathcal{I}) := \frac{1}{2}\{i_0 - 2i_1 + 2i_2 + \cdots + (-1)^{l-1}2i_{l-1} + (-1)^l i_l\} \cdot sgn[q(\omega_0^+)] \tag{9.5.20}$$

Definition 9.5.3. *The set of strings in A with a prescribed imaginary signature* $\gamma = \psi$ *is denoted by* $A(\psi)$. *We also define the set of feasible strings for the constant gain stabilization problem as*

$$F^* = \{\mathcal{I} | \mathcal{I} = \{i_0, i_1, \cdots, i_l\} \in A(n - (Z(N(z)) - P(N(z))))\} \tag{9.5.21}$$

The following example illustrates these definitions.

Example 9.5.2. From (9.5.19), we have

$$\delta(e^{j\omega}, k)N^*(e^{j\omega}) = p(\omega, k) + jq(\omega)$$

where

$$p(\omega, k) = p_1(\omega) + kp_2(\omega)$$

Now suppose, for example, that $\delta(z, k)$ is of degree $n = 6$, and $Z(N(z)) - P(N(z)) = 4$. Let $q(\omega)$ have 4 distinct zeros $0 = \omega_0 < \omega_1 < \omega_2 < \omega_3 = \pi$ with odd multiplicities. Furthermore, suppose that $q(\omega)$ is as shown in Fig. 9.5.

The set A of all the possible strings $\{i_0, i_1, i_2, i_3\}$ is given as follows

$$A = \left\{ \begin{array}{ll} \{-1,-1,-1,-1\} & \{1,-1,-1,-1\} \\ \{-1,-1,-1,1\} & \{1,-1,-1,1\} \\ \{-1,-1,1,-1\} & \{1,-1,1,-1\} \\ \{-1,-1,1,1\} & \{1,-1,1,1\} \\ \{-1,1,-1,-1\} & \{1,1,-1,-1\} \\ \{-1,1,-1,1\} & \{1,1,-1,1\} \\ \{-1,1,1,-1\} & \{1,1,1,-1\} \\ \{-1,1,1,1\} & \{1,1,1,1\} \end{array} \right\}$$

[2] Note that these zeros are independent of k.

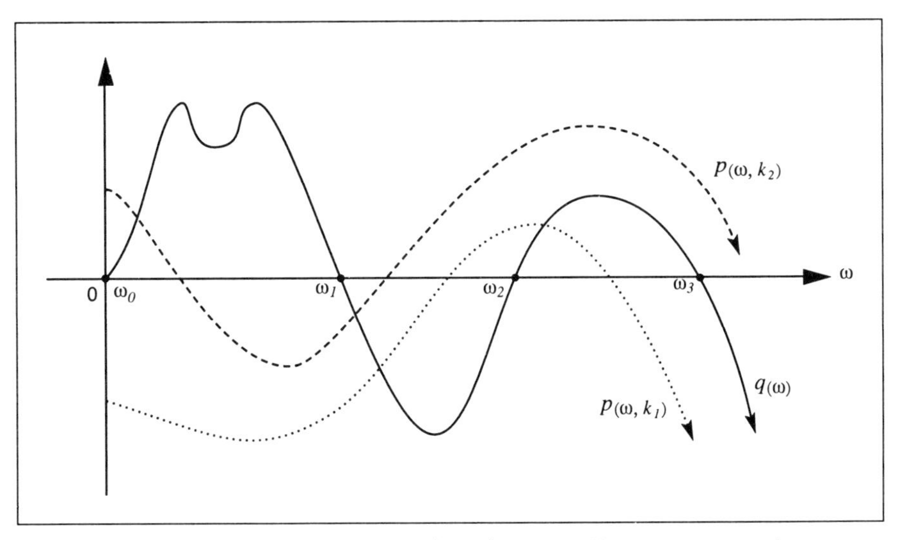

Fig. 9.5. Plots of $q(\omega)$, $p(\omega, k_1)$, and $p(\omega, k_2)$ for the illustrative example

From Lemma 9.5.1, we have $\delta(z,k)$ is Schur if and only if

$$\sigma_i(\delta(z,k)N^*(z)) \quad = \quad n - (Z(N(z)) - P(N(z))) = 2.$$

Also since $\mathrm{sgn}[q(\omega_0^+)] = 1$, it follows using Definition 9.5.3 that every string $\mathcal{I} = \{i_0, i_1, i_2, i_3\} \in F^*$ must satisfy

$$\frac{1}{2}\{i_0 - 2i_1 + 2i_2 - i_3\} = 2 \tag{9.5.22}$$

so that

$$F^* = \{\{-1, -1, 1, -1\}, \{1, -1, 1, 1\}\}.$$

Therefore, the constant gain stabilization problem now reduces to the problem of determining the values of k, if any, such that $\mathrm{sgn}[p(\omega_j, k)] = i_j$, $j = 0,\ 1,\ 2,\ 3$ and $\{i_0, i_1, i_2, i_3\} \in F^*$. For instance, for $k = k_1$, $p(\omega, k_1)$ shown in Fig. 9.5 has

$$\{\mathrm{sgn}[p(\omega_0, k_1)], \mathrm{sgn}[p(\omega_1, k_1)], \mathrm{sgn}[p(\omega_2, k_1)], \mathrm{sgn}[p(\omega_3, k_1)]\} = \{-1, -1, 1, -1\}.$$

On the other hand, for $k = k_2$, $p(\omega, k_2)$ shown in Fig. 9.5 has

$$\{\mathrm{sgn}[p(\omega_0, k_2)], \mathrm{sgn}[p(\omega_1, k_2)], \mathrm{sgn}[p(\omega_2, k_2)], \mathrm{sgn}[p(\omega_3, k_2)]\} = \{1, -1, 1, 1\}.$$

Thus for both $k = k_1$ and k_2, $\delta(z, k)$ is Schur.

We are now ready to state the main result of this section.

Theorem 9.5.1. *(Main Result on Constant Gain Stabilization) The constant gain feedback stabilization problem is solvable for a given plant with transfer function $G(z)$ and no zeros on the unit circle if and only if the following conditions hold:*
(i) F^* is not empty i.e., *at least one feasible string exists*
and
(ii) There exists a string $\mathcal{I} = \{i_0, i_1, \cdots, i_l\} \in F^*$ such that

$$\max_{i_t \in \mathcal{I}, i_t = 1} \left[-\frac{1}{G(e^{j\omega_t})} \right] < \min_{i_t \in \mathcal{I}, i_t = -1} \left[-\frac{1}{G(e^{j\omega_t})} \right]$$

where $\omega_0, \omega_1, \omega_2, \cdots$ are as already defined. Furthermore, if the above condition is satisfied by the feasible strings $\mathcal{I}_1, \mathcal{I}_2, \cdots, \mathcal{I}_s \in F^*$, then the set of all stabilizing gains is given by $K = \cup_{r=1}^{s} K_r$ where

$$K_r = \left(\max_{i_t \in \mathcal{I}_r, i_t = 1} \left[-\frac{1}{G(e^{j\omega_t})} \right], \min_{i_t \in \mathcal{I}_r, i_t = -1} \left[-\frac{1}{G(e^{j\omega_t})} \right] \right), \ r = 1, 2, \cdots, s.$$

Proof. From (9.5.18), we know that $\delta(z, k)$ is Schur if and only if

$$\sigma_i(\delta(z, k)N^*(z)) = n - (Z(N(z)) - P(N(z))).$$

Thus $\delta(z, k)$ is Schur if and only if $\mathcal{I} \in A(n - (Z(N(z)) - P(N(z))))$ where $\mathcal{I} = \{i_0, i_1, \cdots, i_l\}$, $i_j = sgn[p(\omega_j, k)]$, $j = 0, 1, 2, \cdots, l$. We now consider two different cases:
Case 1 $i_j > 0$: If $i_j > 0$, then the stability requirement is $p_1(\omega_j) + kp_2(\omega_j) > 0$. Since $p_2(\omega) = N(e^{j\omega})N(e^{-j\omega})$, we see that $p_2(\omega_j) > 0$, $j = 0, 1, \ldots, l$, and so it follows that

$$k > -\frac{p_1(\omega_j)}{p_2(\omega_j)} \tag{9.5.23}$$

Case 2 $i_j < 0$: If $i_j < 0$, then the stability requirement is $p_1(\omega_j) + kp_2(\omega_j) < 0$. Again since $p_2(\omega_j) > 0$, $j = 0, 1, \ldots, l$, it follows that

$$k < -\frac{p_1(\omega_j)}{p_2(\omega_j)} \tag{9.5.24}$$

Combining (9.5.23) and (9.5.24), we see that

$$-\frac{p_1(\omega_j)}{p_2(\omega_j)} < k \text{ if } i_j > 0 \tag{9.5.25}$$

and

$$k < -\frac{p_1(\omega_j)}{p_2(\omega_j)} \text{ if } i_j < 0 \tag{9.5.26}$$

Thus each i_j in the string $\mathcal{I}$ for which $i_j > 0$ contributes a lower bound on k while each i_j for which $i_j < 0$ contributes an upper bound on k. Thus, if the string $\mathcal{I}$ is to correspond to a stabilizing k then we must have

$$\max_{i_t\in\mathcal{I}, i_t=1}\left[-\frac{p_1(\omega_t)}{p_2(\omega_t)}\right] < \min_{i_t\in\mathcal{I}, i_t=-1}\left[-\frac{p_1(\omega_t)}{p_2(\omega_t)}\right]. \tag{9.5.27}$$

We next proceed to express the above condition (9.5.27) in terms of $G(z)$. Now

$$G(z) = \frac{N(z)}{D(z)}$$

so that

$$\begin{aligned}
\frac{1}{G(e^{j\omega})} &= \frac{D(e^{j\omega})}{N(e^{j\omega})} \\
&= \frac{D(e^{j\omega})N(e^{-j\omega})}{N(e^{j\omega})N(e^{-j\omega})} \\
&= \frac{\frac{1}{2}[D(e^{j\omega})N(e^{-j\omega}) + D(e^{-j\omega})N(e^{j\omega})]}{N(e^{j\omega})N(e^{-j\omega})} \\
&\quad + \frac{\frac{1}{2}[D(e^{j\omega})N(e^{-j\omega}) - D(e^{-j\omega})N(e^{j\omega})]}{N(e^{j\omega})N(e^{-j\omega})} \\
&= \frac{p_1(\omega) + jq(\omega)}{p_2(\omega)}
\end{aligned}$$

Since $q(\omega_t) = 0$, it follows that $-\frac{p_1(\omega_t)}{p_2(\omega_t)} = -\frac{1}{G(e^{j\omega_t})}$. Thus, from (9.5.27), we must have

$$\max_{i_t\in\mathcal{I}, i_t=1}\left[-\frac{1}{G(e^{j\omega_t})}\right] < \min_{i_t\in\mathcal{I}, i_t=-1}\left[-\frac{1}{G(e^{j\omega_t})}\right].$$

which is condition (ii) in the Theorem statement. This completes the proof of the necessary and sufficient conditions for the existence of a stabilizing k. The set of all stabilizing k's is now determined by taking the union of all k's that are obtained from all the strings which satisfy (ii). ♣

We now present an example to illustrate the detailed calculations involved when using Theorem 9.5.1 to analytically determine the set of all stabilizing feedback gains.

Example 9.5.3. Consider the same constant gain stabilization problem as in Example 9.5.1, *i.e.*

$$\begin{aligned}
N(z) &= 100z^3 + 2z^2 + 3z + 11 \\
D(z) &= 100z^5 + 2z^4 + 5z^3 - 41z^2 + 52z + 70.
\end{aligned}$$

The closed loop characteristic polynomial is

$$\delta(z,k) = D(z) + kN(z)$$

and

$$\begin{aligned} N^*(z) &= N(\frac{1}{z}) \\ &= z^{-3}(11z^3 + 3z^2 + 2z + 100). \end{aligned}$$

Therefore

$$\delta(z,k)N^*(z) = [H_r(z) + kN(z)N(\frac{1}{z})] + H_i(z)$$

where

$$\begin{aligned} H_r(z) &= \frac{1}{2}[N(\frac{1}{z})D(z) + N(z)D(\frac{1}{z})] \\ &= \frac{z^{-5}}{2}[1100z^{10} + 322z^9 + 7261z^8 + 14908z^7 - 3127z^6 + \\ & \quad 2688z^5 - 3127z^4 + 14908z^3 + 7261z^2 + 322z + 1100] \\ H_i(z) &= \frac{1}{2}[N(\frac{1}{z})D(z) - N(z)D(\frac{1}{z})] \\ &= \frac{z^{-5}}{2}[1100z^{10} + 322z^9 - 6739z^8 + 4228z^7 + 4445z^6 \\ & \quad -4445z^4 - 4228z^3 + 6739z^2 - 322z - 1100] \\ \text{and } N(z)N(\frac{1}{z}) &= z^{-3}[1100z^6 + 322z^5 + 239z^4 + 10134z^3 + 239z^2 + 322z \\ & \quad +1100]. \end{aligned}$$

Substituting $z = e^{j\omega}$, we obtain

$$\delta(e^{j\omega},k)N^*(e^{j\omega}) = p_1(\omega) + kp_2(\omega) + jq(\omega)$$

where

$$\begin{aligned} p_1(\omega) &:= H_r(e^{j\omega}) \\ p_2(\omega) &:= N(e^{j\omega})N(e^{-j\omega}) \\ q(\omega) &:= \frac{1}{j}H_i(e^{j\omega}). \end{aligned}$$

The distinct upper half plane unit circle zeros of $H_i(z)$ with odd multiplicities and the corresponding arguments are:

$$\begin{array}{rl} z_0 = 1 & \omega_0 = 0 \\ z_1 = 0.8613 + 0.5081j & \omega_1 = 0.5329 \\ z_2 = -0.4860 + 0.8739j & \omega_2 = 2.0784 \\ z_3 = -1 & \omega_3 = \pi \end{array}$$

Now from Definition 9.5.1, the set A becomes

$$\begin{array}{rcl} A & = & \{\{i_0,\ i_1,\ i_2,\ i_3\}\} \\ & = & \left\{ \begin{array}{ll} \{-1,-1,-1,-1\} & \{1,-1,-1,-1\} \\ \{-1,-1,-1,1\} & \{1,-1,-1,1\} \\ \{-1,-1,1,-1\} & \{1,-1,1,-1\} \\ \{-1,-1,1,1\} & \{1,-1,1,1\} \\ \{-1,1,-1,-1\} & \{1,1,-1,-1\} \\ \{-1,1,-1,1\} & \{1,1,-1,1\} \\ \{-1,1,1,-1\} & \{1,1,1,-1\} \\ \{-1,1,1,1\} & \{1,1,1,1\} \end{array} \right\}. \end{array}$$

Since $Z(N(z))-P(N(z)) = 3$ and $\operatorname{sgn}[q(\omega_0^+)] = -1$, it follows using Definition 9.5.3 that every string $\mathcal{I} = \{i_0,\ i_1,\ i_2,\ i_3\} \in F^*$ must satisfy

$$-\frac{1}{2}(i_0 - 2i_1 + 2i_2 - i_3) = 2.$$

Hence $F^* = \{\mathcal{I}_1, \mathcal{I}_2\}$ where

$$\begin{array}{rcl} \mathcal{I}_1 & = & \{-1,1,-1,-1\} \\ \mathcal{I}_2 & = & \{1,1,-1,1\}. \end{array}$$

Furthermore,

$$\begin{array}{rcl} -\dfrac{1}{G(e^{j\omega_0})} & = & -1.6207 \\ -\dfrac{1}{G(e^{j\omega_1})} & = & -0.4178 \\ -\dfrac{1}{G(e^{j\omega_2})} & = & -0.1263 \\ -\dfrac{1}{G(e^{j\omega_3})} & = & -1.4000. \end{array}$$

Hence from Theorem 9.5.1, we have

$$\left\{ \begin{array}{l} K_1 = \emptyset \text{ for } \mathcal{I}_1 \\ K_2 = (-0.4178,\ -0.1263) \text{ for } \mathcal{I}_2. \end{array} \right.$$

Therefore $\delta(z,k)$ is Schur for $k \in (-0.4178,\ -0.1263)$.

9.6 Notes and References

The results in this chapter are based on Ho, Datta and Bhattacharyya [23]. The extension of the constant gain results of this chapter to the case of PI and PID controllers, as also the relaxation of the constraint on the plant zeros for constant gain stabilization, are topics for further investigation.

APPENDIX

ROOT LOCUS IDEAS FOR NARROWING THE SWEEPING RANGE FOR K_P

Consider the problem of determining the root locus of $U(x) + kV(x) = 0$, where $U(x)$ and $V(x)$ are real and coprime polynomials and k varies from $-\infty$ to ∞. Then, we can make the following observations:

(1) The real breakaway points on the root loci of $U(x) + kV(x) = 0$ correspond to a real multiple root and must, therefore, satisfy

$$\frac{d(\frac{V(x)}{U(x)})}{dx} = 0$$

i.e.,

$$\frac{U(x)\frac{dV(x)}{dx} - V(x)\frac{dU(x)}{dx}}{U^2(x)} = 0.$$

The real breakaway points are the real zeros of the above equation.

(2) Let $k_1 < k_2 < \cdots < k_z$ be the distinct, finite, values of k corresponding to the real breakaway points x_i, $i = 1, 2, \cdots, z$ on the root loci of $U(x) + kV(x) = 0$. Also define $k_0 = -\infty$ and $k_{z+1} = \infty$. Then x_i, $i = 1, 2, \cdots, z$ are the multiple real roots of $U(x) + kV(x) = 0$ and the corresponding k's are the k_i's. We note that for $k \in (k_i, k_{i+1})$, the real roots of $U(x) + kV(x) = 0$ are simple and the number of real roots of $U(x) + kV(x) = 0$ is invariant.

(3) If $U(0) + kV(0) \neq 0$ for all $k \in (k_i, k_{i+1})$, then the distribution of the real roots of $U(x) + kV(x) = 0$ with respect to the origin is invariant over this range of k values.

The following example illustrates how the above observations can be used to determine the distribution of the real roots of $U(x) + kV(x) = 0$ with respect to the origin as k varies from $-\infty$ to ∞.

Example A.0.1. Let

$$U(x) = (x+1)^3(x-1)(x^2-x+1)^2$$

and

$$V(x) = (x-2)^2(x+3)(x^2+2x+2)$$

Then

$$\begin{aligned}\frac{U(x)\frac{dV(x)}{dx} - V(x)\frac{dU(x)}{dx}}{U^2(x)} &= [(x^8 - x^6 + 2x^5 - 2x^3 + x^2 - 1)(5x^4 + 4x^3 \\ &\quad -24x^2 - 12x + 8) - (x^5 + x^4 - 8x^3 - 6x^2 \\ &\quad +8x + 24)(8x^7 - 6x^5 + 10x^4 - 6x^2 + 2x)]/ \\ &\quad [(x+1)^6(x-1)^2(x^2 - x + 1)^4] \\ &= [-3x^{12} - 4x^{11} + 41x^{10} + 38x^9 - 82x^8 - 188x^7 \\ &\quad +77x^6 + 82x^5 - 265x^4 + 28x^3 + 160x^2 - 36x \\ &\quad -8]/[(x+1)^6(x-1)^2(x^2 - x + 1)^4]\end{aligned}$$

The breakaway points x_i which are the real zeros of the above expression are:

$$\begin{aligned} x_1 &= 2.96872,\ x_2 = -1,\ x_3 = 0.42142,\ x_4 = 0.66720, \\ &\quad x_5 = -0.14008,\ x_6 = -3.84988 \end{aligned}$$

and the corresponding finite k_i's (arranged in ascending order of magnitude) are:

$$\begin{aligned} k_1 &= -61.44924,\ k_2 = 0,\ k_3 = 0.03689,\ k_4 = 0.03791, \\ &\quad k_5 = 0.04279,\ k_6 = 163.73847 \end{aligned}$$

Furthermore, $U(x) + kV(x)$ has a root at the origin when $k = k^* := 0.04167$.

Now, for $k \in (k_i, k_{i+1})$ and $k^* \notin (k_i,\ k_{i+1})$, the distribution of the real roots of $U(x) + kV(x) = 0$ with respect to the origin is invariant. Thus, we can simply check an arbitrary $k \in (k_i,\ k_{i+1})$ and determine the real root distribution of $U(x) + kV(x) = 0$ with respect to the origin, and this distribution is valid for all k in that interval. In this example, $k^* \in (k_4,\ k_5)$, and so the real root distribution of $U(x) + kV(x) = 0$ with respect to the origin may not be invariant over the entire interval (k_4, k_5). Therefore, we need to split the interval $(k_4,\ k_5)$ into two sub-intervals $(k_4,\ k^*)$, $(k^*,\ k_5)$, and then check the real root distribution for each of these sub intervals.

The real root distribution, with respect to the origin, of $U(x) + kV(x) = 0$ for k belonging to the different intervals, is given below:

$k \in (-\infty,\ -61.44924)$: 3 positive simple real roots
1 negative simple real root

$k \in (-61.44924,\ 0)$: 1 positive simple real root
1 negative simple real root

$k \in (0,\ 0.03689)$: 1 positive simple real root
1 negative simple real root

$k \in (0.03689,\ 0.03791)$: 3 positive simple real roots
1 negative simple real root

$k \in (0.03791,\ 0.04167)$: 1 positive simple real root

		1 negative simple real root
$k \in (0.04167,\ 0.04279)$	:	2 negative simple real roots
$k \in (0.04279,\ 163.73847)$	:	no real roots
$k \in (163.73847,\ \infty)$	:	2 negative simple real roots.

The above example shows how simple root locus ideas can be used to determine the distribution of the real zeros of $U(x) + kV(x)$, with respect to the origin, as k varies from $-\infty$ to ∞.

REFERENCES

1. Astrom K. J. and Hagglund T., "Automatic Tuning of Simple Regulators with Specifications on Phase and Amplitude Margins," *Automatica*, Vol. 20, 645-651, 1984.
2. Astrom K. and Hagglund T., *PID Controllers: Theory, Design, and Tuning*, Instrument Society of America, North Carolina, 1995.
3. Atherton D. P. and Majhi S., "Limitations of PID Controllers," *Proc. of the American Control Conference*, 3843- 3847, 1999.
4. Bellman R. and Cooke K. L., *Differential-Difference Equations*, Academic Press Inc., London, 1963.
5. Bhattacharyya S. P. and deSouza E., "Pole Assignment Via Sylvester's Equation," *Systems and Control Letters*, Vol. 1(4), 261-263, January 1982.
6. Bhattacharyya S. P., *Robust Stabilization Against Structured Perturbations*, Vol. 99, Lecture Notes in Control and Information Sciences, Springer-Verlag, 1987.
7. Bhattacharyya S. P., Chapellat H. and Keel L. H., *Robust Control: The Parametric Approach*, Prentice Hall, 1995.
8. Bhattacharyya S., Keel L. H. and Bhattacharyya S. P., "Robust Stabilizer Synthesis for Interval Plants using H_∞ Methods," *Proceedings of the IEEE Conf. on Decision and Contr.*, 3003-3008, Dec 1993.
9. Bose N. K., *Digital Filters*, Elsevier-Science North-Holland, Krieger Publishing Co., New York, 1993.
10. Boyd S. and Vandenberghe L., *Convex Optimization*, Lecture notes, Electrical Engineering Department, Stanford University.
11. Brasch F. M. and Pearson J. B., "Pole Placement Using Dynamic Compensator," *IEEE Transactions on Automatic Control*, Vol. AC-15, No. 1, 34-43, February 1970.
12. Chapellat H., Mansour M. and Bhattacharyya S. P., "Elementary Proofs of Some Classical Stability Criteria," *IEEE Transactions on Education*, Vol. 33, No. 3, Aug. 1990.
13. Cohen G. H. and Coon G. A., "Theoretical Consideration of Retarded Control," *Trans. ASME*, Vol. 76, 827-834, Jul. 1953.
14. Dahleh M. A. and Diaz-Bobillo I. J., *Control of Uncertain Systems: A Linear Programming Approach*, Prentice Hall, 1995.
15. Doyle J. C., "Guaranteed Margins for LQG Regulators," *IEEE Trans. on Automat. Contr.*, Vol. AC-23, 756-757, Aug. 1978.
16. Doyle J. C., Francis B. A., and Tannenbaum A. R., *Feedback Control Theory*, Macmillan Publishing Company, New York, 1992.
17. Doyle J., Glover K., Khargonekar P. and Francis B., "State Space Solutions to Standard H_2 and H_∞ Control Problems," *IEEE Trans. on Automat. Contr.*, Vol. AC-34, 831-847, Aug. 1989.
18. Dugard L. and Verriest E. I. (Eds), *Stability and Control of Time-delay Systems*, Springer-Verlag, London, 1998.

19. Gantmacher F. R., *The Theory of Matrices*, New York; Chelsea Publishing Company, 1959.
20. Grimble M. J. and Johnson M. A., "Algorithm for PID controller tuning using LQG cost minimization," *Proc. of the American Control Conference*, 4368-4372, 1999.
21. Ho M. T., Datta A. and Bhattacharyya S. P., "A Linear Programming Characterization of All Stabilizing PID Controllers," *Proceedings of the American Control Conference*, 3922-3928, Albuquerque, NM, June 1997.
22. Ho M. T., Datta A. and Bhattacharyya S. P., "Control System Design Using Low Order Controllers: Constant Gain, PI and PID," *Proceedings of the American Control Conference*, 571-578, Albuquerque, NM, June 1997.
23. Ho M. T., Datta A. and Bhattacharyya S. P., "A New Approach to Feedback Stabilization: the Discrete-time Case," *Proceedings of the 36th IEEE Conference on Decision and Control*, 908-914, San Diego, CA, Dec. 1997.
24. Ho M. T., Datta A. and Bhattacharyya S. P., "Design of P, PI and PID Controllers for Interval Plants," *Proceedings of the American Control Conference*, 2496-2501, Philadelphia, PA, June 1998.
25. Ho M. T., Datta A. and Bhattacharyya S. P., "Constant Gain Stabilization with a Specified Damping Ratio and Damped Natural Frequency," *Proceedings of the IFAC World Congress*, Beijing, P.R.C., July 1999.
26. Ho M. T., Datta A. and Bhattacharyya S. P., "Generalizations of the Hermite-Biehler Theorem," *Linear Algebra and its Applications* (to appear).
27. Ho M. T., Datta A. and Bhattacharyya S. P., "An Elementary Derivation of the Routh-Hurwitz Criterion," *IEEE Transactions on Automatic Control*, Vol. AC-43, No. 3, 405-409, March 1998.
28. Ho W., Hang C. and Zhou J., "Self-tuning PID control for a plant with underdamped response with specification on gain and phase margins," *IEEE Trans. Control Syst. Techn.*, Vol. 5, No. 4, 446-452, July 1997.
29. Kalman R. E., "On the General Theory of Control Systems," *Proceedings of the 1st IFAC Congress* (Moscow, USSR, 1960), Vol. 1, Butterworth, London, 1961, pp. 481-492.
30. Keel L. H. and Bhattacharyya S. P., "Robust, Fragile or Optimal ?" *IEEE Trans. on Automat. Contr.*, Vol. AC-42, No. 8, 1098-1105, Aug. 1997.
31. Khalil H. K., *Nonlinear Systems*, Macmillan, New York, 1992.
32. Kharitonov V. L., "Asymptotic Stability of an Equilibrium Position of a Family of Systems of Linear Differential Equations," *Differential'nye Uravneniya*, Vol. 14, 2086-2088, 1978.
33. Kimura H., "Robust Stabilizability For a Class of Transfer Functions," *IEEE Trans. on Automat. Contr.*, Vol. AC-29, No. 9, 788-793, Sept. 1984.
34. Lepschy A., Mian G. A. and Viaro U., "A Geometrical Interpretation of the Routh Test," Journal of the Franklin Institute, Vol. 325, No. 6, 695-703, 1988.
35. Malek-Zavarei M. and Jamshidi M., *Time-delay Systems: Analysis, Optimization and Applications*, Elsevier Science Publishing Company Inc., The Netherlands, 1987.
36. Mansour M., "Robust Stability in Systems Described by Rational Functions," in *Control and Dynamic Systems*, C. T. Leondes, Ed., Vol. 51., 79-128, Academic Press, New York, 1992.
37. Marshall J. E., *Control of Time-delay Systems*, Peter Peregrinus Ltd., Stevenage, UK, 1979.
38. Morari M. and Zafiriou E., *Robust Process Control*, Prentice-Hall, Englewood Cliffs, NJ, 1989.
39. Newton G., Gould L. A. and Kaiser J. F., *Analytical Design of Linear Feedback Controls*, John Wiley & Sons, Inc., New York, 1957.

40. Panagopoulos H., Astrom K. J. and Hagglund T., "Design of PID controllers based on constrained optimization," *Proc. of the American Control Conference*, 3858-3862, 1999.
41. Pessen D. W., "A new look at PID-controller tuning," *Trans. Amer. Soc. Mech. Eng., J. Dynamic Syst., Meas., Contr.*, Vol. 116, 553-557, 1994.
42. Pontryagin, L. S., "On the zeros of some elementary transcendental function," English Translation, *American Mathematical Society Translation*, Vol. 2, pp. 95-110, 1955.
43. Stojic M. R. and Siljak D. D., "Generalization of Hurwitz, Nyquist, and Mikhailov Stability Criteria," *IEEE Trans. on Automat. Contr.*, Vol. AC-10, 250-254, July, 1965.
44. Silva G. J., Datta A. and Bhattacharyya S.P., "Stabilization of First Order Systems with Time Delay," *Department of Electrical Engg., Texas A&M Univ., College Station, TX, Tech. Report*, Sep. 1999.
45. Voda A. A. and Landau I. D., "A method of the auto- calibration of PID controllers," *Automatica*, Vol. 31, No. 1, 41-53, 1995.
46. Youla D. C., Jabr H. A. and Bongiorno J. J., "Modern Wiener-Hopf Design of Optimal Controllers — Part II: The Multivariable Case," *IEEE Trans. on Automat. Contr.*, Vol. AC-21, 319-338, 1976.
47. Zhuang M. and Atherton D. P., "Automatic tuning of optimum PID controllers," *IEE Proceedings-D*, Vol. 140, No. 3, 216-224, 1993.
48. Ziegler J. G. and Nichols N. B., "Optimum Setting for Automatic Controllers," *Trans. ASME*, Vol. 64, 759-768, Nov. 1942.

INDEX